Multivariate
Survival Analysis and
Competing Risks

CHAPMAN & HALL/CRC
Texts in Statistical Science Series

Series Editors

Francesca Dominici, *Harvard School of Public Health, USA*
Julian J. Faraway, *University of Bath, UK*
Martin Tanner, *Northwestern University, USA*
Jim Zidek, *University of British Columbia, Canada*

Texts in Statistical Science

Multivariate Survival Analysis and Competing Risks

Martin Crowder

CRC Press
Taylor & Francis Group
Boca Raton London New York

CRC Press is an imprint of the
Taylor & Francis Group an **informa** business

A CHAPMAN & HALL BOOK

CRC Press
Taylor & Francis Group
6000 Broken Sound Parkway NW, Suite 300
Boca Raton, FL 33487-2742

First issued in paperback 2016

© 2012 by Taylor & Francis Group, LLC
CRC Press is an imprint of Taylor & Francis Group, an Informa business

No claim to original U.S. Government works

Version Date: 20120313

ISBN 13: 978-1-138-19960-6 (pbk)
ISBN 13: 978-1-4398-7521-6 (hbk)

Visit the Taylor & Francis Web site at
http://www.taylorandfrancis.com

and the CRC Press Web site at
http://www.crcpress.com

For Bobbie, Sylvie, Harry, Stanley, and Charlie

Contents

Part IV Counting Processes in Survival Analysis

Preface

This book is an outgrowth of *Classical Competing Risks* (2001). I was very pleased to be encouraged by Rob Calver and Jim Zidek to write a second, expanded edition. Among other things it gives the opportunity to correct the many errors that crept into the first edition at the editorial stage. This edition has been typed in Latex by my own fair hand, so the inevitable errors are now all down to me.

The book is now divided into four parts but I will not go through describing them in detail here since the contents are listed on the next few pages. The book contains a variety of data tables together with R-code applied to them. For your convenience these can be found on the Web site at http://www.crcpress.com/product/isbn/9781439875216.

Survival analysis has its roots in death and disease among humans and animals, and much of the published literature reflects this. In this book, although inevitably including such data, I try to strike a more cheerful note with examples and applications of a less sombre nature. Some of the data included might be seen as a little unusual in the context, but the methodology of survival analysis extends to a wider field. Also, more prominence is given here to discrete time than is often the case.

There are many excellent books in this area nowadays. In particular, I have learned much from Lawless (2003), Kalbfleisch and Prentice (2002), and Cox and Oakes (1984). More specialised works, such as Cook and Lawless (2007 for recurrent events), Collett (2003, for medical applications), and Wolstenholme (1998, for reliability applications) have also been on my list. Hougaard's (2000) book, *Analysis of Multivariate Survival Data*, is devoted to the topic of Part II: its title provides the clue. It is written in a different style and has different emphases to the present book, and the target audience is probably a little more sophisticated than that assumed here; very full discussion and detailed analyses are given to applications, mainly in health and medicine. I apologise to the many authors whose work has not been mentioned here: I know that they will understand that what appears here is just a personal choice in a limited amount of space. It is a mark of the growth of Survival Analysis that in 1993 the journal *LIDA* (*Lifetime Data Analysis*) appeared: this top-quality publication has gone from strength to strength ever since under the energetic editorship of Mei-Ling Ting Lee.

Prerequisites for this book include courses in basic probability and statistics and in mathematics to the level of basic calculus and analysis. I have tried to encompass both probability modelling, along with the necessary mathematical manipulations, and the practicalities of data analysis. It seems to me that the accomplished statistician of the future will need all these skills. A basic introduction to the modern counting-process approach has been left to Part IV. In my opinion, which I confess is not shared by all, the notation and underpinning of counting processes can sometimes get in the way, obscuring what might otherwise be quite transparent.

Some readers might find the style a bit English in places. This cannot be helped: I am part of a nation that pronounces Featherstone-Haugh as Fanshor. I hope such things will be seen as quaint rather than irritating. The jests in the first edition were transcribed from a cuneiform tablet: the ones in this edition are less up-to-date. Dare I whisper that statistics can get a bit dry sometimes. Of course, there is always a risk in trying to be both serious and light, but then one should take risks.

I thank Nick Heard and Niall Adams of Imperial College, London, for their extraordinary patience in guiding a particularly obtuse individual through the tortuous process of setting up and using modern computing facilities such as R. In my defence I would point out that I am familiar with some modern technology: for instance, I regularly speak on the electric telephone. However, I confess that I have to get up at 3 A.M. to record a film from the telly (that's TV to our younger readers).

On a personal note, this is probably my last significant statistical undertaking. Hoots mon, 'Wi' a hundred papers a' more, a' more', and half a dozen books, that'll do — I'm off to play my guitar!

Preface to the First Edition

If something can fail, it can often fail in one of several ways and sometimes in more than one way at a time. This phenomenon is amply illustrated by the MOT (Ministry of Transport) testing records of my cars and bikes over the years. One thing that has long puzzled me is just how much work in Survival Analysis has been accomplished without mentioning Competing Risks. There is always some cause of failure, and almost always more than one possible cause. In one sense, then, Survival Analysis is a lost cause.

The origins of competing risks can be traced back to Daniel Bernoulli's attempt in 1760 to disentangle the risks of dying from smallpox from other causes. Seal (1977) and David and Moeschberger (1977, Appendix A) provide historical accounts. Much of the work in the subject since that time has been demographic and actuarial in nature. However, the results of major importance for statistical inference, and applications of the theory in reliability, are quite recent, the two fields themselves being relatively recent.

One problem with writing a book on a specialized topic is that you have to start somewhere. Different readers will have differing levels of prior knowledge. Too much or too little assumed can be disastrous, and only "just about right" will give the book a fighting chance of serious consideration. What all this is sidling up to is the decision to put certain foundational material into preliminary subsections in some chapters rather than exclude it altogether or bury it in appendices. This decision was taken boldly and decisively, after many months of agonized dithering, shunting the stuff to and fro. Its final resting place will no doubt offend some, known as reviewers, but might be seen by others as a sensible choice, I hope. A related choice is to consider throughout only well-defined risks of real, objectively verifiable, events. I leave it to the philosophers and psychologists to debate how many perceived risks can be balanced on a pinhead.

Another problem with writing a book like this is that one has to stop somewhere. My stopping rule was, appropriately, at martingale methods used for the counting-process approach to Survival Analysis: this requires a discrete jump in the level of mathematical technicality. The last chapter gives a brief introduction, but then the reader who is inspired, and has not expired, is referred to weightier tomes to boldly go where this book has not gone before—hence, the *Classical* in the title of this book.

In short, the target level is for readers with formal statistical training, experience in applied statistics, and familiarity with probability distributions, hazard functions, and Survival Analysis. Competing Risks is a specialized, but widely applicable, topic for the statistician. The book is probably not very suitable for scientists, engineers, biologists, and others, to dip into for a quick fix, nor for pure mathematicians interested only in technical theory.

The plan of the book, besides making a bit on the side for a poor, underpaid, overworked university lecturer, is as follows: In Chapter 1, the bare bones of the Competing Risks setup are laid out, so to speak, and the associated likelihood functions are presented in Chapter 2. Chapter 3 describes the approach to Competing Risks via a joint distribution of latent failure times. Chapter 4 covers more modern approaches to Survival Analysis via hazard functions as a preparation for the extension of these methods to Competing Risks. Chapter 5 carries this work on to discrete failure times in Competing Risks, and Chapter 6 deals likewise with the continuous case. Chapter 7 reports some blockages along the traditional modelling route via latent failure times, in particular, difficulties of identifiability. Chapter 8 gives a brief introduction to the counting-process approach and the associated martingale theory.

Throughout, an attempt has been made to give credit where due for all results, examples, and applications, where known. I apologize if I have missed any innovators in this. The results are often not stated or proved in exactly the original way. This is not claimed to be an improvement on the originals; it just arises through trying to give a coherent, self-contained, sequential account of the subject. Nevertheless, there is the odd refinement: one can often strike more effectively after others have lit up the target. I have been influenced by many authors, though I hope that there is not too much bare-faced plagiarization here, at least not of the sort that would stand up in court. Standard works on which I have drawn include, in chronological order, the books by David and Moeschberger (1978), Kalbfleisch and Prentice (1980), Lawless

(1982), Nelson (1982), and Cox and Oakes (1984); and the papers by Gail (1975) and Elandt-Johnson (1976). Equally excellent, more recent books that should be mentioned are those by Fleming and Harrington (1991), Andersen et al. (1993), Collett (1994), and Bedford and Cooke (2001).

My own involvement in the subject arose from being presented with engineering data and not liking to admit in public that I did not have a clue what to do with it. Eventually, I found out and managed to use the methods in several diverse applications, in both engineering and medical contexts. However, real data always seem to have a sting in the tail in the sense that they are never quite straightforward enough to provide an effective illustration in a textbook. Thus, the experienced statistician will find that the examples here are presented rather cursorily. However, the experienced statistician will also understand that in this way the core subject can be focussed upon without being drawn into side issues, important as they might be in the application context. What this means is that only the central theory and methodology is covered, leaving the practitioner to extend the treatment as necessary for his particular problems. But then, that is what makes the job of statisticians interesting and why we should resist our replacement by computer packages!

Martin John Crowder
Imperial College, London

Part I

Univariate Survival Analysis

To the mighty probabilist, failure times are just positive, or non-negative, random variables, or measurable functions. "Probability is simply a part of measure theory, . . . , and no attempt has been made to sugar-coat this fact," Doob (1953). (Perhaps Doob had been reading Dante: "Abandon all hope ye who enter here.") To the rest of us, humble statisticians, failure times give rise to the kind of data for which there has been a daunting development in methodology over the past 20 or 30 years, the result being a bewildering variety of techniques many of which are pretty hard to come to grips with at the technical level.

In this first part, the basic topic is described in what is now the standard framework. It serves as a brief introduction to the subject, but its main purpose is to set the scene for the subsequent parts. Once the ideas and methods are set out in the simpler framework it is easier to extend them to the more involved situations.

1

Scenario

This first chapter provides an introduction to the kind of data under the microscope, some examples, and some rudiments of processing and simulating such data with R.

1.1 Survival Data

1.1.1 Waiting Times

Survival Analysis is traditionally concerned with the waiting time to the occurrence of a particular event. This might be a terminal incident like death, as the name suggests, but more generally the event can be any well-defined circumstance. For example, the event can be some life-changing occurrence, such as cure from a disease, or winning the hand of the object of one's desire, or finally convincing the editor that one's paper should be accepted. Again, there is a whole parallel universe called *reliability* in which times to breakdown or repair of machines and systems is the subject of study. There are also numerous other applications including stimulus-response times, and performance times in sports and system operation. Response times, in particular, can have wide variation, for example, ranging from (a) a sprinter leaping into action after the starter's gun to (b) a politician asked to apologise for wrecking the economy, education, employment, justice system, and so forth. The old saying, "You can wait till the cows come home" might apply to the latter.

The methods of Survival Analysis can be profitably applied to situations in which *time* is not time at all. For example, in studying construction materials it might be the breaking strength of a concrete block: one can think of the stress being increased steadily until the *time* of fracture. Again, it could be the level attained in some trial: imagine hanging in there as the test proceeds until you are forced to concede defeat. What such measurements do have in common is that they are positive, or at least non-negative, barring ingeniously contrived cases. Also, their values tend to have positively skewed distributions with long upper tails.

1.1.2 Discrete Time

In most conventional applications of survival analysis the times are treated as being continuous, that is, measured on a continuous scale, and almost always as being positive or, at least, non-negative. This reflects a belief that time itself proceeds continuously, though not all physicists are convinced of this nowadays, apparently. Such philosophical puzzles need not detain us here—they just lead to sleepless nights. On the other hand, when one considers carefully the definition of *failure time*, there is plenty of room for consideration of discrete time.

Consider a system in continuous operation that is inspected periodically, manually or automatically. Suppose that, if one or more components are classified as effectively life-expired, the system is taken out of service for renewal or replacement. Then, discounting the possibility of breakdown between inspections, the lifetime of a component is the number of inspections to life expiration. Some systems have to respond to unpredictable demands, a typical task for standby units. A similar situation occurs when a continuously operating system is subject to shocks, peak stresses, loads, or demands. Such stresses can be regular, such as those caused by tides or weekly demand cycles, or irregular, as with storms and earthquakes. The lifetime for such a system is the number of shocks until failure.

1.1.3 Censoring

In Survival Analysis it is almost obligatory to have to deal with *censoring*. The most common case is when the waiting time is *right-censored*. If this occurs at time t, it means that the actual time T is only known to exceed t, that is, the extent of the information is that $T > t$. Such would arise, for example, with a machine still functioning at time t when observation ceased at 5.30 p.m. on Friday, or with an ex-offender still known to be on the right side of the law at time t when probation monitoring finished. Another common example is with patients under observation after some treatment: they could be *lost to follow-up* if they moved away without leaving contact details and were consequently and subsequently lost sight of ("Mum, why do we always move at night?). Yet another situation is where a failure occurs for a reason different to the one under study. So, a patient might die of some condition other than the one being treated. In that case the analysis can be complicated by possible association between the different causes of failure—this is the subject of *competing risks*; see Part III.

Less common is *left-censoring*, when it is only known that failure occurred some time before observation was made. So, the game starts, then you need the loo, and when you return your mates are cheering because your team scored while you were indisposed. (Of course, you would probably have heard the roar of the crowd from your end of the stadium, but please overlook that slight flaw in the example. There is the legend of a supporter whose friends would plead with him to take a break; such was the observed association.) In medicine a left-censored time would result where the patient returns for a scheduled check-up, which then indicates that the disease had recurred sometime previously. Formally, the information is just that $T < t$, where t is observed.

Interval censoring means that the waiting time T is not recorded precisely but can be bracketed, say between times t_1 and t_2. For example, the system might have crashed overnight, between leaving it running in the evening and returning to look at it the following morning. Again, the patient's check-up at three months after treatment might show all clear, but then at six months indicate a recurrence of the disease sometime in between. Even recordings that are nominally exact might be just to the nearest minute or nearest day. It can be argued that virtually all survival times are interval censored, being rounded by the limited accuracy of the recording instrument. However, times are most often treated as exactly equal to the recorded value.

On the other hand, the starting time can also be subject to uncertainty. Switching on a machine is clear enough, but what about patients with chronic conditions? When someone is referred for medical treatment, and subsequently followed up, various initial times can be considered for the final analysis. Should it be when symptoms first appeared, relying on the memory of the patient? Should it be when the doctor was first consulted? Or when treatment started? Or when treatment finished? Such variations of definition can obviously have a significant effect on the analysis. It might be better to recognise that there is often a series of events, producing several inter-event times. Such multivariate data is the topic of Part II.

1.2 Some Small Data Sets

1.2.1 Strengths of Cords

Crowder et al. (1991) gave this data set, shown here in Table 1.1, as Example 2.3. The figures are breaking strengths of parachute rigging lines after certain treament. This is of some interest to those too impatient to wait for the aeroplane to land. There are 48 observations, of which the last 7 are right censored, indicated by adding a + sign.

The figures here just form a random sample from some distribution, possibly the least-structured form of data and more common in textbooks than

TABLE 1.1

Strengths of Cords

36.3	41.7	43.9	49.9	50.1	50.8	51.9	52.1	52.3	52.3
52.4	52.6	52.7	53.1	53.6	53.6	53.9	53.9	54.1	54.6
54.8	54.8	55.1	55.4	55.9	56.0	56.1	56.5	56.9	57.1
57.1	57.3	57.7	57.8	58.1	58.9	59.0	59.1	59.6	60.4
60.7	26.8+	29.6+	33.4+	35.0+	40.0+	41.9+	42.5+		

Source: Crowder, M.J. et al., 1991, *Statistical Analyses of Reliability Data*, Chapman & Hall, London. Reproduced by permission of Chapman & Hall/CRC Press.

TABLE 1.2

Cancer Survival

Time	Group a	Group b	Group c	Group e
0–6	109	98	191	+
6–12	110	83	205	232
12–18	40	42	145	+
18–24	27	20	80	156
24–36	21	22	123	65
36–48	13	9	69	51
48–60	6	11	36	25
60–84	5	6	46	33
84–108	4	2	24	11
>108	1	2	20	3

Source: Boag, J.W., 1949, Maximum Likelihood Estimates of the Proportion
of Patients Cured by Cancer Therapy, *J. R. Statist. Soc.*, B11, 15–44.
Reproduced by permission of Wiley-Blackwell Publishers.

practice. Nevertheless, models can be fitted and assessed and, on the odd
occasion, useful inferences made.

1.2.2 Cancer Survival

Boag (1949) listed the data in his Table II, given in Table 1.2. The groups refer
to different types of cancer, different treatments, and different hospitals. There
were eight groups, listed as *a* to *h* in his Figure 1, but only four appear in the
table. The first column gives survival time in months: the data are grouped
into six-monthly intervals until three years, after which intervals become
wider. In group *e* the count 232 spans interval 0–12 months, and 156 spans
12–24 months, both indicate by a +.

Boag was interested in comparing the fits of lognormal and exponential distributions. He computed expected frequencies to set against those observed,
and found that chi-square tests accepted the lognormal and rejected the exponential for groups *a*, *b*, and *c* but not *e* (for which the opposite was obtained).

1.2.3 Catheter Infection

Collett (2003, Table 4.1) presented some data on 13 kidney dialysis patients,
each of whom had a catheter inserted to remove waste products from the
blood. The original data were used by McGilchrist and Aisbett (1991) to illustrate regression with frailty. If an infection occurs at the entry site, the catheter
has to be removed and the area disinfected. The survival time is the number
of days until a first infection occurs; pre-infection removal of the catheter for
some other reason produces a right-censored time. Among the other variables
recorded were age (years) and sex (1 = male, 2 = female). Collett fitted a Cox
proportional hazards model (Section 5.2) and found that sex, but not age, was
a significant factor; one can only speculate. He then went on to illustrate the
computation and interpretation of various types of residuals.

TABLE 1.3

Catheter Infection

Patient	Age	Sex	tim	cns	Patient	Age	Sex	tim	cns	Patient	Age	Sex	tim	cns
1	23	2	22	0	10	64	1	28	0	19	54	2	11	0
2	21	1	9	1	11	80	1	28	0	20	59	2	28	0
3	34	2	28	0	12	18	2	9	1	21	37	1	23	1
4	55	2	16	1	13	79	2	17	1	22	29	1	28	0
5	24	2	11	1	14	51	2	22	1	23	36	2	3	1
6	71	2	28	0	15	72	1	17	1	24	29	2	4	0
7	78	2	28	0	16	59	1	11	1	25	59	1	18	1
8	19	1	13	1	17	62	2	28	0	26	51	1	28	0
9	38	1	28	0	18	45	1	28	0	27	62	1	10	1

Table 1.3 gives some artificial data on 27 patients of the same general type as Collett's. Here, *tim* is time and *cns* is the censoring indicator (0 for a right-censored time, 1 for an observed time); observation on each patient was terminated at 28 days, so *tim* = 28 entails *cns* = 0. The data will be used for illustration below.

1.3 Inspecting the Data with R

This section and the ones following are for R-novices. If you are among the large number of statisticians more experienced than I am with R, go directly to Chapter 2; do not pass GO; do not collect £200.

I will assume that you have R set up on your computer. Otherwise, and if you do not know how to download it and set it up, phone a friend—I did. If you, like me, grew up in the days before personal computers, when man first stood erect and started to use tools, you will probably need to be guided through abstruse concepts such as working directories. Everitt and Hothorn (2010) tell you how to do it all in plain English that even I can understand. Venables and Ripley (1999) is also highly recommended—when all else fails, read the instructions! (Mrs. Crowder once forced me to stop the car, after driving round in circles for an hour, and ask for directions.)

Incidentally, to perform various data analyses throughout this book, functions have been coded in R; they are available on the Web site referred to in the Preface. There is certainly no claim that they are superior to ones available elsewhere, in the CRAN collection, for instance. But it is often quicker to write your own function than to spend hours searching lists of packages for one that does the job you want. What writing your own code does do, too, is to force you to get to grips with the particular technique better. It also enables you to arrange things as you want them to be arranged. Mainly, it is good practice for tackling data for which there is no off-the-shelf software. How many inappropriate statistical analyses are performed simply because there's a readily available program that does it?

For illustration let us apply R to the catheter data (Table 1.3). First, the data should be set up in a plain text file, say `catheter1.dat`. The standard format is

```
id age sex tim cns
1 23 2 22 0
2 21 1 9 1
...
27 62 1 10 1
```

The data file has a `header` row at the top, giving names to the columns, and then 27 rows of figures. The file must occupy the current *working directory*, as defined in your R setup. Now the data must be loaded into R: in the R-window type

```
dmx=read.table('catheter1.dat',header=T); attach(dmx); dmx; #input and check data
```

The option `header=T` (T means True) indicates that there is a `header` in the data file (use `header=F` if not). The 27 × 5 data matrix will now be stored as dmx: this is created as a *list* variable. (In a moment of weakness I did once look it up in the manual, which has a whole chapter on lists and data frames, but too long to actually read.) The command `attach(dmx)` makes the columns accessible for further processing, for example, `age` is now a numerical vector of length 27. Sometimes you need to force a list to become numeric: this can be done with `dmx=as.numeric(unlist(dmx))`. The # symbol indicates a comment: the rest of the line is ignored by the processor. The semicolon separates commands on the same line: some users prefer to have a new line for each command.

Now try some R commands: type the following, one at a time (pressing the Enter key after each), and see what you get:

```
age; mean(age); avage=mean(age); avage; var(age); summary(dmx);
agf=sort(age); agf; agf[1]; hist(age); plot(age,tim); pairs(dmx);
```

Try variations to see what works and what does not. Incidentally, I just use = in R commands rather than <-   because (a) I am more used to it, (b) it is easier to type, and (c) I cannot rid myself of the feeling that y<-x means that y is less than $-x$. If you come across an unfamilar function, such as y=wotsthisdo(x), look it up online by typing help(wotsthisdo). Many more functions can be found in the online manual and in Venables and Ripley (1999).

You will soon decide to type your commands into a text file and just paste them into the R window: this can save a lot of frustration in retyping to correct minor errors. Throughout this book I will use R for data processing. My listings of R-code are basic and without frills, reflecting my own level in competence. Aficionados will spot slicker ways of doing things. However, there is a case for transparency, hoping to keep down the number of mistakes. One small tip that might come in handy is as follows: after the customary cursing of the computer for daring to produce errors with your code, close the R-window and start again, sometimes previous assignations can corrupt the current run.

1.4 Fitting Models with R

After the data have been inspected comes the next step of analysis, interpretation, inference, drawing conclusions, shaking the data down, or whatever phrase is in current vogue. This rests upon fitting models, fully parametric, semi-parametric, or non-parametric. The first example of fitting a fully parametric model occurs here in Section 3.2, that of a non-parametric model in Section 4.1, and that of a semi-parametric one in Section 4.2. We will make use of the gold-standard R-package `survival`, which includes routines to perform a variety of tasks as well as some data sets to illustrate them. A high priority for the budding survival analyst is to learn how to use `survival` and other packages listed on the CRAN Web site.

In addition to the software available in `survival`, some homegrown programs are used. These are either to make the computations more transparent or to fill minor gaps in the available software. The first such instance occurs in Section 3.2. The R-code used in this book, together with the data sets not subject to copyright restriction, is available on the Web site referred to in the Preface.

1.5 Simulating Data with R

Simulation is a powerful tool in statistics. Many modern techniques using simulation have been developed on the back of fast computing. In this section an introductory exercise in simulating data will be described. It is often useful in assessing the performance of proposed methodology to be able to test its performance on data whose structure is known and controlled. It is a particularly powerful approach in situations where the framework is straightforward to set up but the consequences of interest are analytically intractable.

Let us generate some right-censored survival times whose mean depends on some recorded factors. Sophisticated usage of R is not the point here—just a demonstration of some basic commands. Suppose that T_i, the breakdown time of the ith machine, has an exponential distribution with mean a given function of x_1, a measure of the intensity of usage, and x_2, a measure of the frequency and quality of maintenance. (No prizes for guessing that I have my car in mind here—see the observation with the smallest x_2.) Some basic R code to achieve this is as follows:

```
n=25; b0=1.5; b1=1.2; b2=-2.5; #n= sample size, (b0,b1,b2)= regression coeffs
x1=rep(0,n); x2=x1; tim=x1; #initialise x1,x2,tim as vectors of 0s of length n
for(i in 1:n) { x1[i]=runif(1,min=0,max=1); x2[i]=runif(1);
    lam=exp(b0+b1*x1[i]+b2*x2[i]); ti=rexp(1,rate=lam); tim[i]=min(ti,10); }
for(i in 1:n) {v1=c(x1[i],x2[i],tim[i]); v2=format(v1,width=8,digits=2);
  cat('\n',v2);}
```

Note the exciting variety of brackets: (. . .) to enclose the arguments of a function, [. . .] for indices of a vector or matrix, and {. . .} to group sets of instructions in a `for` loop. The `for` loop runs through the sample elements one by one. The functions `runif` and `rexp` generate samples from uniform and exponential distributions, respectively: look them up, using `help(runif)` and `help(rexp)`, for their full capabilities. For example, `x1=runif(n)` outside the `for` loop would have had the same effect. The function `c(...)` concatenates (look that word up too), for example, `a=c(b,c)` puts b and c together into a, and `cat` prints stuff out. The model here for the mean breakdown time, λ_i^{-1}, is of log-linear form: $\log \lambda_i = b_0 + b_1 x_{i1} + b_2 x_{i2}$, and the signs of the regression coefficients, b_1 and b_2, are meant to reflect the expected effects of x_1 and x_2. The times are right-censored at value 10. The data, printed out, can be copied and pasted into a data file.

2

Survival Distributions

We now get down to developing the mathematical infrastructure for ana-
lysing data of the type described in Chapter 1. The various functions that
need to be manipulated are defined and their application is illustrated.

2.1 Continuous Lifetimes

Let T be the *random variable* representing the lifetime under study. The *distribution function* F and the *survivor function* $\bar{F}$ of T are defined by the probabilities

$$F(t) = \mathrm{P}(T \leq t), \quad \bar{F}(t) = \mathrm{P}(T > t),$$

so $F(t) + \bar{F}(t) = 1$ for all t. Note that $F(t)$ is an increasing, and $\bar{F}(t)$ a decreasing, function of t; normally, $F(t)$ will rise from 0 to 1, and $\bar{F}(t)$ will fall
from 1 to 0, over the range of t. When T is essentially positive, as it is in most
applications, $F(0) = 0$ and $\bar{F}(0) = 1$. The *density function* is defined as

$$f(t) = dF(t)/dt = -d\bar{F}(t)/dt;$$

correspondingly,

$$F(t) = \int_0^t f(s)ds, \quad \bar{F}(t) = \int_t^\infty f(s)ds.$$

(Unless otherwise stated, it will be tacitly assumed that continuous survival
distributions have densities, that is, that they are *absolutely continuous*.)

Modern survival analysis is mostly based around hazard functions. (This
has nothing to do with the over-zealous health-and-safety culture that blights
our lives nowadays.) These functions are concerned with the probability of
imminent failure, that is, that, having got this far, you will get no further. The
formal definition of the *hazard function* h of T is

$$h(t) = \lim_{\delta \downarrow 0} \delta^{-1}\mathrm{P}(T \leq t + \delta \mid T > t).$$

The right-hand side is equal to

$$\lim_{\delta \downarrow 0} \delta^{-1} P(t < T \le t + \delta)/P(T > t) = \lim_{\delta \downarrow 0} \delta^{-1}\{\bar{F}(t) - \bar{F}(t + \delta)\}/\bar{F}(t)$$

$$= -\{d\bar{F}(t)/dt\}/\bar{F}(t) = -d\log\bar{F}(t)/dt.$$

In different contexts $h(t)$ is variously known as the *instantaneous failure rate, age-specific failure rate, age-specific death rate, intensity function*, and *force of mortality* or *decrement*. Integration yields the inverse relationship

$$\bar{F}(t) = \exp\left\{-\int_0^t h(s)ds\right\} = \exp\{-H(t)\},$$

where $H(t)$ is the *integrated hazard function* and the lower limit 0 of the integral is consistent with $\bar{F}(0) = 1$. For a *proper* lifetime distribution, that is, one for which $\bar{F}(\infty) = 0$, $H(t)$ must tend to ∞ as $t \to \infty$.

Both the distribution function F and the hazard function h are concerned with the probability that failure occurs before some given time. The difference is this: with the former, you are stuck at time zero looking ahead to a time maybe a long way into the future (with a telescope); with the latter, you are moving along with time and just looking ahead to the next instant (with a microscope).

2.2 Some Continuous Survival Distributions

A few of the survival distributions in more common use are described here.

2.2.1 The Exponential Distribution

The distribution can be defined by its survivor function $\bar{F}(t) = \exp(-t/\xi)$; the *scale parameter* $\xi > 0$ is actually the mean lifetime $\xi = E(T)$. The primary distinguishing feature of the distribution is arguably its constant hazard function: $h(t) = 1/\xi$. It is the only distribution with such a constant hazard (subject to the usual mathematical provisos). So, as time proceeds, there is no recognition of age: the probability of imminent failure remains at the same level throughout. This makes the distribution both discreditable (in most applications) and compelling (as a prototype model). The *lack-of-memory* property is expressed, if I recall, as

$$P(T > a + b \mid T > a) = P(T > b);$$

your mission, if you decide to accept it, is to prove this—see the Exercises.

2.2.2 The Weibull Distribution

A natural generalisation of the exponential survivor function is $\bar{F}(t) = \exp\{-(t/\xi)^{\nu}\}$, raising t/ξ to a power $\nu > 0$; the exponential distribution is regained when $\nu = 1$. The corresponding hazard function is $h(t) = (\nu/\xi)(t/\xi)^{\nu-1}$, an increasing function of t when the *shape parameter* $\nu > 1$ and decreasing when $\nu < 1$. The mean and variance are expressed in terms of the gamma function (see the Exercises), but with an explicit survivor function, quantiles are more accessible. Thus, the upper qth quantile t_q, for which $P(T > t_q) = q$, is given by $t_q = \xi(-\log q)^{1/\nu}$.

The Weibull distribution is fairly ubiquitous in reliability, even boasting a must-have handbook, commonly referred to as the Weibull Bible. The reasons for this popularity probably have to do with its being an extreme-value distribution with the associated weakest-link interpretation, the variety of shapes that can be accommodated by the hazard function, and the simple form of $\bar{F}(t)$, which facilitates data plotting.

2.2.3 The Pareto Distribution

The Pareto distribution can arise naturally in the survival context as follows. Begin with an exponential distribution for T: $\bar{F}(t) = e^{-t/\xi}$. Suppose that, due to circumstances beyond our control (as they say, whenever I go by train), ξ varies randomly over individual units. Specifically, say $\lambda = 1/\xi$ has a gamma distribution with density

$$f(\lambda) = \Gamma(\gamma)^{-1}\alpha^{\gamma}\lambda^{\gamma-1}e^{-\alpha\lambda}.$$

(This somewhat artificial distributional assumption oils the wheels: a choice always has to be made between hair-shirt realism and mathematical tractability.) Thus, the conditional survivor function of T is $\bar{F}(t \mid \lambda) = e^{-\lambda t}$ and the unconditional one is obtained as

$$\bar{F}(t) = \int_0^{\infty} \bar{F}(t \mid \lambda)\, f(\lambda)\, d\lambda = \Gamma(\gamma)^{-1}\alpha^{\gamma}\int_0^{\infty}\lambda^{\gamma-1}e^{-\lambda(t+\alpha)}\, d\lambda = (1 + t/\alpha)^{-\gamma}.$$

The parameters of this Pareto survivor function are $\alpha > 0$ (scale) and $\gamma > 0$ (shape).

The hazard function, $h(t) = \gamma/(\alpha + t)$, is decreasing in t and tends to zero as $t \to \infty$. Such behaviour is fairly atypical but not altogether unknown in practice. It would be appropriate for units that become less failure-prone with age, ones having ever-decreasing probability of imminent failure, settling in as time goes on. (One might cite as an example humans, who become less gaffe-prone with age—if only that were true!) However, there will certainly be failure at some finite time since $\bar{F}(\infty) = 0$, a reflection of the fact that $h(t)$ here does not decrease fast enough for $\int_0^{\infty} h(s)ds$ to be finite.

2.2.4 Other Distributions

A variety of alternative distributions has been applied in survival analysis and reliability. Essentially, any distribution on $(0, \infty)$ will serve. Thus, one can contemplate distributions such as the gamma, Gompertz, Burr, and inverse Gaussian. And then, when one has finished contemplating them, one can note that distributions on $(-\infty, \infty)$ can be converted, usually via a log-transform: this yields the log-normal and the log-logistic, for example. (And, yes, you can start a sentence with *and*: and one of the best-loved hymns in the English language starts with *And*.) Some properties of these will be set as exercises below.

2.2.5 The Shape of Hazard

For most systems, hazard functions increase with age as the system becomes increasingly prone to crack-up. (No laughing at the back—your professor isn't gaga yet.) The trade description for this is IFR (increasing failure rate). Likewise, DFR stands for decreasing failure rate: this would apply to systems that *wear in* or *learn from experience* (of others, preferably), becoming less liable to failure with age. The bog-standard example is the Weibull hazard, IFR for $v > 1$ and DFR for $v < 1$.

In some cases the legendary *bathtub hazard* shape (down–along–up) crops up. In animal lifetimes this can reflect high infant mortality, followed by a lower and fairly constant hazard, eventually ending up with the ravages of old age. To model such a shape we might simply add together three Weibull possibilities, that is, take the hazard function as

$$h(t) = (v_0/\xi_0) + (v_1/\xi_1)(t/\xi_1)^{v_1-1} + (v_2/\xi_2)(t/\xi_2)^{v_2-1}.$$

The corresponding survivor function is

$$\bar{F}(t) = \exp\left\{-\int_0^t h(s)ds\right\} = \exp\{-t/\xi_0 - (t/\xi_1)^{v_1} - (t/\xi_2)^{v_2}\}$$
$$= \bar{G}_0(t)\,\bar{G}_1(t)\,\bar{G}_2(t),$$

where the $\bar{G}_j$ ($j = 0, 1, 2$) are the survivor functions of the three contributing distributions. The representation in terms of random variables is $T = \min(T_0, T_1, T_2)$, where the T_j are independent with survivor functions $\bar{G}_j$.

In reliability applications the bathtub hazard is relevant for manufactured units when there is relatively high risk of early failure (wear in), a high risk of late failure (wear out), and a lower hazard in between (where it is constant). Unfortunately, in my experience, it is often only the first two characteristics that are apparent in practice. In medical applications the bathtub hazard can reflect treatment with non-negligible operative mortality that is otherwise life preserving; as Boag (1949) put it, it is not the malady but the remedy that can prove fatal. The construction, whereby a specified hazard form is constructed by combining several contributions, is generally applicable, not just confined to the minimum of three independent Weibull variates.

2.3 Discrete Lifetimes

Some systems operate intermittently rather than continuously. The cycles of operation may be regular or irregular, and the system lifetime is the number of cycles until failure. Regular operation is commonly encountered in such areas as manufacturing production runs, cycles of an electrical system, machines producing individual items, and orbiting satellites exposed to solar radiation each time they emerge from the earth's shadow. Another example occurs where certain electrical and structural units on aircraft have to be inspected after each flight: the location of one or more critical faults will lead to failing the system. In such cases, the lifetime, the number of cycles to failure, is a discrete variable.

Let T be a discrete failure time taking possible values $0 = \tau_0 < \tau_1 < \ldots < \tau_m$; m may be finite or infinite, as may τ_m. In many situations it is sufficient to take $\tau_l = l$ ($l = 0, 1, \ldots$), that is, integer-valued failure times. However, when we come on to likelihood functions in later chapters, the generality is necessary to smooth the transition from discrete to continuous time.

The survivor function is $\bar{F}(t) = P(T > t)$ and the corresponding *discrete density function*, or *probability mass function*, is defined as $f(t) = P(T = t)$. They are related by $f(t) = \bar{F}(t-) - \bar{F}(t)$ and $\bar{F}(t) = \sum f(\tau_s)$, where the summation is over $\{s : \tau_s > t\}$. The notation $\bar{F}(t-)$ is a useful abbreviation for $\lim_{\delta \downarrow 0} \bar{F}(t - \delta)$. (For continuous failure times, $\bar{F}(t-) = \bar{F}(t)$.) If t is not equal to one of the τ_s, $f(t) = 0$. Also, we will adopt the convention $\bar{F}(\tau_0) = 1$, that is, $f(0) = 0$, so that zero lifetimes are ruled out. The discrete hazard function is defined as

$$h(t) = P(T = t \mid T \geq t) = f(t)/\bar{F}(t-).$$

(For continuous failure times, the denominator is usually written equivalently as $\bar{F}(t)$.) Note that $0 \leq h(t) \leq 1$ for all t, with $h(t) = 0$ except at the points τ_l ($l = 1, \ldots, m$); also, $h(0) = 0$, and $h(t) = 1$ only at the upper end point τ_m of the distribution, where $\bar{F}(t-) = f(t)$. The inverse relationship, expressing $\bar{F}(t)$ in terms of $h(t)$, can be derived as

$$\bar{F}(t) = \bar{F}(t-) - f(t) = \bar{F}(t-)\{1 - h(t)\} = \prod_{s=1}^{l(t)}\{1 - h(\tau_s)\},$$

where $l(t) = \max\{l : \tau_l \leq t\}$ so that the product is over $\{s : \tau_s \leq t\}$. Also,

$$f(t) = h(t) \prod_{s=1}^{l(t-)}\{1 - h(\tau_s)\}.$$

Otherwise expressed, and writing h_l for $h(\tau_l)$, these representations are

$$\bar{F}(\tau_l) = \prod_{s=1}^{l}(1 - h_s) \text{ for } l \geq 1, \ f(\tau_l) = h_l \prod_{s=1}^{l-1}(1 - h_s) \text{ for } l \geq 2,$$

with $\bar{F}(\tau_0) = 1$, $\bar{F}(\tau_m) = 0$, $f(\tau_0) = 0 = h_0$, and $f(\tau_1) = h_1$. If we interpret $\prod_{s=1}^{0}$ as simply contributing a factor 1, then the product formulae here hold for all l.

The *integrated hazard function* is

$$H(t) = -\log \bar{F}(t) = -\sum_{s=1}^{l(t)} \log(1 - h_s).$$

If the h_s in the summation are small, then $\log(1 - h_s) \approx -h_s$ and $H(t) \approx \sum_{s=1}^{l(t)} h_s$, which can justifiably be called the *cumulative hazard function*.

2.4 Some Discrete Survival Distributions

Continuous distributions can be segmented to form discrete ones on a specified set of times τ_l. Thus, starting with the survivor function $\bar{F}$ of some continuous variate, we take $P(T = \tau_l)$ as $\bar{F}(\tau_{l-1}) - \bar{F}(\tau_l)$. However, there are some "genuine" discrete waiting-time distributions and we now look at a couple (or just one, depending on how you look at it).

2.4.1 The Geometric Distribution

This is just about the simplest discrete waiting-time distribution. It arises as the time to failure in a sequence of Bernoulli trials, that is, independent trials each with probability ρ of success. In the present context the process carries on while it is still winning. The τ_l here are the non-negative integers: $\tau_l = l$ for $l = 0, 1, \ldots, m = \infty$. We have $f(0) = 0$ and, for $l \geq 1$,

$$f(l) = \rho^{l-1}(1 - \rho), \quad \bar{F}(l) = \rho^l, \quad h_l = (1 - \rho).$$

Note that the hazard function is constant, independent of l, as for the exponential among continuous distributions. Further, the famed *lack of memory* property of the exponential is also shared by the geometric:

$$P(T > t + s \mid T > s) = \bar{F}(t + s)/\bar{F}(s) = \rho^t = \bar{F}(t) = P(T > t).$$

Actually, of course, this is really only another way of saying that the hazard function is constant, as can be seen by considering the general identity

$$\bar{F}(t + s)/\bar{F}(s) = \prod_{r=s+1}^{s+t} (1 - h_r):$$

if h_r is independent of r, this expression is equal to $\bar{F}(t)$; conversely, if the expression is equal to $\bar{F}(1)$ for $t = 1$ and every s, then $\bar{F}(1) = 1 - h_{s+1}$, and so h_{s+1} is independent of s.

A standard textbook connection between the geometric and exponential distributions is as follows: Suppose that the continuous time scale is divided into equal intervals of length δ, and that independent Bernoulli trials are performed with probability $\pi = \lambda\delta$ of a failure event within each interval. Let $M = T/\delta$, the number of intervals survived without failure. Then M has the geometric survivor function $P(M > m) = (1 - \pi)^m$ for $m = 1, 2, \ldots$; hence, $P(T/\delta > t/\delta) = (1 - \lambda\delta)^{t/\delta}$. As $\delta \downarrow 0$, $P(T > t) \to e^{-\lambda t}$, that is, an exponential distribution.

A slightly extended version can be based on the discrete survivor function $P(M > m) = (1 - \pi)^{m^\phi}$, with $\phi > 0$; π is the probability of failure on the first trial, and $(1 - \pi)^{m^\phi - (m-1)^\phi}$ is the probability of failure on the mth, given survival that far. This leads to $(1 - \lambda\delta^\phi)^{(t/\delta)^\phi}$ and thence, allowing $\delta \downarrow 0$, to a Weibull distribution with $P(T > t) = \exp(-\lambda t^\phi)$.

2.4.2 The Negative Binomial Distribution

The negative binomial is often introduced as a waiting-time distribution. Let M be the number of independent Bernoulli trials performed up to and including the κth failure, where κ is a given positive integer. The probability that there are $\kappa - 1$ failures in the first $\kappa + m - 1$ trials is given by the binomial expression $\binom{\kappa + m - 1}{\kappa - 1}(1 - \rho)^{\kappa-1}\rho^m$. This is to be multiplied by $1 - \rho$, the probability of failure on the $(\kappa + m)$th trial. Thus,

$$P(M = \kappa + m) = \binom{\kappa + m - 1}{\kappa - 1}\rho^m(1 - \rho)^\kappa \quad (m = 0, 1, \ldots).$$

More generally, the expression can be taken to define a probability distribution on the non-negative integers for any positive real number κ. In the Exercises you are encouraged to verify that these probabilities sum to 1, and also to find out why it is called *negative binomial*.

When κ is an integer, M can be represented as the sum of κ consecutive waiting times to a first failure, that is, as the sum of κ independent geometric variates. Then the limiting process described in the preceding example yields the sum of κ independent exponential variates with rate parameter ρ, that is, a gamma distribution for T with density $\rho^\kappa t^{\kappa-1}e^{-\rho t}/\Gamma(\kappa)$.

2.5 Mixed Discrete-Continuous Survival Distributions

Suppose that T has a mixed discrete–continuous distribution, with atoms of probability at points $0 = \tau_0 < \tau_1 < \tau_2 < \ldots$, together with a density $f^c(t)$ on $(0, \infty)$. An all-too-familiar example is the waiting time in a queue: T is then either zero (rarely) or capped at closing time τ (when the shutter comes down just as you reach the counter), or continuous on $(0, \tau)$. (My wife always seems to beat the queue, although she maintains that when she first met me there was no queue to beat.)

If $\bar{F}$ is continuous at t, $\bar{F}(t-) = \bar{F}(t)$, whereas if $t = \tau_l$, $\bar{F}(\tau_l-) = \bar{F}(\tau_l) + P(T = \tau_l)$. Then,

$$\bar{F}(t) = P(T > t) = \sum_{\tau_l > t} P(T = \tau_l) + \int_t^\infty f^c(s)ds;$$

the density component $f^c(t)$ is defined as $-d\bar{F}(t)/dt$ at points between the τ_l. Now,

$$\bar{F}(t) = \bar{F}(0) \prod_{l=1}^{l(t)} \left[\{\bar{F}(\tau_l)/\bar{F}(\tau_l-)\}\{ \bar{F}(\tau_l-)/\bar{F}(\tau_{l-1})\} \right] \{\bar{F}(t)/\bar{F}(\tau_{l(t)})\},$$

where $l(t) = \max\{l : \tau_l \le t\}$. But $\bar{F}(0) = 1 - h_0$, where $h_0 = P(T = 0)$, and

$$\bar{F}(\tau_l)/\bar{F}(\tau_l-) = P(T > \tau_l)/P(T > \tau_l-) = 1 - P(T \le \tau_l \mid T > \tau_l-) = 1 - h_l,$$

say. Also,

$$\bar{F}(\tau_l-)/\bar{F}(\tau_{l-1}) = \exp\left\{ -\int_{\tau_{l-1}}^{\tau_l-} h^c(s)ds \right\},$$

where

$$h^c(s) = f^c(s)/\bar{F}(s) = -d \log \bar{F}(s)/ds.$$

The h_l are the discrete hazard contributions at the discontinuities, and h^c is the continuous component of the hazard function. Last,

$$\bar{F}(t)/\bar{F}(\tau_{l(t)}) = \exp\left\{ -\int_{\tau_{l(t)}}^{t} h^c(s)ds \right\},$$

which equals 1 if $t = \tau_{l(t)}$. Substituting into the expression given for $\bar{F}(t)$ a few lines above yields the well-known formula (e.g., Cox, 1972, Section 1):

$$\bar{F}(t) = \left\{ \prod_{s=1}^{l(t)} (1 - h_s) \right\} \exp\{- \int_0^t h^c(s)ds\}.$$

2.5.1   From Discrete to Continuous

Consider now a purely discrete distribution in which the density component is absent, so that

$$h_l = P(\tau_{l-1} < T \le \tau_l \mid T > \tau_{l-1}).$$

Let $\delta_l = \tau_l - \tau_{l-1}$ and $g(\tau_l) = h_l/\delta_l$, so that, in the limit $\delta_l \downarrow 0$, $g(\tau_l)$ is defined as the hazard function at τ_l of a continuous variate. Now,

$$\log \bar{F}(t) = \log \prod_{\tau_l \leq t}(1 - h_l) = \sum_{\tau_l \leq t} \log\{1 - g(\tau_l)\delta_l\}$$

$$= -\sum_{\tau_l \leq t} g(\tau_l)\delta_l + O\left\{\sum_{\tau_l \leq t} g(\tau_l)^2 \delta_l^2\right\}.$$

In the limit $\max(\delta_l) \to 0$ we obtain

$$\bar{F}(t) = \exp\left\{-\int_0^t g(s)ds\right\}.$$

This illustrates the transition from an increasingly dense discrete distribution to a continuous one. This is just an informal sketch of material dealt with in much greater depth by Gill and Johansen (1990, Section 4.1). The reverse transition, obtained by dividing up the continuous time scale into intervals (τ_{l-1}, τ_l), is accomplished simply by defining

$$h_l = 1 - \exp\left\{-\int_{\tau_{l-1}}^{\tau_l} g(s)ds\right\}.$$

Then,

$$\bar{F}(\tau_k) = \exp\left\{-\int_0^{\tau_k} g(s)ds\right\} = \exp\left\{-\sum_{l=1}^k \int_{\tau_{l-1}}^{\tau_l} g(s)ds\right\} = \prod_{s=1}^k(1 - h_s).$$

2.5.2 Rieman–Stieltjes Integrals

We describe here a convenient notation, which can be used for discrete, continuous, and mixed distributions alike. Suppose first, that T is continuous with distribution function $F(t) = P(T \leq t)$. Its mean is then calculated as $E(T) = \int_0^\infty t f(t)dt$, where f is its density function. But $f(t) = dF(t)/dt$, so we can write $E(T) = \int_0^\infty t \, dF(t)$. Now suppose that T is discrete, taking values t_j with probabilities p_j ($j = 1, 2, \ldots$): then $E(T) = \sum_j t_j p_j$. But $dF(t) = F(t+dt) - F(t)$, so $dF(t) = 0$ if the interval $(t, t+dt]$ does not include one of the t_j, and $dF(t) = p_j$ if $t_j \in (t, t+dt]$. In that case, $\int_0^\infty t \, dF(t)$ reduces to $\sum_j t_j p_j$ since $dF(t)$ is only non-zero at the t_j. In either case, continuous or discrete, and also when T has a mixed discrete–continuous distribution, the form $\int_0^\infty t \, dF(t)$ serves to define $E(T)$. More generally, we can define $\int_0^\infty g(t)dF(t)$ in the same way, where g is some function of t. This style of integral is known as *Rieman–Stieltjes*. (Of course, there is a more formal argument for all this, but here is not the place to be pedantic. It is sufficient that g be continuous and F of bounded variation—look it up if you feel the need.)

2.6 Reliability Topics

We give a brief introduction here to some areas of application in Reliability. Later, in Part II, some of the material will be covered in more detail.

2.6.1 Reliability of Systems

Consider a system, comprising components $c_1, \ldots, c_n$, and let p_j be the reliability of component c_j, that is, its probability of being *up* (in working order). We wish to calculate the reliability R of the whole system from the component reliabilities. It is assumed throughout that the components operate independently.

Example

1. Series system: this is shown in Figure 2.1a. Since the system can operate only if both components are up, the reliability is

$$R = \mathrm{P}(c_1 \text{ up and } c_2 \text{ up}) = \mathrm{P}(c_1 \text{ up}) \times \mathrm{P}(c_2 \text{ up}) = p_1 p_2,$$

   using the independence of c_1 and c_2.

2. Parallel system: this is shown in Figure 2.1b. The system can operate if either c_1 or c_2 is up, so

$$R = 1 - \mathrm{P}(c_1 \text{ down and } c_2 \text{ down})$$
$$= 1 - \mathrm{P}(c_1 \text{ down}) \times \mathrm{P}(c_2 \text{ down}) = 1 - (1 - p_1)(1 - p_2).$$

This can easily be extended to n components: for a series system $R = p_1 p_2 \ldots p_n$, and for a parallel system $R = 1 - (1 - p_1)(1 - p_2) \ldots (1 - p_n)$. Further, *components* can be replaced by subsystems in more complex systems and networks.

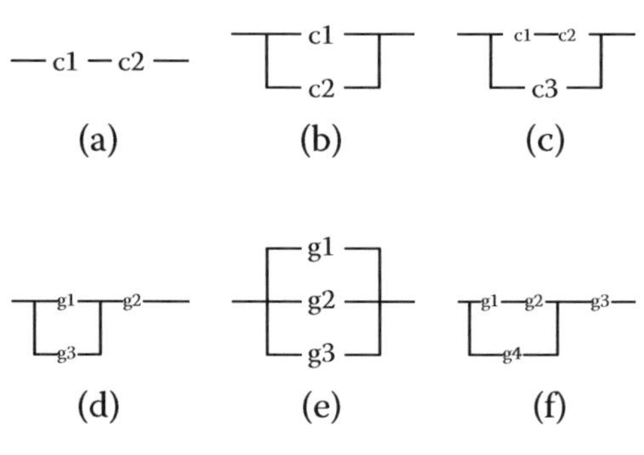

| (a) | (b) | (c) |
| (d) | (e) | (f) |

FIGURE 2.1
Some systems of components.

Example

The system is shown in Figure 2.1c. The subsystems (c_1, c_2) and c_3 here are in parallel, so $R = 1 - (1 - R_{12})(1 - R_3)$, where $R_{12} = p_1 p_2$ for subsystem (c_1, c_2) and $R_3 = p_3$ for subsystem c_3.

2.6.2 k-out-of-n Systems

This title refers to a system containing a certain type of redundancy, namely, that it can continue to operate as long as any k of its n components are up.

Example

For a 2-out-of-3 system,

$$R = \text{P}(c_1, c_2 \text{ up, } c_3 \text{ down}) + \text{P}(c_1, c_3 \text{ up, } c_2 \text{ down}) + \text{P}(c_2, c_3 \text{ up, } c_1 \text{ down})$$
$$+ \text{P}(c_1, c_2, c_3 \text{ up})$$
$$= p_1 p_2 (1 - p_3) + p_1 p_3 (1 - p_2) + p_2 p_3 (1 - p_1) + p_1 p_2 p_3.$$

2.6.3 Survival Aspect

In the foregoing there has been no specific mention of failure times, which is what this book is supposed to be about. But this is easily rectified—just interpret p_j as $\text{P}(T_j > t)$, where T_j is the lifetime of component j. Then the equations give $R(t)$, the system *reliability function* (also known as the *survivor function*) in terms of the component reliability functions.

2.6.4 Degradation Processes

Suppose that D_t measures the amount of wear or deterioration in a system at time t and that, as time goes on, D_t increases monotonically until it reaches some critical level, say d^*, at which point the system fails. The probability distribution of the time T to failure depends on that of the D_t-curve via the fundamental relation

$$\text{P}(T > t) = \text{P}(D_t < d^*).$$

Example

Suppose that the D_t-curve is linear, starting at $(0,0)$ with slope b, that is, $D_t = bt$, and that b varies across systems with a Weibull distribution: $\text{P}(b > x) = \exp\{-(x/\xi)^v\}$ for $x > 0$. Then,

$$\text{P}(T > t) = \text{P}(D_t < d^*) = \text{P}(b < d^*/t) = 1 - \exp[-\{d^*/(t\xi)\}^v].$$

Thus, T has a *Frechet distribution*, which is also known as a *Type-II extreme-value distribution*.

2.6.5 Stress and Strength

Suppose that a system (electronic, mechanical, chemical, biological) has strength X on some scale that measures its resistance to failure. The system will have been constructed or assembled to some nominal strength specification, say μ_X; but in practice, X will be randomly distributed around μ_X. Once in operation, suppose that the system will be exposed to stress Y randomly distributed around some level μ_Y. The system will fail if $X < Y$, and we are interested in the probability of this event.

Let $f(x, y)$ be the joint density of X and Y, then

$$P(failure) = P(X < Y) = \int_{x<y} f(x, y)dxdy = \int_0^\infty dx \int_x^\infty dy\{f(x, y)\}.$$

If, as is usual, X and Y are independent, $f(x, y) = f_X(x)f_Y(y)$, and then

$$P(X < Y) = \int_0^\infty dx \int_x^\infty dy\{f_X(x)f_Y(y)\} = \int_0^\infty f_X(x)\{1 - F_Y(x)\}dx,$$

where F_Y is the distribution function of Y. Equivalently,

$$P(X < Y) = \int_0^\infty dy \int_0^y dx\{f(x, y)\} = \int_0^\infty f_Y(y)F_X(y)dy.$$

In the survival version of the situation, X and Y may vary over time, and then we have $X(t)$ and $Y(t)$. The system will fail at time s if $X(t) \geq Y(t)$ for $0 \leq t < s$, and $X(s) < Y(s)$. The forms taken by $X(t)$ and $Y(t)$ can vary widely between different applications; for example, $X(t)$ might be constant or steadily decreasing over time, and $Y(t)$ might be constant, steadily increasing, randomly fluctuating, or result from intermittent shocks to the system. The analysis in such cases can be quite difficult: T is the first time at which the *stochastic process* $X(t) - Y(t)$ becomes negative.

2.7 Exercises

2.7.1 Survival Distributions

1. Derive the density $f(t) = \xi^{-1}e^{-t/\xi}$ of the exponential distribution and show that its mean is ξ and that its variance is ξ^2. Prove the *lack-of-memory* property. What is the distribution of $(T/\xi)^v$ for $v > 0$? What about $v < 0$?

2. Derive the mean and variance of the Weibull distribution. Verify that the upper qth quantile is $t_q = \xi(-\log q)^{1/v}$.

3. Suppose that Y has distribution $N(\mu, \sigma^2)$ and that $T = e^Y$: then T has a log-normal distribution. Confirm that $E(T^r) = e^{r\mu + r\sigma^2/2}$. Derive the mean and variance of T as $\mu_T = E(T) = e^{\mu + \sigma^2/2}$ and

$\sigma_T^2 = \mathrm{var}(T) = (e^{\sigma^2} - 1)\mu_T^2$. Note that the survivor and hazard functions are not explicit, only expressible in terms of Φ, the standard normal distribution function.

4. Another generalisation of the exponential distribution is the gamma, which has density $f(t) = \Gamma(v)^{-1}\xi^{-v}t^{v-1}e^{t/\xi}$. The parameters are $\xi > 0$ (scale) and $v > 0$ (shape); the exponential is recovered with $v = 1$. Derive the survivor function as $\bar{F}(t) = 1 - \Gamma^{(r)}(v; t/\xi)$, where $\Gamma^{(r)}(v; z) = \Gamma(v)^{-1}\int_0^z y^{v-1}e^{-y}dy$ is the *incomplete gamma function ratio*.

5. Calculate the mean, variance, and upper $100q\%$ quantile of the Pareto distribution. Take care regarding the size of γ.

6. As a generalisation of the Pareto distribution, step up the Burr: this has survivor function $\bar{F}(t) = \{1 + (t/\alpha)^\rho\}^{-\gamma}$, just replacing t/α by $(t/\alpha)^\rho$ with $\rho > 0$. Derive its hazard function: is it IFR, DFR, or what? You might suspect that, by analogy with the derivation of the Pareto given above, the Burr can be mocked up as some sort of Weibull–gamma combination. Can it? The special case $\gamma = 1$ gives the log-logistic distribution.

7. Prove the following:
   a. If $\int_0^\infty h(t)\, dt < \infty$, then $P(T = \infty) > 0$.
   b. If $h(t) = h_1(t) + h_2(t)$, where $h_1(t)$ and $h_2(t)$ are the hazard functions of independent failure time variates T_1 and T_2, then T has the same distribution as $\min(T_1, T_2)$.

8. Calculate $P(T > t_0)$ and $P(T > 2t_0)$ when T has hazard function $h(t) = a$ for $0 < t < t_0$, $h(t) = b$ for $t \geq t_0$.

9. Show that, for continuous T, $E(T) = \int_0^\infty \bar{F}(t)dt$, provided that $t\bar{F}(t) \to 0$ as $t \to \infty$. For discrete T, taking values $0, 1, \ldots$ with probabilities $p_0, p_1, \ldots$, let $q_j = P(T > j)$: show that $E(T) = \sum_{j=1}^\infty q_j$.

10. Negative binomial distribution: verify that the probabilities $p_r = \binom{\kappa + r - 1}{\kappa - 1} \rho^r (1 - \rho)^\kappa$ $(r = 0, 1, 2, \ldots)$ sum to 1.

11. Sometimes survival data are reduced to binary outcomes. Thus, all that is recorded is whether $T > t^*$ or not, where t^* is some threshold, for example, five-year survival after cancer treatment. For Weibull lifetimes $p^* = P(T > t^*) = \exp\{-(t^*/\xi)^v\}$, so $\log(-\log p^*) = v \log t^* - v \log \xi$. A log-linear regression model for ξ then gives a *complementary log–log* form for p^*: $\log(-\log p^*) = \beta_0 + \mathbf{x}^T\beta$. What lifetime distribution corresponds to a logit-linear model for binary data?

2.7.2 Reliability of Systems

1. A system has n components in parallel, that is, it functions as long as at least one component functions. If component j functions with probability p_j, independently of the other components, what is the probability that the system functions?

2. Diagrams (d), (e), and (f) in Figure 2.1 show paths through a network with independently functioning gates. The probability that the jth gate gj will be open is p_j. For each diagram, calculate the probability of an open path through.

2.7.3 Degradation Processes

1. Calculate $P(T \leq t)$ when the degradation curve D_t has the following forms, d^* being the critical failure level:

   a. $D_t = bt$, where b has a Pareto distribution with $P(b > x) = (1 + x/\xi)^{-\nu}$ on $(0, \infty)$;

   b. $D_t = e^{bt} - 1$, where b has an exponential distribution with $P(b > x) = e^{-\lambda x}$ on $(0, \infty)$.

2. Suppose that D_t follows a logistic curve, $D_t = a(1 - e^{-bt})/(1 + e^{-bt})$, where a is exponentially distributed with $P(a > x) = e^{-\lambda x}$ and b is a fixed constant. Calculate $P(T \leq t)$ and $P(T = \infty)$.

2.7.4 Stress and Strength

1. Calculate the probability of failure of a system with strength X uniformly distributed on $(0, a)$ and stress Y exponentially distributed on $(0, \infty)$ with parameter λ, X and Y being independent.

2. Calculate the failure probability of systems in which the strength X and stress Y are jointly distributed with density proportional to $c^2 - xy$ on $(0, c)^2$.

3. Suppose that X has density $f_X(x) = \lambda_X e^{-\lambda_X(x-\mu_X)}$ on (μ_X, ∞) and Y has density $f_Y(y) = \lambda_Y e^{-\lambda_Y y}$ on $(0, \infty)$. Thus, X can vary from μ_X to ∞ with exponentially diminishing probability density as x increases, μ_X being a guaranteed minimum value, and Y can vary similarly from 0 to ∞. Assuming that X and Y are independent, calculate $P(X < Y)$.

2.8 Hints and Solutions

2.8.1 Survival Distributions

1. For lack of memory, start with $P(T > a + b \mid T > a) = P(T > a + b, T > a)/P(T > a)$. Second part: consider $P\{(T/\xi)^\nu > a\}$.

2. Begin with $E(T^r) = \int_0^\infty t^r f(t)dt$, in which the density $f(t) = -d\bar{F}(t)/dt$, and use the definition $\Gamma(a) = \int_0^\infty x^{a-1}e^{-x}dx$ of the gamma function. You should obtain $\mu_T = E(T) = \xi\Gamma(1 + 1/\nu)$ and $\sigma_T^2 = \text{var}(T) = \xi^2\Gamma(1 + 2/\nu) - \mu_T^2$.

3. $\quad \mathrm{E}(T^r) = \displaystyle\int_{-\infty}^{\infty} e^{ry}(2\pi\sigma^2)^{-1/2}e^{-(y-\mu)^2/2\sigma^2}dy$

$$= (2\pi\sigma^2)^{-1/2}\int_{-\infty}^{\infty} e^{-\{(y-\mu-r\sigma^2)^2-(\mu+r\sigma^2)^2+\mu^2\}/2\sigma^2}dy$$

$$= e^{\{(\mu+r\sigma^2)^2-\mu^2\}/2\sigma^2}(2\pi\sigma^2)^{-1/2}\int_{-\infty}^{\infty} e^{-t^2/2\sigma^2}dy = e^{r\mu+r^2\sigma^2/2}.$$

$\bar{F}(t) = \mathrm{P}(e^Y > t) = \Phi(\log t).$

4. $\bar{F}(t) = 1 - \int_0^t f(s)ds = 1 - \Gamma(\nu)^{-1}\int_0^{t/\xi} y^{\nu-1}e^{-y}dy = 1 - \Gamma(\nu; t/\xi).$

5. The density is $-d\bar{F}(t)/dt = (\gamma/\alpha)(1+t/\alpha)^{-\gamma-1}$. Then,

$$\mathrm{E}(1+T/\alpha) = (\gamma/\alpha)\int_0^{\infty}(1+t/\alpha)^{-\gamma}dt$$

$$= \gamma/(\gamma-1) \text{ for } \gamma > 1, \text{ so } \mathrm{E}(T) = \alpha/(\gamma-1)$$

Likewise, $\mathrm{E}\{(1+T/\alpha)^2\} = \ldots$, giving $\mathrm{var}(T) = \mathrm{E}(T^2) - \mathrm{E}(T)^2 = \alpha^2\gamma/\{(\gamma-1)^2(\gamma-2)\}$ for $\gamma > 2$.
For quantile, $q = \bar{F}(t_q) = (1+t_q/\alpha)^{-\gamma}$ yields $t_q = \alpha(q^{-1/\gamma}-1)$.

6. Hazard function $h(t) = (\gamma\rho/\alpha)(t/\alpha)^{\rho-1}/\{1+(t/\alpha)^{\rho}\}$. The function $t^{\rho-1}/(1+t^{\rho})$ has derivative $(\rho-1)t^{\rho-2}(1+t^{\rho})^{-1}\{1-(\frac{\rho}{\rho-1})(\frac{t^{\rho+1}}{1+t^{\rho}})\}$. For $\rho > 1$ this is negative, so DFR; for $\rho > 1$, the derivative is positive for small t, zero when t solves $(\frac{\rho}{\rho-1}) = (\frac{t^{\rho+1}}{1+t^{\rho}})$, and thereafter negative, so neither uniformly IFR nor DFR. Weibull–gamma mixture: suppose that T has survivor function, conditional on λ, $\mathrm{P}(T > t \mid \lambda) = e^{-\lambda t^{\nu}}$ and that λ has a gamma distribution with density $f(\lambda) = \Gamma(\gamma)^{-1}\alpha^{\gamma}\lambda^{\gamma-1}e^{-\alpha\lambda}$. Then, the unconditional survivor function of T is

$$\bar{F}(t) = \int_0^{\infty} e^{-\lambda t^{\nu}}f(\lambda)d\lambda = \Gamma(\lambda)^{-1}\int_0^{\infty} s^{\gamma-1}e^{s(1+t^{\nu}/\alpha)}ds = (1+t^{\nu}/\alpha)^{-\gamma}.$$

7. a. Let $t \to \infty$ in $\mathrm{P}(T > t) = \bar{F}(t) = \exp\{-\int_0^t h(t)\,dt\}$.

  b. $\mathrm{P}(T > t) = \exp\left\{-\int_0^t h(t)\,dt\right\} = \exp\left[-\int_0^t \{h_1(t) + h_2(t)\}\,dt\right]$

$$= \exp\left\{-\int_0^t h_1(t)\,dt\right\} \times \exp\left\{-\int_0^t h_2(t)\,dt\right\}$$

$$= \mathrm{P}(T_1 > t) \times \mathrm{P}(T_2 > t)$$

$$= \mathrm{P}(T_1 > t, T_2 > t) = \mathrm{P}\{\min(T_1, T_2) > t\}.$$

8. $\mathrm{P}(T > t) = \bar{F}(t) = \exp\{-\int_0^t h(t)dt\} = e^{-at}$ for $0 < t < t_0$, $e^{-at_0-b(t-t_0)}$ for $t > t_0 \Rightarrow \mathrm{P}(T > t_0) = e^{-at_0}$, $\mathrm{P}(T > 2t_0) = e^{-(a+b)t_0}$.

9. Continuous $T: \mathrm{E}(T) = \int_0^{\infty} tf(t)dt = (by\ parts)\ \left[-t\bar{F}(t)\right]_0^{\infty} + \int_0^{\infty}\bar{F}(t)dt$.
   Discrete $T: \mathrm{E}(T) = \sum_{j=0}^{\infty} jp_j = p_1 + 2p_2 + 3p_3 + \cdots = (p_1 + p_2 + p_3 + \cdots) + (p_2 + p_3 + \cdots) + (p_3 + \cdots) + \cdots = \sum_{j=1}^{\infty} q_j.$

10. For $0 < \rho < 1$ and $\kappa > 0$,

$$(1 - \rho)^{-\kappa} = \sum_{r=0}^{\infty} \binom{-\kappa}{r}(-\rho)^r = \sum_{r=0}^{\infty}(r!)^{-1}(-\kappa)(-\kappa - 1)\ldots(-\kappa - r + 1)(-\rho)^r$$

$$= \sum_{r=0}^{\infty}(r!)^{-1}\kappa(\kappa + 1)\ldots(\kappa + r - 1)\rho^r = \sum_{r=0}^{\infty}\binom{\kappa + r - 1}{\kappa - 1}\rho^r.$$

11. The logistic survivor function has $p^* = \{1 + (t^*/\alpha)^\rho\}^{-1}$, so

$$\log\{p^*/(1 - p^*)\} = -\rho \log t^* + \rho \log \alpha = \beta_0 + \mathbf{x}^T\beta,$$

taking a log-linear model for α.

2.8.2 Reliability of Systems

1. prob $= 1 - $ P(all compts down) $= 1 - \prod_{j=1}^{n}$ P(compt j down) $= 1 - \prod_{j=1}^{n}(1 - p_j)$

2. (i) $\{1 - (1 - p_1)(1 - p_3)\}p_2$; (ii) $1 - (1 - p_1)(1 - p_2)(1 - p_3)$; (iii) $\{1 - (1 - p_1 p_2)(1 - p_4)\}p_3$;

2.8.3 Degradation Processes

1. a. $P(T \leq t) = P\{D_t \geq d^*\} = \{1 + d^*/(\xi t)\}^{-\nu}$.
   b. $P(T \leq t) = P\{b \geq t^{-1}\log(1 + d^*)\} = \exp\{-\lambda t^{-1}\log(1 + d^*)\} = (1 + d^*)^{-\lambda/t}$.

2. $P(T \leq t) = P\{D_t \geq d^*\} = P\{a > d^*(1 + e^{-bt})/(1 - e^{-bt})\} = \exp\{-\lambda d^*(1 + e^{-bt})/(1 - e^{-bt})\}$, $P(T = \infty) = 1 - \lim_{t \to \infty}\exp\{-\lambda d^*(1 + e^{-bt})/(1 - e^{-bt})\} = 1 - e^{-\lambda d^*}$.

2.8.4 Stress and Strength

1. $P(\text{failure}) = P(X < Y) = \int_0^\infty f_X(x)\{1 - F_Y(x)\}dx = \int_0^a a^{-1}e^{-\lambda x}dx = (1 - e^{-\lambda a})/(\lambda a)$.

2. Take $f(x, y) = k(c^2 - xy)$ with k determined by

$$1 = \int_0^c dx \int_0^c dy\{k(c^2 - xy)\} = k\int_0^c [c^2 y - xy^2/2]_0^c$$

$$= k\int_0^c (c^3 - c^2 x/2)dx = k[c^3 x - c^2 x^2/4]_0^c = 3kc^4/4$$

$$\Rightarrow k = 4/(3c^4).$$

$$P(\text{failure}) = P(X < Y) = \int_0^c dx \int_x^c dy\{k(c^2 - xy)\} = k \int_0^c [c^2 y - xy^2/2]_x^c$$

$$= k \int_0^c (c^3 - c^2 x/2 - c^2 x + x^3/2) dx$$

$$= k[c^3 x - 3c^2 x^2/4 + x^4/8]_0^c = 3kc^4/8 = 1/2.$$

3.
$$P(X < Y) = \int_0^\infty f_X(x)\{1 - F_Y(x)\} dx = \int_0^\infty f_X(x)\{e^{-\lambda_Y x}\} dx$$

$$= \int_{\mu_X}^\infty \lambda_X e^{-\lambda_X (x - \mu_X)} \{e^{-\lambda_Y x}\} dx = \ldots$$

$$= \left(\frac{\lambda_X}{\lambda_X + \lambda_Y}\right) e^{-\lambda_Y \mu_X}.$$

3

Continuous Time-Parametric Inference

We will use likelihood-based methods predominantly in this book. That is not to say that there are no useful ad hoc methods for obtaining answers to particular questions, but we will stick to the general approach, which is capable of giving answers to any well-formulated parametric question. However valid the methodology, of course, the data need to be sufficiently informative and the model has to be sufficiently well fitting. How many seminars have you attended where the conclusion was, "Nice model; shame about the data?"

Traditional parametric statistical methods for survival data are covered in this chapter. Data in both discrete and continuous time frameworks are considered.

3.1 Parametric Inference: Frequentist and Bayesian

There are historical arguments about which came first, the chicken (Bayesian approach) or the egg (Frequentist approach). Some of the more vocal proponents of the different approaches to inference have been shouting at each other for years from their respective hilltops. Personally, I cannot raise much enthusiasm for the debate since both approaches have their merits and drawbacks. That said, I do think that the broad differences should be appreciated by the statistician—it is a bit depressing nowadays to hear research students say that they are Bayesian because they do McMC or because they do Bayesian modelling (meaning statistical modelling).

Let us define a parameter, say θ, here as an unknown constant (maybe a vector) occurring in the expression for the statistical model under consideration. The likelihood function, based on data D, is $p(D \mid \theta)$, where p is just used to represent a probability or a density. Both Frequentist and Bayesian will use the likelihood, when it is accessible, to make inferences, but in different ways.

3.1.1 Frequentist Approach

The routine Frequentist approach is to maximise the likelihood over θ to obtain the *maximum likelihood estimate* (*mle*), $\hat{\theta}$ (in regular likelihood cases). Then the machinery of *asymptotics* can be brought to bear: as the sample size (or the information content of the data) increases, the distribution of $\hat{\theta}$ tends toward normal with mean θ and covariance matrix estimated as $-l''(\hat{\theta})^{-1}$, where $l''(\theta)$ is the second derivative (*Hessian*) matrix of the log-likelihood function, $l(\theta) = \log p(D \mid \theta)$. Standard errors, and the resulting confidence intervals, for component parameters can now be obtained. For hypothesis tests, appropriate likelihood ratio tests can be applied, or asymptotic equivalents such as those based on the score function (*score statistics*) and the *mle* (*Wald statistics*). The latter are less well recommended, though, in view of their lack of invariance under parametric transformation (e.g., Cox and Hinkley, 1974, Section 9.3[vii]). The asymptotic normal approximation to the distribution of the *mle* can sometimes be usefully improved by transformation of the parameters (e.g., Cox and Hinkley, 1974, Section 9.3[vii]).

3.1.2 Bayesian Approach

The general literature in this area is not sparse. O'Hagan and Forster (2004) give a comprehensive, general treatment. For reliability and survival analysis, in particular, the book by Martz and Waller (1982) contains much detail and gives many references to applications. *Lifetime Data Analysis* (*LIDA*) published a special issue in 2011: "Bayesian Methods in Survival Analysis."

I know that the distance from where I am sitting to Tipperary is a long way, because the old song says so, but I don't know exactly how far. However, I do believe that it is constant, subject to a few earthquakes and my not stirring from this armchair. I would be prepared to say it is about 100 miles, give or take, though geography was never my strong point. Adopting the Bayesian approach, I would have to elaborate on this by specifying a probability that the distance does not exceed 150 miles: in fact I would have to think up a whole probability distribution for the distance. In practice, life is too short for such navel-gazing (as it has been called), and one usually adopts a convenient distribution with suitable attributes, such as an appropriate mean and variance. This is called a *prior distribution* for the parameter, being the aforesaid distance in this case.

Commonly, it is said that because a parameter is endowed with a probability distribution, it becomes a random variable. To my mind that is a lazy way of looking at it. A random variable, notwithstanding all the measurable function stuff, is a quantity that can take different values on different occasions. How can that be true of an unknown constant? I know that geographical areas are sometimes described as being "on the move," but I do not think that this applies to Tipperary in quite that way.

Note that the prior distribution gives probabilities that are not the usual coin-tossing, die-rolling, card-shuffling types of probabilities—those are frequency based. It gives *subjective probabilities*, based on beliefs held by the subject. The crux of the matter is whether such probabilities can be combined with frequency probabilities, that is, whether the prior and the likelihood can be validly multiplied together to form a *posterior* distribution for θ. Mr. Bayesian, he says yes; Mr. Frequentist, he says no; not sure about Mr. Del Monte.

3.1.3 Proceed with Caution

The first jibe thrown at the Bayesian by the Frequentist is that his analysis is not "objective." One retort is that of course it isn't; the clue is in the phrase *subjective probabilities*. The argument is that there is no such thing as objectivity or that the data can speak for themselves. Inference from data is always channelled through an interpreter, different interpreters start with different background information, and this is bound to influence the inference. The Frequentist will then say that scientific statements should not depend on the opinions of the speaker, particularly the opinions of one of those dodgy Bayesians. He will add, more seriously, that the subjective approach is a vehicle for individual decision making, but it is not appropriate for objective scientific reporting.

A more practical criticism of the Bayesian approach is the difficulty of creating a prior distribution. When the data are extensive, we know that the posterior is mainly determined by the likelihood, the prior having little impact. However, particularly in multiparameter cases, an apparently harmless prior can hide unsuspected and undesirable features. Further, even if you want your input to be negligible compared with that of the data, there is no such thing as an *uninformative prior*, though time and again you will see this claimed in published work. The classic example is taking a uniform prior for a probability, say π, to express no preference for any one value over any other in the range (0,1). Unfortunately, the unintended consequence is that this choice expresses preference for smaller values of π^2 over larger ones on (0,1).

Let us now turn our fire on the Frequentist approach. A hypothesis test produces a p-value: if this is very small, doubt is thrown on the hypothesis. By *doubt* we must mean that, in the light of the data D, we view the hypothesis H as being dubious, unlikely, and improbable. But the p-value arises from $p(D \mid H)$, not $p(H \mid D)$, and for "H improbable" we need the second one. So, the p-value does not do the job that we might like it to do (but see DeGroot, 1973).

Confidence intervals are open to similar criticism. They do not give probabilities: the carefully calculated interval (0.19,0.31), for example, either spans θ or does not, which could have been said without getting out of bed. The argument that this interval has been randomly selected from a population of such intervals, 95% of which do span θ, sounds like a cunning attempt to persuade the listener that this one spans θ with probability 0.95. But the latter statement is invalid to the Frequentist because it confers a probability on the parameter. If you want probabilities out (posterior), you have to put probabilities in (prior).

So, what do I think, you ask? Or maybe you don't, but I will tell you anyway. Well, I cannot really say, even though this fence has rather sharp spikes. On this, I am not able to be dogmatic—I can see both sides of the argument. The Bayesian approach has some attractive features: it is logically consistent; nuisance parameters are not a nuisance, you just integrate them out; you do not have to rely on sometimes-dubious asymptotic approximations, as you do with the Frequentist approach; the computational problems, which inhibited the application of the Bayesian approach in years gone by, are now mainly solved. On the other hand, the assignment of appropriate priors is tricky, and the Frequentist stand against subjectivism in scientific inference is eminently reasonable.

In putting this book together, as an outgrowth of *Classical Competing Risks*, much new material needed to be introduced. Some of the former things had to go, and one of them was the use of McMC to produce Bayesian posteriors in some of the applications. Nevertheless, all the methodology is still based on likelihood functions, which lies from the Bayesian approach but a short step or a long stretch, depending on your point of view. Also, currently the major R package, `survival`, is mainly Frequentist, and I did feel the need to base things around it and other freely available R programs.

3.2 Random Samples

We have a sample $(t_1, \ldots, t_n)$ of observed lifetimes. Strictly speaking, no two observed values should be equal when they arise from a continuous distribution. In practice, though, rounding will often produce such ties.

The likelihood contributions are $f(t_i; \theta)$ for an observed failure time t_i and $\bar{F}(t_i; \theta)$ for one right-censored at t_i. The latter give information from events that have not yet occurred. It is sometimes not appreciated that such non-events, like unobserved failures, can provide useful information. Sheerluck Holmes was well aware of this: he gained a vital clue from the "curious incident" that the dog did not bark in *Silver Blaze* (Doyle, 1950). The overall likelihood function is

$$L = \prod_{obs} f(t_i; \theta) \times \prod_{cens} \bar{F}(t_i; \theta),$$

where $\prod_{obs}$ and $\prod_{cens}$ are the products over the observed and right-censored times, respectively. The appearance of $\bar{F}(t_i; \theta)$ in the expression for L assumes that the censoring tells us nothing further about the failure time than that it is beyond t_i. It is not always the case that censoring is *non-informative*; for example, in some circumstances censoring is associated with imminent failure.

Let $c_i = I(t_i \ observed)$, in terms of the indicator function. So, c_i is the *censoring indicator*, $c_i = 1$ if t_i is observed and $c_i = 0$ if t_i is right-censored. The likelihood function can then be written as

$$L = \prod_{i=1}^{n} \{ f(t_i; \theta)^{c_i} \bar{F}(t_i; \theta)^{1-c_i} \} = \prod_{i=1}^{n} \{ h(t_i; \theta)^{c_i} \bar{F}(t_i; \theta) \}.$$

Different symbols are used for the censoring indicator by different authors. Some use δ_i, but we will mostly reserve Greek letters for parameters here; further, I prefer to spell *censoring* with a c. Perhaps we should use C_i instead of c_i, adhering to the convention that capitals are used for random variables. However, that looks a bit odd, against most usage. Finally, the term *censoring indicator* should, strictly speaking, be replaced by *non-censoring*, or *observation*, indicator; but let's not be too fussy.

3.2.1 Type-I Censoring

Consider a random sample from an exponential distribution with mean ξ. The observations are right-censored at fixed time $a > 0$, that is, we only observe $t_a = \min(a, t)$: this is known as *Type-I censoring*. Thus,

$$E(t_a) = \int_0^a t(\xi^{-1}e^{-t/\xi})dt + aP(t > a) = \xi(1 - e^{-a/\xi}).$$

Suppose that the data comprise $t_1, \ldots, t_r$ (observed values, all $\leq a$) and $t_{r+1}, \ldots, t_n$ (right-censored, all $= a$). Then,

$$E(r) = nP(t \leq a) = n(1 - e^{-a/\xi}).$$

The log-likelihood function is given by

$$l(\xi) = \log \left\{ \prod_{i=1}^r (\xi^{-1}e^{-t_i/\xi}) \times \prod_{i=r+1}^n e^{-a/\xi} \right\} = -r \log \xi - \xi^{-1}t_+,$$

where $t_+ = \sum_{i=1}^r t_i + (n-r)a$ is the *Total Time on Test*, a term from reliability engineering. The *score function* is $l'(\xi) = -r\xi^{-1} + t_+\xi^{-2}$, and the information function is $-l''(\xi) = -r\xi^{-2} + 2t_+\xi^{-3}$. The *mle*, obtained as the solution of $l'(\xi) = 0$, is $\hat{\xi} = t_+/r$, and its variance is approximated by $-l''(\hat{\xi})^{-1} = \hat{\xi}^2/r$ (Appendix B).

3.2.2 Type-II Censoring

Consider a random sample from an exponential distribution with mean ξ. However, this time we observe only the r smallest t_is, where r is a predetermined number: this is known as *Type-II censoring*. Let $t_{(1)}, \ldots, t_{(n)}$ be the sample *order statistics* (the t_is rearranged in ascending order). To calculate the likelihood function we use (a) the density $\xi^{-1}e^{-t/\xi}$ for $t_{(1)}, \ldots, t_{(r)}$ (since their values are observed) and (b) the survivor function $e^{-t/\xi}$ evaluated at $t = t_{(r)}$ for $t_{(r+1)}, \ldots, t_{(n)}$ (since we know only that their values exceed $t_{(r)}$). The log-likelihood function is now

$$l(\xi) = \log \left\{ \prod_{i=1}^r (\xi^{-1}e^{-t_{(i)}/\xi}) \times \prod_{i=r+1}^n (e^{-t_{(r)}/\xi}) \right\} = -r \log \xi - \xi^{-1}t_+,$$

which looks much the same as for Type-I censoring though now r is non-random and $t_+ = \sum_{i=1}^r t_{(i)} + (n-r)t_{(r)}$. The score function is $l'(\xi) = -r\xi^{-1} + t_+\xi^{-2}$ and the information function is $-l''(\xi) = -r\xi^{-2} + 2t_+\xi^{-3}$. The *mle* is $\hat{\xi} = t_+/r$, and its variance is approximated by $-l''(\hat{\xi})^{-1} = \hat{\xi}^2/r$.

3.2.3 Left Truncation

Suppose that a unit comes under scrutiny sometime after switching on, meaning the time at which its inexorable downward spiral toward its demise begins. Specifically, suppose that observation on the unit starts at random time Z after switch-on and its lifetime is T: let $S = T - Z$. If $S < 0$, the unit will fail before it comes under observation; otherwise, it will be under observation for time $S \geq 0$. Let $\bar{F}(t \mid z) = P(T > t \mid Z = z)$ and let Z have density function $g(z)$ on $(0, \infty)$. Then,

$$P(S > s \mid Z = z) = P(T > s + z \mid Z = z) = \begin{cases} \bar{F}(s + z \mid z) & \text{if } s + z \geq 0 \\ 1 & \text{if } s + z < 0 \end{cases}$$

Hence,

$$P(S > s) = \int_{-s}^{\infty} \bar{F}(s + z \mid z) g(z) dz + \int_{0}^{-s} g(z) dz :$$

for $s \geq 0$ the second integral is zero, and for $s < 0$ it is equal to $P(Z \leq -s)$.

Assume now that T has an exponential distribution with mean ξ, and that Z has an exponential distribution with mean ν. Then, for $s \geq 0$,

$$P(S > s) = \int_{0}^{\infty} e^{-(s+z)/\xi} \nu^{-1} e^{-z/\nu} dz = \left(\frac{\xi}{\xi + \nu}\right) e^{-s/\xi},$$

and for $s < 0$,

$$P(S > s) = \left(\frac{\xi}{\xi + \nu}\right) e^{s/\nu} + (1 - e^{s/\nu}) = 1 - \left(\frac{\nu}{\xi + \nu}\right) e^{s/\nu}.$$

The proportion of units that survive to become observed is $P(S > 0) = \frac{\xi}{\xi + \nu}$, which is greater than 50% if $\xi > \nu$.

Lawless (2003, Section 2.4) gave a general treatment of this sort of situation from a slightly different standpoint.

3.2.4 Probabilities of Observation versus Censoring

Consider the situation where a unit is liable to failure, at time T^f, or censoring (being lost to observation), at time T^c. Assume that T^f and T^c are independent with probability functions $(f^f, \bar{F}^f)$ and $(f^c, \bar{F}^c)$, respectively. The probabilities of observed failure and censoring at time t are

$$P(T^f = t, T^c > t) = f^f(t)\bar{F}^c(t) \text{ and } P(T^c = t, T^f > t) = f^c(t)\bar{F}^f(t).$$

The overall probabilities of observed failure and of censorship are obtained by integration over t. The likelihood function for a random sample is

$$L = \prod_{obs}\{f^f(t)\bar{F}^c(t)\} \times \prod_{cens}\{f^c(t)\bar{F}^f(t)\}$$

$$= \left\{\prod_{obs} f^f(t)\right\}\left\{\prod_{cens}\bar{F}^f(t)\right\} \times \left\{\prod_{obs} f^c(t)\right\}\left\{\prod_{cens}\bar{F}^c(t)\right\}.$$

The first product here contains the probabilities determining the T^f distribution, which is usually the one of primary interest. Suppose that the T^c probabilities are not linked in any way to those of T^f, through having parameters in common, for example. Then we may focus upon the first term, which is the likelihood function shown at the beginning of this section. Of course, we can likewise use the second product to estimate the T^c probabilities if we want to.

If the censoring process is not independent of the failure process, we have *dependent competing risks*, which is covered in Part III.

3.2.5 Weibull Lifetimes

Consider a random sample of Weibull-distributed lifetimes. The survivor function has form $\bar{F}(t;\theta) = \exp\{-(t/\xi)^\nu\}$, with $\theta = (\xi, \nu)$. This can be recast as

$$\log\{-\log\bar{F}(t;\theta)\} = \nu \log t - \nu \log \xi.$$

Given a random sample of uncensored times, $(t_1, \ldots, t_n)$, we can order them as $t_{(1)} < t_{(2)} < \ldots < t_{(n)}$; the $t_{(i)}$ are the *order statistics*. An estimate of $\bar{F}(t_{(i)};\theta)$ can then be extracted as $(n - i)/n$, the sample proportion of times beyond $t_{(i)}$ ($i = 1, \ldots, n$); to avoid the extreme and unlikely value 1 for $\bar{F}(t_{(n)};\theta)$, in practice we use a slightly modified version such as $a_i = (n - i + 1)/(n + 1)$. Now, consider plotting the points $(\log t_{(i)}, \log(-\log a_i))$. According to the equation above, with Weibull-distributed times, the points plotted should lie near a straight line with slope ν and intercept $-\nu \log \xi$. This is the basic *Weibull probability plot*, widely used by engineers, for example. When there is right censoring a more sophisticated approach to estimating the survivor function is needed—see Section 4.1.

For random samples that may include right-censored observations, we can write down the likelihood function as follows, using the Weibull form of $\bar{F}(t;\theta)$ given above together with the corresponding hazard function $h(t) = (\nu/\xi)(t/\xi)^{\nu-1}$:

$$\log L(\xi, \nu) = \log \prod_{i=1}^{n}\left[\{(\nu/\xi)(t_i/\xi)^{\nu-1}\}^{k_i}\exp\{-(t_i/\xi)^\nu\}\right]$$

$$= n_c(\log \nu - \nu \log \xi) + (\nu - 1)\sum_{cens}\log t_i - \sum_{i=1}^{n}(t_i/\xi)^\nu,$$

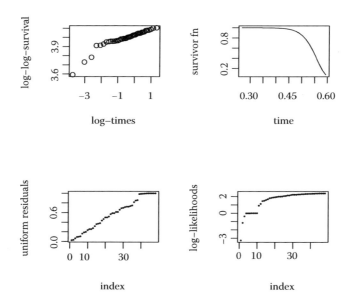

FIGURE 3.1
Weibull plots for the 41-cord data.

where n_c is the number of right-censored times. Setting $\partial \log L / \partial \xi$ to zero produces $\xi^\nu = \sum_{i=1}^{n} t_i^\nu / n_c$, which yields the *mle* $\hat{\xi}_\nu$ in terms of ν. So, ξ in L can be replaced by $\hat{\xi}_\nu$ to produce the *profile likelihood* $L(\hat{\xi}_\nu, \nu)$ for ν. This can then be maximised to compute $\hat{\nu}$ using a simple one-dimensional search.

3.2.6 Strengths of Cords

Let us apply the Weibull distribution to these data (Section 1.2). To begin with, Figure 3.1 (top left panel) shows a Weibull probability plot for 41 cords, omitting the censored values: the plot was drawn using R-function wblplot1, which is listed on the Web site noted in the Preface. There are three rogue points at the bottom end, but the rest look fairly close to a straight line (if one does not look too hard).

We will now fit a Weibull model to the data, this time including the censored values. The data can be set up using

```
#cords data (Sec I.3.2)
nd=48; nobs=41; ncens=7;
strengths=
c(36.3,41.7,43.9,49.9,50.1,50.8,51.9,52.1,52.3,52.3,
  52.4,52.6,52.7,53.1,53.6,53.6,53.9,53.9,54.1,54.6,
  54.8,54.8,55.1,55.4,55.9,56.0,56.1,56.5,56.9,57.1,
  57.1,57.3,57.7,57.8,58.1,58.9,59.0,59.1,59.6,60.4,
  60.7,26.8,29.6,33.4,35.0,40.0,41.9,42.5);
cens=c(rep(1,nobs),rep(0,ncens));
```

We apply the routine `survreg` from the `survival` package as follows:

```
#Weibull fit
library(survival);
sobj=Surv(strengths,cens,type='right',origin=0); #create 'survival object' sobj
sfit=survreg(sobj~1,dist='weibull'); summary(sfit);
```

Note the capital S for `Surv` and the formula `sobj`~ 1 for a model fit with no covariates, just a constant *intercept* term. Unfortunately, the parametrisation in `survreg` is different from that used here. To transform back to (ξ, ν), as in Section 2.2, take $\nu = 1/scale$ and $\xi = \exp(-scale * intercept)$, where *scale* and *intercept* are printed out in the R-summary. The output from `survreg` is

```
Call:
survreg(formula = sobj ~ 1, dist = "weibull")
            Value Std. Error    z        p
(Intercept)  4.03      0.010 402.9  0.00e+00
Log(scale)  -2.79      0.125 -22.3 3.19e-110
Scale= 0.0613
Weibull distribution
Loglik(model)= -115.8   Loglik(intercept only)= -115.8
Number of Newton-Raphson Iterations: 11
n= 48
```

The model can also be fitted by applying homegrown R-functions `fit-model`, `rsampc1`, and `wblb`: `fitmodel` is just a convenient way of using the R-optimisation routine `optim` with a user-supplied function; `rsampc1` returns the negative log-likelihood for random samples in continuous time, and `wblb` specialises to the Weibull distribution (listed on the Web site). The R-code is

```
#fit Weibull model
cdl=matrix(0,nd,2); cdl[1:nd,1]=strengths/100; cdl[1:nd,2]=cens;
jt=1; jc=2; opt=0; iwr=0; adt=c(nd,jt,jc,opt,iwr);
np=2; par0=c(1,2); adt=c(nd,jt,jc,opt,iwr);
par1=fitmodel(rsampc1,np,par0,cdl,adt);
adt=c(nd,jt,jc,1,2); rs1=rsampc1(par0,cdl,adt);
```

In the code, `jt` and `jc` identify the columns of the data matrix that contain the times (here, strengths/100) and censoring indicators; `opt` is an option that tells `rsampc1` to return either the log-likelihood or other information; `iwr` tells it how much to print out (nothing while `optim` is working its magic); `np` is the number of parameters and `par0` is the initial estimate.

The estimated survivor function is plotted in Figure 3.1 (top right panel): the x axis, time, represents breaking strength/100. The lower two panels are residual plots, which will be discussed in more detail below. Suffice it to say here that if the fitted model is correct, the left plot should resemble a line from (0,0) to (1,1) and the right one should show a fairly smooth curve. The bottom left panel is a plot of the uniform residuals: here the censored strengths are taken at face value; the bunch of points at the top end arises from the seven censored values and the three rogue points. The bottom right panel shows the ordered log-likelihood contributions: the seven censored and three rogue

TABLE 3.1

Cancer Survival Times

Group 1														
0.3	5.0	5.6	6.2	6.3	6.6	6.8	7.5	8.4	8.4	10.3	11.0	11.8	12.2	12.3
13.5	14.4	14.4	14.8	15.7	16.2	16.3	16.5	16.8	17.2	17.3	17.5	17.9	19.8	20.4
20.9	21.0	21.0	21.1	23.0	23.6	24.0	24.0	27.9	28.2	29.1	30	31	31	32
35	35	38	40	41	41	42	44	48	51	51	52	54	56	60
78	80	89	90	126	39	40	46	48	78	87	131	84	97	98
100	114	174												

Group 2														
0.3	4.0	7.4	15.5	23.4	46	46	51	65	68	83	88	96	111	132
110	112	162												

Group 3														
111	112	113	114	114	117	121	123	129	131	133	134	134	136	141
143	167	177	179	189	201	203	203	213						

Group 4		
228		

Source: Boag, J.W., 1949, Maximum Likelihood Estimates of the Proportion of Patients Cured by Cancer Therapy, *J. R. Statist. Soc.*, B11, 15–44. Reproduced by permission of Wiley-Blackwell Publishers.

points appear at the lower end of the curve. These last two panels were drawn using R-function indxplot.

3.2.7 Survival of Breast Cancer Patients

The survival times (months) of 121 breast cancer patients are given in Table 3.1. They were taken from clinical records of a hospital from the 1930s and appear in Boag's (1949) Table IX. The times up to 30 months are given to the nearest tenth, and after that to the nearest month: they are measured from commencement of treatment. The groups referred to in the table are as follows.

Group 1: death with the cancer present ($n_1 = 78$)

Group 2: death without the cancer from other causes ($n_2 = 18$)

Group 3: still alive without the cancer ($n_3 = 24$)

Group 4: still alive but with the cancer ($n_4 = 1$)

Boag pointed to previous findings that such survival times were well fitted by the log-normal distribution. However, he also pointed out that there might be an inflated frequency in the lower tail due to "operative mortality": as he says, "the primary cause of death in such cases is not the malady but the remedy" (p. 16, op cit). He compared the fits of the log-normal distribution with others, for example, the exponential. In addition, he went on to fit the data, by maximum likelihood, with a model containing three parameters, μ

and σ for the log-normal distribution, and c, the probability of permanent cure. Much of the paper is taken up with ingenious methods, giving tables and nomograms, for numerical solution of the likelihood equations in that pre-computer age; it is humbling to be made aware of how much easier it is for us nowadays.

This was a paper read to the Royal Statistical Society in the post-war years. Interestingly, the proposer of the vote of thanks (the chairman, no less) saw "no great advantage of maximum likelihood here" over a simple head count of survival after a given period of time, citing the relative difficulty of obtaining the *mle*. However, according to the seconder, "We must all be grateful to Mr. Boag and his collaborators for introducing these methods into this new field in a very ingenious manner," citing the replacement of previous *ad hoc* methods. So, the score was one-all so far. The third discussant referred to the "arithmetical examples of the working which could be followed by intelligent people who had not a passion for algebra": one in the eye for those pesky mathematicians. Other discussants made additional points: the problem, even arbitrariness, with defining time zero (appearance of symptoms, start or end of treatment, imputed initiation of disease); the often-difficult assignment of cause of death—was it the cancer or something else to which resistance had been lowered by the cancer; does "death with the cancer present" in Group 1 equate to death from the cancer; the actuarial method (several speakers) for use when other causes of death are present (see Part III, "Competing Risks"); unobserved treatment selection correlated with perceived disease severity.

There is much to study in these data. Does the log-normal fit well? Should one mix in an early component to reflect operative mortality? Is there a permanently cured component (with $P(T = \infty) > 0$) as considered by Boag? How about other survival distributions? Boag considered densities $\alpha t^{-\alpha t}$ (exponential), $\alpha^2 t e^{-\alpha t}$ and $\frac{1}{2}\alpha^3 t^2 e^{-\alpha t}$ (gamma), and $2\alpha t e^{-\alpha t^2}$ (Weibull). I leave it to you to investigate.

3.3 Regression Models

In most applications, the data are not independently identically distributed, and a regression model is called for to relate the basic parameters to *covariates* or *explanatory variables* or *design, classification*, and *grouping indicators*. Suppose that to the ith case in the sample is attached a vector $\mathbf{x}_i$ of such covariates. Thus, $\mathbf{x}_i$ might contain components recording the age, height, weight, and health history of a patient, or the age, dimensions, specification, and usage history of an engineering system component, or binary indicators of the levels of factors in some experimental design.

Some models will be described in this section, with applications to come later. The likelihood function is basically the same as that given for random samples in the previous section. We just have to replace $f(t_i; \theta)$ and $\bar{F}(t_i; \theta)$ by $f(t_i; \mathbf{x}_i, \theta)$ and $\bar{F}(t_i; \mathbf{x}_i, \theta)$, these forms incorporating a regression model

TABLE 3.2

Times to Profitability

Sector 1												Sector 2										
5	1	5+	1	1	2	1+	2	11+	2	12+	2	11	15	12	10+	17	4	8	5			
4	2	1	3	1	2+	1	2	3	2	30	4+	3	8	4	14	6+	2	1	6			
13+	1	2	2	1																		

involving covariate x_i and parameter θ. So, for the record,

$$L(\theta) = \prod_{obs} f(t_i; \mathbf{x}_i, \theta) \times \prod_{cens} \bar{F}(t_i; \mathbf{x}_i, \theta).$$

Example

A regression model for Weibull failure times (Section 2.2) has a log-linear model for the scale parameter ξ: $\log \xi = \mathbf{x}^T \beta$. Typically, the shape parameter ν is left as an unknown constant, not related to $\mathbf{x}$ (Smith, 1991).

3.3.1 Business Start-Ups

The data listed in Table 3.2 represent times (in quarters) for start-up businesses to achieve profitability. The 45 firms are from two industry sectors, and right-censored times (at which the businesses were removed from the survey for reasons unconnected with t) are indicated by $+$.

The data can be set up as $\{(t_i, x_i) : i = 1, \ldots, n\}$, where $n = 25 + 20$, $x_i = -1$ for sector 1 and $x_i = 1$ for sector 2. We will model the times as exponentially distributed with means specified by a log-linear form, $\log E(t_i) = \beta_0 + \beta_1 x_i$: the difference between sectors is then quantified as $2\beta_1$. Denote the means by $\xi_1 = e^{\beta_0 - \beta_1}$ (sector 1) and $\xi_2 = e^{\beta_0 + \beta_1}$ (sector 2). The likelihood function reduces to

$$L(\beta_0, \beta_1) = \prod^{(1)} \left\{ \prod_{obs} (\xi_1^{-1} e^{-t_i/\xi_1}) \prod_{cens} e^{-y_i/\xi_1} \right\} \times \prod^{(2)} \left\{ \prod_{obs} (\xi_2^{-1} e^{-t_i/\xi_2}) \prod_{cens} e^{-y_i/\xi_1} \right\},$$

where $\prod^{(j)}$ refers to sector j. To fit the model in R we code as follows.

```
#Business start-ups (Sec I.3.3)
library(survival);
#set up time and censoring vectors for sectors
tv1=c(5,1,5,1,1,2,1,2,11,2,4,2,1,3,1,2,1,2,3,2,13,1,2,2,1);
cv1=c(1,1,0,1,1,1,0,1,0,1,1,1,1,1,1,0,1,1,1,1,0,1,1,1,1);
tv2=c(12,2,11,15,12,10,17,4,8,5,30,4,3,8,4,14,6,2,1,6);
cv2=c(0,1,1,1,1,0,1,1,1,1,1,0,1,1,1,1,0,1,1,1);
#vectors of times, censoring and covariates for both sectors
tv=c(tv1,tv2); cv=c(cv1,cv2); xv=c(rep(-1,25),rep(1,20));
sobj=Surv(tv,cv,type='right');
sfit1=survreg(sobj~1+xv,dist='exponential'); summary(sfit1);
```

Note the formula, sobj~1+xv, which specifies an intercept plus covariate; you can omit the 1 or, if the model is to have no intercept, replace it by -1. The output is

```
Call:
survreg(formula = sobj ~ 1 + xv, dist = "exponential")
            Value Std. Error    z        p
(Intercept)  1.83     0.168 10.89 1.25e-27
xv           0.56     0.168  3.34 8.45e-04
Scale fixed at 1
Exponential distribution
Loglik(model)= -99.5   Loglik(intercept only)= -105
        Chisq= 11.03 on 1 degrees of freedom, p= 9e-04
Number of Newton-Raphson Iterations: 4
n= 45
```

The ratio $\hat{\beta}_1/se(\hat{\beta}_1) = 3.34$, suggesting a real difference between sectors. The ξ-parameters for the two sectors are estimated as $\hat{\xi}_1 = \exp(1.83 - 0.56) = 3.55$ and $\hat{\xi}_2 = \exp(1.83 + 0.56) = 10.88$. Assuming that you didn't fall asleep halfway through, you will have realized that ξ_1 and ξ_2 could have been estimated as in Section 3.2 from the two random samples, and thence the β_j. The reason for going the long way round here is to illustrate the fitting of a regression model.

A goodness-of-fit test can be made by comparing the exponential fit with a Weibull fit, the former distribution being a special case of the latter. The R-code is the same as before except for dist='weibull'. The Weibull fit yields β estimates similar to those above. The maximised log-likelihood values are -99.5 for the exponential fit and -98.8 for the Weibull, yielding a log-likelihood ratio $2(99.5 - 98.8) = 1.4$, which gives $p > 0.10$ when referred to χ_1^2. Evidently, the Weibull distribution does not fit the data significantly better than the exponential.

3.3.2 Proportional Hazards (PH)

The univariate version is $h(t; \mathbf{x}) = \psi_x h_0(t)$, where h_0 is some baseline hazard function and ψ_x is a positive function, for example, log-linear with $\psi_x = \exp(\mathbf{x}^T \beta)$. The model decrees that, as time proceeds, the hazard profiles of different individuals mirror each other in the sense that their ratios are fixed. Thus, for individuals i and j,

$$h(t; \mathbf{x}_i)/h(t; \mathbf{x}_j) = \exp\{(\mathbf{x}_i - \mathbf{x}_j)^T \beta\},$$

a function of $\mathbf{x}_i$ and $\mathbf{x}_j$ but not of t. The consequent for the survivor function is

$$\bar{F}(t; \mathbf{x}) = \exp\left\{-\int_0^t \psi_x h_0(s)ds\right\} = \bar{F}_0(t)^{\psi_x}.$$

The new family, obtained by raising a given survivor function to a power, is known as a set of *Lehmann alternatives*.

3.3.3 Accelerated Life (AL)

This model specifies the form $\bar{F}(t; \mathbf{x}) = \bar{F}_0(\psi_x t)$ with $\mathbf{x}$ and ψ_x as for proportional hazards and $\bar{F}_0$ some baseline survivor function. The effect of $\mathbf{x}$ is to *accelerate* the time scale t by a factor ψ_x in the baseline model: so, $T_x = T_0/\psi_x$, where failure time T_x has covariate x and T_0 has the baseline survivor function $\bar{F}_0$. If $\psi_x > 1$ the individual's progress along the time scale is "speeded" up, and if $\psi_x < 1$ progress is slowed down. The model is often cast in the form

$$\log T_x = \psi'_x + \log T_0,$$

where $\psi'_x = -\log \psi_x$. The hazard function correspondingly satisfies

$$h(t; \mathbf{x}) = -d \log \bar{F}(t; \mathbf{x})/dt = \psi_x h_0(\psi_x t).$$

3.3.4 Proportional Odds (PO)

The odds of a random event E is defined as $P(E)/\{1 - P(E)\}$. I would place a bet that certain sections of society are well versed in the art of interpreting and manipulating odds. The odds on the event $\{T \leq t\}$ are constrained by this model to satisfy

$$\{1 - \bar{F}(t; \mathbf{x})\}/\bar{F}(t; \mathbf{x}) = \psi_x\{1 - \bar{F}_0(t)\}/\bar{F}_0(t).$$

The consequent constraint on the hazard function defies simple expression unless you can dissect

$$h(t; \mathbf{x})/\bar{F}(t; \mathbf{x}) = \psi_x\{h_0(t)/\bar{F}_0(t)\}.$$

If so, "You're a better man than I am, Gunga Din."

3.3.5 Mean Residual Life (MRL)

The *mean residual life (mrl)* at age t is defined as

$$m(t) = E(T - t \mid T > t);$$

this is the (expected) time left to an individual who has survived to age t; obviously, $m(0) = E(T)$. The corresponding *life expectancy* is $E(T \mid T > t) = m(t) + t$. The *mrl* is the answer to that age-old question, "How long have I got, Doc?" (The answer, in England, is seven minutes, the average time allotted following, "The doctor will see you now," that rather condescending summons from the receptionist.) The *mrl* can be evaluated as

$$m(t) = \int_t^\infty (y - t) f(y \mid y > t) dy = \int_t^\infty (y - t)\{f(y)/\bar{F}(t)\} dy$$
$$= \bar{F}(t)^{-1} \int_t^\infty \bar{F}(y) dy,$$

where $f(y \mid y > t)$ is the indicated conditional density of T and an integration by parts has been performed under the assumption that $y\bar{F}(y) \to 0$ as $y \to \infty$.

For regression, the *mrl* can be modelled as follows, following Oakes and Desu (1990) and Maguluri and Zhang (1994):

$$m(t; \mathbf{x}) = \psi_x m_0(t),$$

where m_0 is some baseline *mrl* function and ψ_x is some positive function of the vector $\mathbf{x}$ of explanatory variables, for example, $\psi_x = \exp(\mathbf{x}^T \beta)$.

3.3.6 Catheter Infection

For a small illustration let us consider an accelerated-life model for these data (Section 1.2). There are 27 patients, the response variable is the time for a catheter infection to appear, and the covariates are age and sex. Let us consider a log-linear model: for patient i,

$$\log T_i = x_i^T \beta + e_i.$$

This is of accelerated-life form with $\psi_x = \exp(-x_i^T \beta)$. We'll take the e_i as independent $N(0, \sigma^2)$ residuals and ignore the fact that some of the times are right-censored. The data can be assembled in a file `catheter1.dat` with a header containing the variable names followed by 27 rows each containing five numbers (Section 1.3). The standard function `lm` (linear model) can be used. Note that the variables `tim`, `cns`, `age`, and `sex` have been made available by the preceding `attach(dx1)`. The R-code for performing the fit is

```
#catheter data (Sec I.3.3)
dx1=read.table('catheter1.dat',header=T); attach(dx1);
fit1=lm(log(tim)~age+sex,dx1); summary(fit1);
```

The output includes the following Anova table:

	Estimate	Std. Error	t value	Pr(>|t|)	
(Intercept)	2.594865	0.456686	5.682	7.48e-06	***
age	0.011867	0.005758	2.061	0.0503	.
sex	-0.228139	0.222462	-1.026	0.3153	

The factor `age` is nudging significance at 5%, but `sex` makes no impact. For the plots shown in Figure 3.2 add

```
par(mfrow=c(2,2)); plot(fit1);
```

It would take us too far afield to start explaining what these plots do exactly—details can be found in the R-manual and in books on diagnostics, for example, Cook and Weisburg (1982) and Atkinson (1985).

For comparison, here is a Weibull regression fit to the data.

```
#Weibull regn
library(survival);
sobj=Surv(tim,cns,type='right',origin=0); #create 'survival object' sobj
fit2=survreg(sobj~age+sex,dist='weibull'); summary(fit2);
```

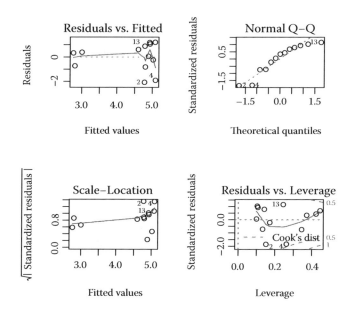

FIGURE 3.2
Goodness-of-fit plots for catheter data.

The output for this is

```
Call:
survreg(formula = sobj ~ age + sex, dist = "weibull")
              Value Std. Error       z        p
(Intercept)  2.83286     0.6552  4.3235 1.54e-05
age          0.01325     0.0097  1.3670 1.72e-01
sex         -0.00811     0.3408 -0.0238 9.81e-01
Log(scale)  -0.49529     0.2426 -2.0414 4.12e-02

Scale= 0.61

Weibull distribution
Loglik(model)= -58.6   Loglik(intercept only)= -59.6
        Chisq= 1.96 on 2 degrees of freedom, p= 0.37
Number of Newton-Raphson Iterations: 5
n= 27
```

The results bear some resemblance to the normal linear regression output—
the Weibull is an accelerated-life model (see Section 3.7, Exercises).

3.4 Goodness of Fit

A huge variety of methods has been developed over the years for this problem.
The widely accepted view in statistics is that no proposed model is "true"
but might be useful for the purpose at hand. In some cases it isn't even the

"fit" of the model that matters, however that is defined, but only whether it performs the job required (e.g., Crowder et al., 2002). What follows is a brief overview: much more detail can be found in Collett (2003, Chapters 4 and 7) and Lawless (2003, Chapter 10). Also left to other works is the extensive topic of *model selection*. And, while we are on the subject of missing topics, those of *missing data* and *measurement error* must be found elsewhere, for example, in Little and Rubin (2002), Wu (2010), and Buonaccorsi (2010).

3.4.1 Enhanced Models

A well-worn approach to goodness of fit, which has worn well, is to compare the model used with an extended, more general one. For example, a Weibull model might be used as an enhanced version of an exponential fit. Regression models can be enhanced by including additional explanatory variables.

A standard method is to employ a so-called *Lehmann alternative*. When the model is based on a particular distribution function $F(t)$, a modified model, based on $F(t)^\phi$, where ϕ is an extra parameter, can be entertained. If $F(t)$ already has a power law form, the Lehmann alternative model will be no more general than the original one; in that case the transformation can be performed on the survivor function, $\bar{F}(t)$. The original model will be supported if, when the alternative is fitted and tested, $\phi = 1$ is acceptable.

Another standard method is to apply a *Neyman smooth test*. Here the density function, say $f(t; \theta)$, is multiplied by a factor $\exp\{g(t; \theta)^T \beta\}$, where the components of the vector $g(t; \theta)$ are specified functions. The result needs to be rescaled by the reciprocal of $\int_0^\infty f(t; \theta) \exp\{g(t; \theta)^T \beta\} dt$ to make it a proper density. The test for adequacy of the unadorned model is then of the parametric hypothesis that $\beta = 0$. See Rayner and Best (1990) and the references therein.

An extreme case of an extended model is a non-parametric one, one that assumes no particular parametric form. In cases where non-parametric estimation is possible, this can provide the enhanced model to be used for the comparison. For example, a fitted survivor function can be plotted in the frame along with the Kaplan–Meier estimate (Section 4.1).

3.4.2 Uniform Residuals

Residuals can be computed via the *probability integral transform* (see Section 3.7, Exercises). Thus, if $\bar{F}$ is the common survivor function of independent random variables $(t_1, \ldots, t_n)$, the $u_i = \bar{F}(t_i)$ form a random sample from $U(0, 1)$, the uniform distribution on the interval $(0, 1)$. Denote by $(u_{(1)}, \ldots, u_{(n)})$ the *order statistics* of the u_is. Then the $u_{(i)}$ form an ordered sample from $U(0, 1)$, for which the means are given by $\mathrm{E}(u_{(i)}) = i/(n+1)$ (see Section 3.7, Exercises). This suggests plotting the points $(\frac{i}{n+1}, u_{(i)})$: if the correct form of $\bar{F}$ is used, the points should land near the 45° line from (0,0), to (1,1). An index plot of the points $(i, u_{(i)})$ will serve just as well; one such appears in Figure 3.1. In practice, the estimate of $\bar{F}$ from the fitted model is used to compute the u_i.

The transform can be extended to the non-iid case simply by replacing $\bar{F}$ with $\bar{F}_i$. For example, in a model incorporating covariates $\bar{F}_i(t) = \bar{F}_0(t; x_i)$.

3.4.3 Cox–Snell Residuals

An alternative transformation of the t_i is based on the *integrated hazard function*, $H(t) = -\log \bar{F}(t)$. The $H(t_i)$ form a random sample from $E(1)$, the exponential distribution with unit mean. So, a plot of points $(t_i, H(t_i))$ should show a near-line of slope 1. Alternatively, a Weibull plot (Section 3.2), which is just a plot of points $(\log t_i, \log H(t_i))$, should show a near-line of points of slope 1 and intercept $-\log \xi$, where $\xi = E(T)$. This second one allows some assessment of departure from $E(1)$ towards the Weibull distribution if the latter is indeed more appropriate. As usual, an estimate of H is used in practice, and the extension to the non-iid case replaces H by H_i.

3.4.4 Right-Censored Times

As usual, censoring introduces complications. If the ith time is right-censored at t_i, then $H_i(t_i)$ should be adjusted by adding 1, since this is the mean residual lifetime for an $E(1)$ variate (see Section 3.7, Exercises). The form $v_i = H_i(t_i) + (1 - c_i)$, where c_i is the non-censoring indicator, covers both censored and uncensored cases; these v_i are often referred to as *modified Cox–Snell residuals*. The corresponding uniform residuals can then be obtained as $-\log v_i$.

3.4.5 Other Residuals

More sophisticated approaches have been developed in recent years. Barlow and Prentice (1988) and Therneau, Grambsch, and Fleming (1990) defined and employed so-called *martingale residuals*. These cover a wide class of residuals applicable to both parametric and semi-parametric survival analysis and also cater for time-dependent covariates. More generally, consider $v_i(t) = H_i\{\min(t_i, t)\} + 1 - d_i(t)$, where t represents evolving time and $d_i(t) = I(t < t_i, d_i = 1)$ indicates whether unit i is still under observation and unfailed at time t. The process $\{v_i(t)\}$ can be shown to be a continuous-time martingale (see Part IV), and the papers cited show how to use such martingale residuals for investigations of goodness of fit, regression diagnostics, influence, and outlier detection.

Many other types of residuals and diagnostics are used in applied work. The books by Cook and Weisburg (1982) and Atkinson (1985) contain a wealth of material, and an overview is given by Davison and Snell (1991). Among the other types of residual are *leave-one-out residuals*, where each observation is compared with its value predicted from the model fitted to the rest (a type of cross-validation). The *deviance residuals* of *generalised linear models* fame are essentially signed square roots of shifted log-likelihood contributions (see Figure 3.1).

3.4.6 Tests on Residuals

Besides plotting, residuals can be used to construct test statistics. A useful one, based on uniform residuals, is *Moran's test* (Cheng and Stevens, 1989). An important property of this statistic is that its large-sample distribution is not distorted by using an estimate of $\bar{F}$ in place of $\bar{F}$. A more general treatment of this asymptotic theory is given by Crowder (2001).

3.5 Frailty and Random Effects

Statistical models are often constructed with a sense of capturing the broad behaviour of the observed data but with an uncomfortable recognition that there can be a great deal of unexplained variation. In some cases this irritating misconduct on the part of the individuals under study can be ascribed to variability in particular aspects of the basic model. Thus, we have *random effects*, in which certain parameters of the model are deemed to vary over individuals with some probability distribution. An example of this was given in Section 2.2, where the basic model is an exponential distribution but the rate parameter (reciprocal of the mean) varies with a gamma distribution: the model produced by averaging over the rate parameter is a Pareto distribution.

In survival analysis, random effects have gained widespread acclaim under the name *frailty*. This is a random factor by which the hazard function is multiplied to reflect variation over the population of units under study. If the frailty is large, the individual is more prone to imminent failure; and if it is small, then the individual is less, er, frail. We will assume here, as is common, that the frailty of a unit does not change over time—that it is "fixed at birth."

In the univariate case the process of introducing a random effect can look like just replacing one distribution with another. That said, in some cases there are good reasons for doing it: this is so when the original distribution is firmly supported, say by some theoretical underpinning, and the modification by the random effect has a clear contextual interpretation. The procedure really comes into its own in the multivariate case, to be described in Part II. There, a dependence structure can be created for a set of separate components.

3.5.1 Frailty

Let $h_0(t)$ be a common baseline hazard function, and let the hazard for a particular unit be $Zh_0(t)$, where the Zs are independently generated from some probability distribution across the population of units. The idea is that each unit has a latent frailty Z that cannot be observed directly but affects its hazard function multiplicatively. The survivor function for the unit, conditional on $Z = z$, is then

$$\bar{F}(t \mid z) = \exp\left\{-\int_0^t zh_0(s)ds\right\} = \exp\{-zH_0(t)\},$$

where $H_0(t) = \int_0^t h_0(s)ds$ is the baseline integrated hazard function. Assuming that z is unknown (because Z is unobserved), $\bar{F}(t \mid z)$ is unusable for computation and so we have to average over z. The unconditional survivor function is then

$$\bar{F}(t) = \int \bar{F}(t \mid z) \, dK(z) = \int e^{-zH_0(t)} \, dK(z),$$

where K is the distribution function of Z. (See Section 2.5 for this type of integral.) So, $\bar{F}(t)$ is the *Laplace transform*, equivalently the *moment generating function*, of the distribution K evaluated at $H_0(t)$.

Example

Let $h_0(t) = \xi^{-1}$ on $(0, \infty)$ (constant hazard corresponding to an exponential distribution) and $K(z) = 1 - e^{-z/\phi}$ on $(0, \infty)$ (exponentially distributed frailty with mean ϕ). Then $H_0(t) = \int_0^t \xi^{-1}ds = t/\xi$ and $dK(z) = \phi^{-1}e^{-z/\phi}dz$, so the unconditional lifetime survivor function is

$$\bar{F}(t) = \int e^{-zH_0(t)} \, dK(z) = \int_0^\infty e^{-zt/\xi}\phi^{-1}e^{-z/\phi}dz = (1 + \phi t/\xi)^{-1}.$$

This is a Pareto distribution with infinite mean. Its hazard function is $h(t) = (\phi/\xi)(1 + \phi t/\xi)^{-1}$, which is decreasing in t, unlike the originally constant hazard. The usual explanation for this phenomenon is that, later, only units with small frailty will be left and they are not about to give up the ghost so easily. This is really just a re-run of the example in Section 2.2 with $\gamma = 1$, but here presented in terms of the hazard function.

3.5.2 Recovering the Frailty Distribution

Suppose that, in the formula

$$\bar{F}(t) = \int e^{-zH_0(t)} \, dK(z).$$

the functions $\bar{F}(t)$ and $H_0(t)$ are known but $K(z)$ is not. So, can we recover the form of $K(z)$? The problem is essentially one of inverting a Laplace transform, and numerical methods have been much investigated, for example, Abate et al. (1996). However, we do not usually have complete knowledge of $\bar{F}(t)$, only data on which to base estimates. If $H_0(t)$ were completely specified, and with sufficient data to estimate $\bar{F}(t)$ reliably, one might consider proceeding in this way. However, the problem has been tackled in a respectably statistical way.

We have data $(t_1, \ldots, t_n)$, and let $p_i(t \mid z)$ be either (a) the conditional density function of the ith time when it is observed to be equal to t, or (b) its conditional survivor function when it is right-censored at t; for the *iid* case we could write $p(t \mid z)$, without the subscript i. Then,

$$p_i(t; K) = \int p_i(t \mid z) \, dK(z),$$

writing $p_i(t; K)$ for the marginal density or survivor function of t_i evaluated under distribution K. The likelihood function, to be maximised over distribution K, is then

$$L(K) = \prod_{i=1}^{n} p_i(t_i; K).$$

It turns out that the *mle* of K is a discrete distribution with positive probability masses assigned to at most n z-points, often considerably fewer.

Laird (1978) set out the problem and described an algorithm for computing the NPML (*non-parametric maximum likelihood*) estimate of K based on *self-consistency* and showed it to be a case of the EM (*expectation maximisation*) algorithm. Since then a number of papers have proposed more efficient algorithms, for example, Lesperance and Kalbfleisch (1992) and Wang (2007). Theoretical properties of the *mle* of K in various settings have been presented by Simar (1976), Hill et al. (1980), Jewell (1982), and Lindsay (1983a,b).

3.5.3 Discrete Random Effects and Frailty

In some circumstances, it is appropriate to consider discretely distributed random effects or frailty. In the general formula,

$$\bar{F}(t) = \int \bar{F}(t \mid z) \, dK(z),$$

K is now the distribution function of a discrete distribution. In this case, the integral becomes a sum (Section 2.5):

$$\bar{F}(t) = \sum_{j} p_j \bar{F}(t \mid z_j),$$

when Z takes values z_j with probabilities p_j ($j = 1, 2, \ldots$). We then have a mixture distribution with a number of components.

Examples

1. Suppose that a general population is sampled but is known to contain two sub-populations, which are unidentified in the sample. The survivor function for the observed units is then

$$\bar{F}(t) = p_1 \bar{F}(t \mid subpop\ 1) + p_2 \bar{F}(t \mid subpop\ 2),$$

where p_1 and p_2 are the proportions of the sub-populations in the whole population. Extension to three and more sub-populations works similarly. The standard undergraduate coursework example is where the units are manufactured items from different sources.

2. Some people seem, annoyingly, to be immune to the coughs and colds that affect the rest of us. (I believe I have also mentioned this elsewhere, so you can infer that I find it doubly annoying.) So, if

the proportion of super-beings is p, and the survivor function (until taking to one's bed) is $\bar{F}_{mm}(t)$ for mere mortals, the overall survivor function is

$$\bar{F}(t) = p + (1 - p)\bar{F}_{mm}(t).$$

In this example and the next z takes two values only.

3. The opposite case to (2), in a sense, is when some units never enter the race—their lifetime is zero. So, in a batch of components there might be hiding some dud ones that only reveal themselves after you have taken hours setting the equipment up and further hours tracking down the fault. If the proportion of duds is p, then we have

$$\bar{F}(t) = (1 - p)\bar{F}_{go}(t),$$

where $\bar{F}_{go}(t)$ is the survivor function of the good ones.

Other examples of discretely distributed frailty include exposure of a unit to a random number of damage incidents, giving rise to units of varying vulnerability to applied stress, and the presence of a random number of flaws in fibres, causing heterogeneity in their breaking strengths. Again, with the spread of disease in a closed population a variable number of contacts might take place for an unsuspecting individual minding his own business. Caroni et al. (2010) pursued the matter remorselessly: here we just give a very brief summary.

Assume that Z has a discrete distribution on the non-negative integers with $P(Z = k) = p_k$. The proportional hazards model then gives the hazard function as $Zh_0(t)$ and, correspondingly, $\bar{F}(t \mid z) = \bar{F}_0(t)^z$. The unconditional survivor function of T is then given by

$$\bar{F}(t) = \sum_{k=0}^{\infty} p_k \bar{F}_0(t)^k = G_z\{\bar{F}_0(t)\},$$

where G_z is the *probability generating function* of Z. The case $Z = 0$, which entails $P(T > t \mid Z = 0) = 1$ for all t, will be addressed presently (if you are English) or momentarily (if you are American).

Example

Suppose that Z has a geometric distribution with parameter $\rho \in (0, 1)$, so $p_k = \rho^k(1 - \rho)$ for $k = 0, 1, \ldots$. This gives

$$\bar{F}(t) = (1 - \rho)/\{1 - \rho\bar{F}_0(t)\}.$$

Let $\bar{F}_0(t_q) = q$, so that t_q is the upper-tail q-quantile of the baseline distribution. Then $\bar{F}(t_q) = (1 - \rho)/(1 - \rho q)$, which exceeds q if and only if $\rho < (1+q)^{-1}$. In this case, when ρ is sufficiently small, the $\bar{F}$-quantiles lie to the right of those of $\bar{F}_0$; when $\rho > (1 + q)^{-1}$, they lie to the left.

3.5.4 Accommodating Zero Frailty

Frailty distributions that allow $p_0 > 0$ can generate units with zero frailty. For such units, the proportional hazards model entails zero hazard and, in consequence, $\bar{F}_0(t)^0 = 1$ for all t. These units are long-term survivors, ones that never die (but only fade away, like old golfers). In the medical context, they are individuals immune from or cured of the illness in question; in the product reliability context, er, no, can't think of any. Various alternative strategies were proposed by Caroni et al. (2010) for addressing this aspect; I won't bore you with the details (too late?).

3.6 Time-Dependent Covariates

The procedures described above can be extended to cope with covariates that change over time. Let the value of the covariate vector for the ith individual at time t be $x_i(t)$, denote by $x_i(0, t)$ the covariate history over the time period $(0, t)$, and let $x_i = x_i(0, \infty)$. (In the previous development $x_i(t)$ was constant over time.) More generally, some components of $x_i(t)$ might be constant and others not. A distinction is made between *external covariates* and *internal covariates* (Kalbfleisch and Prentice, 1980, Section 5.3).

An external covariate is determined independently of the progress of the individual under observation. Thus, x_i can represent a prespecified regime of conditions to which the ith individual is subjected, some function of clock time, or a non-predetermined stochastic process whose probability distribution does not involve the parameters of the failure time model; environmental conditions might be a case in point. In such cases it is natural to condition the inferences on the observed realisation of x_i.

An internal covariate is generated as part of the individual's development over time. An example is where $x_i(t)$ gives information on the state of the individual or system at time t and therefore can have a very close bearing on the probable failure time. This case is often more difficult to interpret because, for instance, the value of the internal covariate might essentially determine whether failure has occurred or not and hence effectively remove some or all of the stochastic structure of T. On the other hand, $x_i(t)$, being much more relevant to imminent failure than $x_i(0)$, might be an essential ingredient of a useful model.

When there are time-dependent covariates, the continuous-time hazard function becomes

$$h(t; x_i) = h\{t; x_i(0, t)\} = \lim_{\delta \downarrow 0} P\{T \le t + \delta \mid T > t, x_i(0, t)\}$$

and the corresponding survivor function is given formally as

$$\bar{F}(t; x_i) = \exp\left[-\int_0^t h\{s; x_i(0, s)\}ds \right].$$

A natural extension of the PH model (Section 3.3) for time-dependent co-variates is

$$h(t; \mathbf{x}_i) = h_0(t)\, \psi\{\mathbf{x}_i(0, t); \beta\}.$$

This more general class is known as *relative risk models* (Kalbfleisch and Prentice, 2002, Section 4.1). The hazard formally depends on the process $\mathbf{x}_i$ only through its history $\mathbf{x}_i(0, t)$ up to time t; one has to specify which functions of $\mathbf{x}_i$ are to be represented, for example, an integral over $(0, t)$ for some cumulative effect. The relation in Section 3.3, expressing $\bar{F}(t; \mathbf{x})$ as $\bar{F}_0(t)^{\psi_x}$, no longer holds when ψ varies with t; this is simply because ψ cannot be taken outside the integral that relates $\bar{F}$ to h. Also, the hazards might no longer be proportional: the hazard ratio between cases i and j is

$$h(t; \mathbf{x}_i)/h(t; \mathbf{x}_j) = \psi\{\mathbf{x}_i(0, t); \beta\}/\psi\{\mathbf{x}_j(0, t); \beta\},$$

and this is likely to vary with t.

Example

Take $h(t; x_i) = h_0(t)\psi\{x_i(0, t); \beta\}$ with scalar x_i, $h_0(t) = \alpha t^{\gamma-1}$ (Weibull form), $\psi\{x_i(0, t); \beta\} = \exp\{\beta x_i(0, t)\}$, and $x_i(0, t) = x_i \log t$. Then,

$$h(t; x_i) = \alpha t^{\gamma-1} \exp\{\beta x_i \log t\} = \alpha t^{\gamma-1+\beta x_i},$$

and

$$\bar{F}(t; x_i) = \exp(\alpha_i t^{\gamma+\beta x_i}),$$

where $\alpha_i = \alpha/(\gamma + \beta x_i)$. The baseline hazard obtains when $x_i = 0$, and the corresponding survivor function is $\bar{F}_0(t) = \exp(-\alpha t^\gamma/\gamma)$. However, unlike in Section 3.4, $\bar{F}(t; x_i) \neq \bar{F}_0(t)^{\psi_i}$ here. Also, $h(t; x_i)/h(t; x_j) = t^{\beta(x_i-x_j)}$ varies with t, so the hazards are not proportional.

Time-dependent covariates can be put to use as a general approach to detecting departures from proportionality: the suspected departure is built into the model, and then the corresponding coefficients are parametrically tested. For example, there might be some intervention, like treatment, repair, or maintenance, which is supposed to change the hazard function for an individual (advantageously).

Example

Suppose that the hazard for individual i is $h_0(t)\psi_1(\mathbf{x}_{it})$ before intervention and $h_0(t)\psi_2(\mathbf{x}_{it})$ afterwards; here, $\mathbf{x}_{it}$ is an abbreviation for $\mathbf{x}_i(0, t)$. Then, with intervention taking place at time τ_i, we may take the full hazard function to be $h_0(t)\psi(\mathbf{x}_i; t)$ with

$$\psi(\mathbf{x}_i; t) = I(t < \tau_i)\psi_1(\mathbf{x}_{it}) + I(t \geq \tau_i)\psi_2(\mathbf{x}_{it}),$$

$I(.)$ representing the indicator function. An assessment of the effect of the intervention can now be made by comparing the estimates of $\psi_1(\mathbf{x})$ and $\psi_2(\mathbf{x})$ over the sample.

Suppose now that, instead of the sudden change from $\psi_1(\mathbf{x})$ to $\psi_2(\mathbf{x})$ at time τ_i, the transition is smoother: the covariate function is ψ_1 initially and then, as time goes on, moves over to ψ_2 in a continuous manner. This might reflect a cumulative effect of the covariates, or steadily changing conditions. A suitable model would be

$$\psi(\mathbf{x}_i; t) = w(t)\psi_1(\mathbf{x}_{it}) + \{1 - w(t)\}\psi_2(\mathbf{x}_{it}),$$

in which $w(t)$ is a function continuously decreasing from 1 at $t = 0$ to 0 at $t = \infty$, for example, $w(t) = \mathrm{e}^{-\gamma t}$ for some $\gamma > 0$. More generally, $w(t)$ might also depend on $\mathbf{x}_{it}$.

3.7 Exercises

3.7.1 Regression Models

1. The *Gompertz distribution* has survivor function $\bar{F}(t) = \exp\{\psi \xi (1 - \mathrm{e}^{t/\xi})\}$ on $(0, \infty)$, with $\xi > 0$ and $\psi > 0$. Calculate its hazard function. Show that it has the PH property under a log-linear regression model $\log \psi_x = \mathbf{x}^T \beta$.

2. Let T have a Weibull distribution with survivor function $\exp\{-(t/\xi)^\nu\}$ and let $\bar{F}_0(t) = \exp\{-t^\nu\}$. Show that, under a log-linear model $\log \xi_x = \mathbf{x}^T \beta$, the Weibull has both PH and AL properties.

3. The *log-logistic distribution* has survivor function $\bar{F}(t) = \{1 + (t/\alpha)^\rho\}^{-1}$ (see Section 7.2, Exercises). Show that, under a log-linear model $\log \alpha_x = \mathbf{x}^T \beta$, the distribution is both AL and PO.

4. Mean residual life: show how $\bar{F}(t)$ can be obtained from $m(t)$.

5. Construct a log-logistic plot for random samples along the lines of the Weibull plot. The log-logistic distribution is a useful alternative to the Weibull in that it has a non-monotonic hazard, one that rises to a maximum then tails off to zero eventually (see Section 2.7, Exercises).

6. Suppose that t^* is a significant milestone such that $T > t^*$ represents a success of some sort. For example, $t^* = 5$ years of survival after cancer treatment. Let $p_x^* = \bar{F}(t^*; \mathbf{x})$ be the probability of success associated with covariate value $\mathbf{x}$, and let $p_0^* = \bar{F}_0(t^*)$. Discover how p_x^* and p_0^* are related under PH and PO. Suggest plots, based on data comprising only success and failure counts, for checking these regression models.

3.7.2 Residuals

1. The *probability integral transform*. Suppose that T has continuous, monotone increasing distribution function $F(t)$. Verify that $F(T)$ has distribution $U(0, 1)$.

2. The *integrated hazard transform*. Show that $H(T)$, the integrated hazard function, has distribution $E(1)$.

3. *Order statistics.* Let $t_1, \ldots, t_n$ be *iid* with common distribution function F. Then the distribution function of the kth order statistic $t_{(k)}$ may be calculated as follows: let $p_x = P(t_i \leq x)$, then

$$P(t_{(k)} \leq x) = P(at\ least\ k\ of\ the\ unordered\ t_is \leq x)$$

$$= \sum_{j=k}^{n} P(exactly\ j\ of\ the\ t_is \leq x)$$

$$= \sum_{j=k}^{n} \binom{n}{j} p_x^j (1 - p_x)^{n-j}.$$

Now, assuming that the density $f(x) = dp_x/dx$ exists, differentiate to obtain the density of $t_{(k)}$ as

$$f_k(x) = k\binom{n}{k} f(x) p_x^{k-1}(1 - p_x)^{n-k}.$$

4. Specialise the previous problem to show that, for the uniform distribution, the kth order statistic has density $B(n-k+1, k)^{-1} x^k (1-x)^{n-k}$. Hence, the ordered uniform residual $u_{(k)}$ has a beta distribution with mean $k/(n+1)$ and variance $k(n-k+1)/\{(n+1)^2(n+2)\}$. (For the beta distribution see Section 5.6.)

5. Verify that the mean residual life of $H(t)$, where t is the right-censored value, is 1.

3.7.3   Discrete Frailty

1. Take z to be binary with $P(z = 1) = p_1$. Derive $\bar{F}(t) = (1 - p_1) + p_1 \bar{F}_0(t)$. This is a mixture model allowing a proportion $1 - p_1$ of immortals.

2. Take the z distribution to be Poisson with parameter λ: $p_k = e^{-\lambda}\lambda^k/k!$. Verify that

$$\bar{F}(t) = \exp\left[-\lambda\{1 - \bar{F}_0(t)\}\right].$$

3. Take the z distribution as negative binomial with parameters $\nu > 0$ and $\pi \in (0, 1)$: $p_k = \binom{k+\nu-1}{k}\pi^k(1-\pi)^\nu$. Verify that

$$\bar{F}(t) = \left[(1 - \pi)/\{1 - \pi \bar{F}_b(t)\}\right]^\nu = \left\{1 - \left(\frac{\pi}{1 - \pi}\right)\bar{F}_b(t)\right\}^{-\nu}.$$

(The second formula does agree with the first—look again.)

4. In Question 2 note that taking $\nu = 1$ gives the geometric distribution, and the Poisson form is recovered when $\nu \to \infty$ with $\pi = \lambda/\nu$.

3.8 Hints and Solutions

3.8.1 Regression Models

1. $h(t) = \psi e^{t/\xi}$, $h(t; \mathbf{x}) = e^{\mathbf{x}^T \beta} e^{t/\xi} = \psi(\mathbf{x}; \beta) h_0(t)$.

2. PH: $h(t; \mathbf{x}) = \nu \xi_x^{-\nu} t^{\nu-1} = e^{-\nu \mathbf{x}^T \beta} h_0(t)$, where $h_0(t) = \nu t^{\nu-1}$.
   AL: $\bar{F}(t; \mathbf{x}) = \bar{F}_0(t/\xi_x)$.

3. AL: $\bar{F}(t; \mathbf{x}) = \{1 + (t/\alpha_x)^\rho\}^{-1} = \bar{F}_0(t/\alpha_x)$, where $\bar{F}_0(t) = (1 + t^\rho)^{-1}$.
   PO: $\bar{F}(t; \mathbf{x})/\{1 - \bar{F}(t; \mathbf{x})\} = (t/\alpha_x)^{-\rho} = \alpha_x^\rho \bar{F}_0(t)/\{1 - \bar{F}_0(t)\}$.

4. Differentiate with respect to t,

$$m'(t)\bar{F}(t) + m(t)\bar{F}'(t) = -\bar{F}(t),$$

which yields

$$\{m'(t) + 1\}/m(t) = -\bar{F}'(t)/\bar{F}(t) = h(t),$$

and thence, by integration,

$$\log\{m(t)/m(0)\} + \int_0^t m(y)^{-1} dy = -\log \bar{F}(t) = H(t);$$

note that a constant of integration has been included to match the two sides at $t = 0$. Hence, we get the inverse relationship, expressing $\bar{F}$ in terms of m, as

$$\bar{F}(t) = \{m(t)/m(0)\}^{-1} \exp\left\{-\int_0^t m(y)^{-1} dy\right\}.$$

5. $\log\left[\bar{F}_0(t)/\{1 - \bar{F}_0(t)\}\right] = \rho \log \alpha - \rho \log t$, so plot the estimated log-odds *versus* $\log t$.

6. PH: $p_x^* = (p_0^*)^{\psi_x}$, so $\log p_x^* = \psi_x \log p_0^*$ and a plot of $\log p_x^*$ *versus* ψ_x should look like a straight line through the origin of slope $\log p_0^*$.
   PO: $p_x^*/(1 - p_x^*) = \psi_x p_0^*/(1 - p_0^*)$, yielding a straight-line plot of the odds ratios. Sufficient replication at each $\mathbf{x}$-value would be required to estimate p_x^*.

3.8.2 Residuals

1. $P\{F(T) \le a\} = P\{T \le F^{-1}(a)\} = F\{F^{-1}(a)\}$. Note that, by the same token, $\bar{F}(T)$ has distribution $U(0, 1)$.

2. Start with $P\{H(T) \le a\} = P\{-\log \bar{F}(T) \le a\}$.

3. There is a lot of cancelling of terms in the summation.

4. See Section 5.6, Exercises.

5. If x has distribution $E(1)$, then $E(x - t \mid x > t) = \int_t^\infty (x - t) f(x \mid x > t) dx$, and $f(x \mid x > t) = e^{-(x-t)}$.

4

Continuous Time:
Non- and Semi-Parametric Methods

The continuous-time framework has been much developed over the past couple of decades. In the absence of explanatory variables there is only the hazard function to estimate. This can be done parametrically, with a fully specified model, or non-parametrically, using the Kaplan–Meier estimator. To deal with explanatory variables the options are, broadly, (a) a fully parametric model for the hazards $h(t; \mathbf{x})$ or their equivalents, and (b) a model that separates the regression and hazard components, that is, the $\mathbf{x}$ and t parts, leaving the latter parametrically unspecified. In (b) it might be possible to find a combination of model and estimating function that allows the unspecified baseline hazards to be eliminated, as in Cox's combination of proportional hazards and partial likelihood (Section 4.2). The full likelihood function needed for (a), and for (b) when the baseline hazards are to be estimated, is as given in Section 3.2.

4.1 Random Samples

4.1.1 The Kaplan–Meier Estimator

Consider a random sample (i.e., without explanatory variables) of continuous failure-time observations generated from some unknown underlying survivor function $\bar{F}(t)$. A parametric model for $\bar{F}$ is *not* assumed. The derivation here follows that of Kalbfleisch and Prentice (2002, Section 1.4.1).

Let the distinct observed failure times be $t_1 < t_2 < \ldots < t_m$. Suppose that r_l failures are observed at time t_l, and that s_l cases are right-censored during the interval $[t_l, t_{l+1})$ at times t_{ls} (known or unknown), where $t_{ls} \leq t_{l,s+1}$ ($s = 1, \ldots, s_l; l = 0, \ldots, m; t_0 = 0, t_{m+1} = \infty$). Strictly speaking, for continuous time r_l should be 1 for all l. In practice, however, because of rounding, tied failure times do occur. The likelihood function can then be written as

$$L = \prod_{l=1}^{m} \{\bar{F}(t_l-) - \bar{F}(t_l)\}^{r_l} \times \prod_{l=0}^{m} \prod_{s=1}^{s_l} \bar{F}(t_{ls}).$$

The terms in the first product would normally be written as $f(t_l)^{r_l}$, where $f(t)dt = \bar{F}(t-) - \bar{F}(t)$. However, the reason for adopting this odd representation will now become apparent.

The maximum likelihood estimator $\hat{\bar{F}}$ of $\bar{F}$ must be discontinuous on the left at t_l; otherwise $\bar{F}(t_l-) = \bar{F}(t_l)$, which makes $L = 0$. Also, since $t_{ls} \geq t_l$, $\hat{\bar{F}}(t_{ls}) \leq \hat{\bar{F}}(t_l)$, and so L is maximised with $\hat{\bar{F}}(t_{ls}) = \hat{\bar{F}}(t_l)$. So, in the expression for L we may replace the second term by $\prod_{l=0}^{m} \bar{F}(t_l)^{s_l}$. The expression does not now involve $\bar{F}(t)$ for $t_l < t < t_{l+1}$ and we can take $\hat{\bar{F}}(t) = \hat{\bar{F}}(t_l)$ on this interval. Hence, $\hat{\bar{F}}$ is a discrete survivor function with steps at $t_1, \ldots, t_m$ and so we can express $\hat{\bar{F}}(t_l)$ in terms of discrete hazards (Section 2.3) as $\prod_{s=1}^{l}(1 - \hat{h}_s)$ and $\hat{\bar{F}}(t_l-) - \hat{\bar{F}}(t_l)$ as $\hat{h}_l \prod_{s=1}^{l-1}(1 - \hat{h}_s)$. The $\hat{h}_s$ are chosen to maximise

$$L = \prod_{l=1}^{m} \left\{ h_l \prod_{s=1}^{l-1}(1 - h_s) \right\}^{r_l} \times \prod_{l=0}^{m} \left\{ \prod_{s=1}^{l}(1 - h_s) \right\}^{s_l};$$

the terms $\prod_{s=1}^{0}(.)$ are interpreted as 1 because the first is $\bar{F}(0) - \bar{F}(t_1) = h_1$ and the second is $\prod_{s=1}^{0} \bar{F}(0) = 1$. Collecting terms,

$$L = \prod_{l=1}^{m} \{ h_l^{r_l}(1 - h_l)^{q_l - r_l} \},$$

where

$$q_l = (r_l + s_l) + \cdots + (r_m + s_m) \text{ for } l = 1, \ldots, m;$$

taking $r_0 = 0$, q_0 is the overall sample size and q_l for $l > 0$ is the number of cases at risk at time t_l-. The likelihood function L attains its maximum when $h_l = r_l/q_l$. The *product limit estimator* of $\bar{F}(t)$ is then

$$\hat{\bar{F}}(t) = \prod(1 - \hat{h}_l) = \prod(1 - r_l/q_l),$$

where the product is taken over $\{l : t_l \leq t\}$. The formula is also known as the *Kaplan–Meier estimator* after their much-referenced 1958 paper. It is a non-parametric estimator of the survivor function.

If $s_m > 0$, so that at least one failure time is known to exceed t_m, $r_m/q_m < 1$ and so $\hat{\bar{F}}(t_m) > 0$. The downward progress of $\hat{\bar{F}}(t)$ beyond t_m to its eventual demise at value 0 is usually taken to be undefined in this case. The censoring times t_{ls} have no effect on $\hat{\bar{F}}(t)$, which might seem a bit strange at first sight. However, as far as the lifetime distribution is concerned, nothing happens at time t_{ls}—it is only the observation process that is interrupted there. When there is no censoring, $s_l = 0$ for all l, so $\hat{h}_l = r_l/(r_l + \cdots + r_m)$ and $1 - \hat{h}_l = q_{l+1}/q_l$, and then

$$\hat{\bar{F}}(t) = q_{l+1}/q_1 = (r_{l+1} + \cdots + r_m)/(r_1 + \cdots + r_m) \text{ for } t_l \leq t < t_{l+1},$$

which is the usual *empirical survivor function*. If, in addition, the t_i are all distinct, then $r_l = 1$ for all l, $q_1 = m$, $q_{l+1} = m - l$, and $\hat{\bar{F}}(t_l) = 1 - l/m$.

TABLE 4.1

Strengths of Cords

τ_l	r_l	s_l	q_l	$\hat{h}_l$	$\hat{F}_l$
25	0	2	48	0.0000	1.000
35	1	2	46	0.0022	0.978
45	3	3	43	0.0698	0.910
52	14	0	37	0.3784	0.566
56	16	0	23	0.6957	0.172
60	7	0	7	1.0000	0.000

An asymptotic variance calculation can be made using an informal argument based on *Greenwood's formula* (Section 5.2):

$$\text{var}\{\hat{F}(t)\} \approx \hat{F}(t)^2 \sum \hat{h}_l/\{q_l(1-\hat{h}_l)\} = \hat{F}(t)^2 \sum r_l/\{q_l(q_l-r_l)\}.$$

More rigorous asymptotic theory has been given by Meier (1975), for the case of censorship at fixed, predetermined times, and by Breslow and Crowley (1974), for the case of random censoring.

4.1.2 Strengths of Cords

Here we use the data from Section 1.2 to illustrate the Kaplan–Meier estimate. The figures have been grouped into bands $20-30-40-50-54-58-62$, taking the mid-points as the representative τ-values. Table 4.1 gives details for computing the estimates to a fussy accuracy to enable checking. For example, $\hat{h}_3 = 3/43 = 0.0698$ and $\hat{F}_3 = 0.978(1-0.0698) = 0.910$.

If the last s_l had not been zero, it would reveal that $m > 6$ and $\tau_m > 60$. Suppose that s_6 were 1, say. Then, $q_6 = 8$, $\hat{h}_6 = 7/8 = 0.8750$, $\hat{F}_l = 0.1756(1-0.8750) = 0.0219$ and $\hat{F}_6$ would be undefined beyond τ_6. You can also check the arithmetic with the function `survfit` from the R-package `survival`. Note the use of `weights`.

```
#cords data (Sec 4.1)
library(survival);
tv2=c(25,35,45,52,56,60, 25,35,45,52,56,60); #times
cv2=c(rep(1,6), rep(0,6)); #censoring indicators
freqs=c(0,1,3,14,16,7, 2,2,3,0,0,0); #frequencies
sobj=Surv(tv2,cv2,type='right',origin=0); #create 'survival object' sobj
sfit1=survfit(sobj~1,weights=freqs); summary(sfit1);
```

The output appears as

```
time n.risk n.event survival std.err lower 95% CI upper 95% CI
  35     46       1    0.978  0.0215       0.9370       1.000
  45     43       3    0.910  0.0429       0.8296       0.998
  52     37      14    0.566  0.0773       0.4328       0.739
  56     23      16    0.172  0.0592       0.0878       0.338
  60      7       7    0.000     NaN           NA           NA
```

In the last line, the NaN and NA codes just reflect the end of time (don't panic, this is only for the data).

4.1.3 The Integrated and Cumulative Hazard Functions

The integrated hazard function is $H(t) = -\log \bar{F}(t)$. This suggests the estimator

$$\hat{H}(t) = -\log \hat{\bar{F}}(t) = -\sum \log(1 - \hat{h}_l) = -\sum \log(1 - r_l/q_l),$$

with summation over $\{l : t_l \leq t\}$. Applying the expression given above for $\text{var}\{\hat{\bar{F}}(t)\}$, together with the *delta method*,

$$\text{var}\{\hat{H}(t)\} \approx \hat{\bar{F}}(t)^{-2}\text{var}\{\hat{\bar{F}}(t)\} \approx \sum \hat{h}_l/\{q_l(1 - \hat{h}_l)\}.$$

If the $\hat{h}_l$ are small the alternative *Nelson–Aalen estimator*,

$$\tilde{H}(t) = \sum \hat{h}_l = \sum r_l/q_l,$$

is suggested by the *cumulative hazard* approximation to $H(t)$ given in Section 2.3. In fact, the r_l/q_l will tend to be small early on, that is, for smaller values of l, when the q_l are larger. So $\hat{H}(t)$ and $\tilde{H}(t)$ will often not differ much for the smaller values of t. The corresponding alternative estimator of the survivor function is $\tilde{S}(t) = \exp\{-\tilde{H}(t)\}$. The variance of $\tilde{H}(t)$ can be approximated as

$$\text{var}\{\tilde{H}(t)\} \approx \sum \hat{h}_l(1 - \hat{h}_l)/q_l;$$

this is similar to that of $\hat{H}(t)$ when the $\hat{h}_l$ are small. Plots of $\hat{H}(t)$ or $\tilde{H}(t)$ versus t are useful for both exploratory analysis and assessment of goodness of fit (Nelson, 1970). That "a picture paints a thousand words" was even immortalised in song—are you old enough to remember Kojak? (If so, you have probably got a similar hairstyle.)

4.1.4 Interval-Censored Data

In this case the failure times are only partially observed: what is recorded for t_i is an interval $(L_i, R_i]$ such that $L_i < t_i \leq R_i$. The setup covers the following special cases:

1. Exact observation ($L_i = R_i-$)
2. Interval censoring ($L_i < R_i$)
3. Right-censoring at L_i ($R_i = \infty$)
4. Left-censoring at R_i ($L_i = -\infty$)

The likelihood contribution for the ith observation is $p_i = F(R_i) - F(L_i)$, where F is the common distribution function of the t_i. The overall likelihood function is then $\prod_{i=1}^{n} p_i$. Note that a maximum likelihood estimator of F

will assign no probability to intervals not overlapping any of the $(L_i, R_i]$; otherwise, one or more of the p_i would be correspondingly reduced. So, the *mle* $\hat{F}$ is flat on such intervals.

Construct disjoint intervals $(a_j, b_j]$ whose union is the union of the $(L_i, R_i]$ and whose end points are all Ls and Rs. This can be done by arranging the whole set of Ls and Rs in ascending order, which gives $2n$ points, some of which may coincide. Then take the as and bs as the resulting cut-points to produce $a_1 \leq b_1 = a_2 \leq b_2 = a_3 \ldots$; note that some $a - b$ intervals might be "empty," in the sense that they do not overlap any L–R intervals (though this can be corrected with a bit more trouble). For computation, zero-width intervals (where the observed value is exact) can be treated by taking $a_j = b_j - \epsilon$ for some small value ϵ. If the number of zero-width L–R intervals is $n_1 < n$, then the number of $a - b$ intervals is $m = n_1 + 2(n - n_1) - 1$, which is at most $2n - 1$; to see this, start with one L–R interval of positive width and then add L-R intervals one at a time to observe the effect.

Let $q_j = F(a_j)$. Then, with the construction above, $b_j = a_{j+1}$ and so $p_i = \sum_{j \in i}(q_{j+1} - q_j)$, with summation over just those $a - b$ intervals that partition $(L_i, R_i]$; this can be written as $p_i = \sum_{j=1}^{m} \alpha_{ij}(q_{j+1} - q_j)$, where $\alpha_{ij} = I\{(a_j, b_j] \subseteq (L_i, R_i]\}$ in terms of the indicator function. Computation of $\hat{F}$ now reduces to maximising the log-likelihood

$$l(\mathbf{q}) = \sum_{i=1}^{n} \log \left\{ \sum_{j=1}^{m} \alpha_{ij}(q_{j+1} - q_j) \right\}$$

over $\mathbf{q} = (q_1, \ldots, q_m)$ subject to constraints $0 \leq q_1 \leq q_2 \leq \ldots \leq q_{m+1} = 1$. The corresponding score and information functions have components

$$\partial l / \partial q_k = \sum_{i=1}^{n} (\alpha_{i,k-1} - \alpha_{ik}) \left\{ \sum_{j=1}^{m} \alpha_{ij}(q_{j+1} - q_j) \right\}^{-1},$$

and

$$\partial^2 l / \partial q_k \partial q_l = -\sum_{i=1}^{n} (\alpha_{i,k-1} - \alpha_{ik})(\alpha_{i,l-1} - \alpha_{il}) \left\{ \sum_{j=1}^{m} \alpha_{ij}(q_{j+1} - q_j) \right\}^{-2}.$$

The information matrix is thus positive definite, so $-l(\mathbf{q})$ is a convex function: this is good news for numerical maximisation (Appendix D).

Peto (1973) set out the framework described above, and Turnbull (1974; 1976) proposed a "self-consistent" algorithm for computation of the *mle*. The asymptotic properties of the *mle* depend on the censoring mechanism, that is, how L_i and R_i are generated in relation to t_i, and are not generally straightforward: see, for example Groenboom and Wellner (1992), Huang and Wellner (1995), and Huang (1996).

4.2 Explanatory Variables

The observational setting here comprises observed failure times t_l ($l = 1, \dots,$ m), taken to be in increasing order, and individual cases of whom the ith is accompanied by a vector $\mathbf{x}_i$ of explanatory variables (covariates, design, grouping or classification indicators, etc.).

4.2.1 Cox's Proportional Hazards Model

You might not have heard about it but in 1972 a paper was read to the Royal Statistical Society in which Cox proposed the *semi-parametric* PH (*proportional hazards*) model. This specifies the hazard function for the ith individual as

$$h(t; \mathbf{x}_i) = \psi_i h_0(t);$$

here, $h_0(t)$ is some unspecified *baseline hazard function* and $\psi_i = \psi(\mathbf{x}_i; \beta)$ is some positive function, often taken to be of the form $\psi(\mathbf{x}; \beta) = \exp(\mathbf{x}^T \beta)$, incorporating a vector β of regression coefficients.

The survivor function corresponding to $h(t; \mathbf{x})$ is given by

$$\bar{F}(t; \mathbf{x}) = \exp \left\{ - \int_0^t h(s; \mathbf{x}) ds \right\},$$

and then the PH form yields

$$\bar{F}(t; \mathbf{x}_i) = \bar{F}_0(t)^{\psi_i},$$

where $\bar{F}_0(t)$ is the *baseline survivor function* corresponding to $h_0(t)$.

The PH model is *semi-parametric* in that the ψ_i are parametrically specified but $h_0(t)$ is not. It will be seen that this specification, together with the partial likelihood, yields a major benefit: one can estimate and test β without having to do anything about the unknown $h_0(t)$. However, as always, such an assumption should not be taken in too carefree a manner. It requires that the only effect of the covariate is to multiply the hazard by a factor that is constant over time. For instance, it would not be strictly valid in the case of a transient medicinal effect, where the hazard is reduced for a while but later resumes its untreated level, the reduction depending on the dosage x. Cox and Oakes (1984, Section 5.7) discussed such an example.

4.2.2 Cox's Partial Likelihood

Let $R(t)$ be the *risk set* at time $t-$, that is, the set of individuals still at risk just before time t of eventually being recorded as failures. Then the probability that individual $i \in R(t)$ fails in the time interval $(t, t + dt]$ is $h(t; \mathbf{x}_i)dt$. We assume for the moment that the individuals have distinct failure times, that is, there are no ties. Given the events up to time t_l-, and given that there is a failure at time t_l, the conditional probability that, among those still at risk, it is individual i_l at whom the famous fickle finger of fate and fortune points is

$$h(t_l; \mathbf{x}_{i_l}) \bigg/ \sum{}^l h(t_l; \mathbf{x}_i) = \psi_{i_l} \bigg/ \sum{}^l \psi_i;$$

here the PH assumption has been applied, and the summation $\sum^l$ is over $i \in R(t_l)$, that is, over all the individuals still biding their time in the risk set at time t_l. Notice that $h_0(t)$ has been cancelled out: this nuisance parameter, or function, has been eliminated; it is deceased; this nuisance parameter is no more; it is an ex nuisance parameter. By analogy with ordinary likelihood, take the product over $l = 1, \ldots, m$ of these terms: the result is the so-called *partial likelihood function*

$$P(\beta) = \prod_{l=1}^{m} \left\{ \psi_{i_l} \bigg/ \sum{}^l \psi_i \right\}.$$

The censored failure times make their presence felt only in the denominators of the factors of $P(\beta)$, the individuals concerned being included in the appropriate risk sets. Cox (1972) argued that there is no further information in the data about β, the parameter of interest. This is because $h_0(t)$ could be very small between failure times and not small at the observed t_l. It is only at the times t_l that β can affect the outcome through the *relative* chances of failure between different individuals. Note also that the times t_l do not appear explicitly in $P(\beta)$, it is only their ordering that has been used. This makes partial likelihood a method based on *ranks*, and this was shown formally by Kalbfleisch and Prentice (1973).

4.2.3 Inference

The function $P(\beta)$ provides estimates and tests for β analogous to those based on ordinary likelihoods. Thus, the *mple* (maximum partial likelihood estimator) is found by maximising $P(\beta)$ or, equivalently, $\log P(\beta)$. We have

$$\log P(\beta) = \sum_{l=1}^{m} \left(\log \psi_{i_l} - \log \sum{}^l \psi_i \right)$$

with first and second derivatives

$$\mathbf{U}(\beta) = \partial \log P(\beta)/\partial \beta = \sum_{l=1}^{m} \mathbf{U}_l(\beta) \text{ and } \mathbf{V}(\beta) = -\partial^2 \log P(\beta)/\partial \beta^2 = \sum_{l=1}^{m} \mathbf{V}_l(\beta),$$

where the vector $\mathbf{U}_l(\beta)$ has jth component $\partial \log(\psi_{i_l}/\sum^l \psi_i)/\partial \beta_j$ and the matrix $\mathbf{V}_l(\beta)$ has (j, k)th element $-\partial^2 \log(\psi_{i_l}/\sum^l \psi_i)/\partial \beta_j \partial \beta_k$. With the usual choice, $\psi_i = \exp(x_i^T \beta)$, we have $U_l(\beta) = x_{i_l} - v_l(\beta)$ and $V_l(\beta) = X_l(\beta) - v_l(\beta)v_l(\beta)^T$, where

$$v_l(\beta) = \left(\sum^l \psi_i \right)^{-1} \sum^l \psi_i x_i \text{ and } X_l(\beta) = \left(\sum^l \psi_i \right)^{-1} \sum^l \psi_i x_i x_i^T.$$

Under some regularity conditions $P(\beta)$ has asymptotic properties similar to those of ordinary likelihoods. Thus, $V(\hat{\beta})^{-1}$ provides an estimate for the covariance matrix of $\hat{\beta}$, and so follow hypothesis tests and confidence regions for β. Inferences can also be based on $P(\beta)$ in the same way as ordinary likelihood-ratio tests.

4.2.4 Computation

A computationally useful aspect of partial likelihood with $\psi_i = \exp(x_i^T \beta)$ is that $P(\beta)$ is a convex function of β, and so has a unique maximum, $\hat{\beta}$ (see Section 4.6, Exercises). The numerical implications of this for maximising such functions are discussed in Appendix D. The R-function coxph, part of the R survival package, does the work—an illustration is given below.

4.2.5 Catheter Infection

Again we use these data for illustration. Assume that the data have been loaded and that the survival library has been called up, as previously in Section 3.3. The following R-code will produce Kaplan–Meier estimates.

```
#catheter data (Sec I.4.2)
dx1=read.table('catheter1.dat',header=T); attach(dx1);
library(survival); sobj=Surv(tim,cns,type='right',origin=0);
#Kaplan-Meier estimate
sfit1=survfit(sobj~sex); summary(sfit1);
plot(sfit1,conf.int=T,lty=3:2,xlab='time',ylab='survival prob');
title(main='Catheter data: KM plot',sub=' ');
```

The output contains the following lines, and the plot is given as Figure 4.1.

```
                  sex=1
time n.risk n.event survival std.err lower 95% CI upper 95% CI
   9     13       1    0.923  0.0739        0.789        1.000
  10     12       1    0.846  0.1001        0.671        1.000
  11     11       1    0.769  0.1169        0.571        1.000
  13     10       1    0.692  0.1280        0.482        0.995
  17      9       1    0.615  0.1349        0.400        0.946
  18      8       1    0.538  0.1383        0.326        0.891
  23      7       1    0.462  0.1383        0.257        0.830
                  sex=2
time n.risk n.event survival std.err lower 95% CI upper 95% CI
   3     14       1    0.929  0.0688        0.803        1.000
   9     12       1    0.851  0.0973        0.680        1.000
  11     11       1    0.774  0.1152        0.578        1.000
  16      9       1    0.688  0.1306        0.474        0.998
  17      8       1    0.602  0.1397        0.382        0.949
  22      7       1    0.516  0.1438        0.299        0.891
```

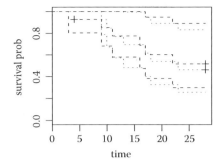

FIGURE 4.1
Kaplan–Meier survival curves for catheter data.

A proportional hazards model may be fitted with R-code.

```
#Cox PH fit
sfit2=coxph(sobj~age+sex); summary(sfit2);
```

and the output contains

```
n= 27
          coef exp(coef)  se(coef)      z Pr(>|z|)
age -0.020772  0.979442  0.015655 -1.327    0.185
sex -0.006843  0.993180  0.559272 -0.012    0.990

    exp(coef) exp(-coef) lower .95 upper .95
age    0.9794      1.021    0.9498     1.010
sex    0.9932      1.007    0.3319     2.972

Rsquare= 0.065    (max possible= 0.941 )
Likelihood ratio test= 1.82  on 2 df,   p=0.4033
Wald test             = 1.77  on 2 df,   p=0.4123
Score (logrank) test = 1.82  on 2 df,   p=0.4021
```

which may be compared with the previous fits in Section 3.3.

4.3 Some Further Aspects

4.3.1 Stratification

A useful extension to the analysis can be made via stratification. Suppose that there are groups of individuals in the data arising from different sub-populations or strata, between which $h_0(t)$ is likely to differ. One can allow for this by constructing a separate partial likelihood function for each group and then taking the overall $P(\beta)$ as their product. The regression parameter β may be common to all groups or may differ between groups in any respect.

4.3.2 Tied Lifetimes

When tied values are among the observed failure times, the expression for $P(\beta)$ is more complicated. Thus, if there are r_l observed failure times equal to t_l, the lth term in the product defining $P(\beta)$ should allow for all distinct subsets of r_l individuals from the risk set $R(t_l)$. This can lead to an unwieldy computation, and an approximation due to Peto (1972) is to replace this lth term by $\prod(\psi_{i_l}/\sum^l \psi_i)$, where the product is taken over the r_l tied cases. Since $\sum^l \psi_i$ is the same for all these cases, the replacement term can be written as $(\prod \psi_{i_l})/(\sum^l \psi_i)^{r_l}$. This approximation will give reasonable results provided that the proportion of tied values in each $R(t_l)$ is small. Otherwise, one must compute the partial likelihood contributions taking full account of all the combinatorial possibilities of r_l individuals from $R(t_l)$ (see Section 5.3).

4.3.3 The Baseline Survivor Function

The baseline survivor function $\bar{F}_0$ can be estimated non-parametrically by an extension of the Kaplan–Meier estimator. Let t_l ($l = 1, \ldots, m$) be the distinct failure times in increasing order, with $t_0 = 0$ and $t_{m+1} = \infty$; let R_l be the set of individuals failing at time t_l, and let S_l likewise comprise the cases right-censored during $[t_l, t_{l+1})$. The likelihood function analogous to that in Section 4.1 becomes, applying $\bar{F}(t; \mathbf{x}_i) = \bar{F}_0(t)^{\psi_i}$,

$$L = \prod_{l=1}^{m} \left[\prod_{i \in R_l} \{\bar{F}_0(t_l-)^{\psi_i} - \bar{F}_0(t_l)^{\psi_i}\} \times \prod_{i \in S_l} \{\bar{F}_0(t_i)^{\psi_i}\} \right].$$

To maximise L, $\hat{\bar{F}}_0$ must jump at t_l- for each l, and $\hat{\bar{F}}_0(t) = \hat{\bar{F}}_0(t_l)$ for $t_l \leq t < t_{l+1}$. Hence,

$$\hat{\bar{F}}_0(t_l) = \prod_{s=1}^{l} (1 - \hat{h}_{0s}),$$

where the $\hat{h}_{0s}$ maximise

$$L = \prod_{l=1}^{m} \left[\prod_{i \in R_l} \left\{ \prod_{s=1}^{l-1}(1 - h_{0s})^{\psi_i} - \prod_{s=1}^{l}(1 - h_{0s})^{\psi_i} \right\} \times \prod_{i \in S_l} \prod_{s=1}^{l}(1 - h_{0s})^{\psi_i} \right].$$

Collecting terms, as was done in Section 4.1, we obtain

$$L = \prod_{l=1}^{m} \left[\prod_{i \in R_l} \{1 - (1 - h_{0l})^{\psi_i}\} \times \prod_{i \in Q_l} (1 - h_{0l})^{\psi_i} \right],$$

where $Q_l = R(t_l) - R_l$. Thus, the $\hat{h}_{0l}$ may be computed by maximisation of L, after substituting the maximum partial likelihood estimate $\hat{\beta}$ for β. Note that L is of orthogonal form, being a product of m factors each involving just one h_{0l}. Thus, the maximisation problem can be reduced to a set of m one-dimensional searches (Appendix D).

Alternatively, L could be maximised jointly over β and the h_{0l} (Kalbfleisch and Prentice, 1980, Section 4.3). However, the resulting estimator for β might not be as well behaved computationally as the maximum partial likelihood estimator.

4.3.4 The Log-Rank Test

Suppose that we have k random samples, the jth arising from a population with survivor function $\bar{F}_j$. Express $\bar{F}_j(t)$ as $\bar{F}_0(t)^{\exp(\beta_j)}$, in terms of some baseline survivor function $\bar{F}_0$. Now let $\beta = (\beta_1, \ldots, \beta_k)^T$ and, for unit i arising from F_j, let $\mathbf{x}_i$ be the jth column of the $k \times k$ unit matrix. Then, $\mathbf{x}_i^T \beta = \beta_j$ and the survivor function for unit i can be written in PH form as $\bar{F}_0(t)^{\exp(\mathbf{x}_i^T \beta)}$. So, to determine whether the k survivor functions are equal, we can fit the Cox PH model and test the null hypothesis that $\beta = \mathbf{0}$.

To perform a score test we calculate the various quantities as follows, in the notation of Section 4.2. For unit i arising from $\bar{F}_j$, $\psi_i(\beta) = \exp(x_i^T \beta) = \exp(\beta_j)$. Thus, $\sum^l \psi(\mathbf{0}) = q_l$, the size of the risk set $R(t_l)$, and

$$\sum^l \psi(\mathbf{0}) \mathbf{x}_i = \mathbf{q}_l = (q_{l1}, \ldots, q_{lk})^T,$$

where q_{lj} is the number of F_j units at risk at time t_l, so $q_{l+} = q_l$. We can now obtain

$$\mathbf{v}_l(\mathbf{0}) = \mathbf{q}_l / q_l \quad \text{and} \quad \mathbf{U}_l(\mathbf{0}) = \mathbf{x}_{i_l} - \mathbf{q}_l / q_l;$$

note that the binary vector $\mathbf{x}_{i_l}$ does have expected value $\mathbf{q}_l / q_l$ when $\beta = \mathbf{0}$. Lastly,

$$\mathbf{X}_l(\mathbf{0}) = q_l^{-1} \operatorname{diag}(q_{l1}, \ldots, q_{lk}) \quad \text{and} \quad \mathbf{v}_l(\mathbf{0}) \mathbf{v}_l(\mathbf{0})^T = \mathbf{q}_l \mathbf{q}_l^T / q_l^2 = q_l^{-2} \{q_{lj} q_{lj'}\}_{jj'},$$

which yields

$$\{\mathbf{V}_l(\mathbf{0})\}_{jj'} = \delta_{jj'} q_{lj} / q_l - q_{lj} q_{lj'} / q_l^2.$$

The score test statistic would now normally be based on the quadratic form $\mathbf{U}(\mathbf{0})^T \mathbf{V}(\mathbf{0})^{-1} \mathbf{U}(\mathbf{0})$. But $\mathbf{V}(\mathbf{0})$ is singular because its rows all sum 0; this just reflects the lack of identiability in the β_j. The usual cure is to take one of the β_j equal to 0, say the last, which entails lopping off the last element of $\mathbf{U}(\mathbf{0})$ and the last row and column of $\mathbf{V}(\mathbf{0})$. The resulting statistic has asymptotic distribution χ^2_{k-1} under the null hypothesis.

Finally, why *log-rank* test? Buried in the annals of statistics are derivations of the form of test statistic under various guises, nothing to do with the PH score. That $\mathbf{x}_{i_l}$ has expected value $\mathbf{q}_l / q_l$ when the F_j are equal is true in the wider context. Names associated with the test include Mantel, Haenszel, Peto, and Cox; and it can be derived from a rank test statistic based on the logarithm of the Nelson–Aalen estimator. The *Wilcoxon test*, also known as the *Breslow test*, is an alternative to the log-rank test—see Collett (2003, Sections 2.6 and 2.7).

4.3.5 Schoenfeld Residuals

The source paper is Schoenfeld (1982): the *Schoenfeld residual vector* attached to failure time t_l is

$$\mathbf{r}(Sch)_l = \mathbf{x}_l - \hat{\mathbf{v}}_l,$$

where $\mathbf{v}_l$ defined in Section 4.2 as

$$\mathbf{v}_l = \left(\sum^l \psi_i \right)^{-1} \sum^l \psi_i \mathbf{x}_i;$$

$\hat{\mathbf{v}}_l$ means that the ψ_i are evaluated at $\beta = \hat{\beta}$. The $\mathbf{r}(Sch)_l$ are the contributions to $\mathbf{U}(\beta)$, the first derivative, with respect to parameter vector β, of the log-partial-likelihood. Since $\mathbf{U}(\hat{\beta}) = \mathbf{0}$ they sum to zero. Note also that there are just m of these residuals, one for each observed failure time t_l, but none for censored times. So that all shall have prizes, they are often defined, for $i = 1, \ldots, n$, as

$$\mathbf{r}(Sch)_i = c_i(\mathbf{x}_i - \hat{\mathbf{v}}_i),$$

incorporating the non-censoring indicator c_i. This is a bit disingenuous because now the censored units just get a zero residual. This reminds me of the football team whose goal score was zero but, in general opinion, was lucky to get that.

Schoenfeld residuals are useful for checking that the hazards are in fact proportional in a fitted PH model. Grambsch and Therneau (1994) proposed various diagnostics for the Cox PH model. They showed how the residuals can be used to throw light on the constancy or otherwise of the regression coefficients β_j over time: if β_j is not constant the hazards are not proportional. See Therneau and Grambsch (2000) for an extended account.

4.3.6 Time-Dependent Covariates

When $\mathbf{x}$ varies over time, the partial likelihood becomes

$$P(\beta) = \prod_{l=1}^{m} \left\{ \psi_{i_l l} \Big/ \sum^l \psi_{il} \right\},$$

where ψ_{il} is written for $\psi\{\mathbf{x}_i(t_l); \beta\}$ and $\mathbf{x}_i(t_l)$ is the covariate value for unit i at time t_l. This extended form of $P(\beta)$ can be employed in the same way as described above to obtain an estimate for β and associated asymptotic inference. Such time-varying $\mathbf{x}$s can be introduced artificially to check the PH assumption: if the associated regression coefficient is significantly non-zero, it suggests that the hazard ratios are not constant over time. However, a baseline survivor function cannot be estimated in the manner described above because, as noted previously, the fundamental relation equating $\bar{F}(t; \mathbf{x})$ to $\bar{F}_0(t)^{\psi_x}$ fails here.

Notice that in the *l*th term of the product defining $P(\beta)$, the covariate vectors for all individuals involved are to be evaluated at time t_l. In practice, when covariate information is less complete, most-recent values or ones estimated by some sort of interpolation or extrapolation have to be used. Continuously varying $\mathbf{x}$s are easier to handle than categorical ones in this respect. For example, it might only be possible to determine the state or condition of a unit by destructive investigation after failure, for example, a post-mortem examination.

Time-varying covariates can be handled computationally by ascribing multiple identities to each unit. Thus, at each time t_l before its final curtain call, when it will be eventually observed to fail or be censored, the unit appears as a censored case with its current covariate value $\mathbf{x}(t_l)$, and so contributes correctly to the risk set in the denominator of $P(\beta)$. One by one its identities are eliminated as each t_l is passed, until its number is up, like the nine lives of a cat.

4.3.7 Interval-Censored Data

The estimation described in Section 4.1 for random samples can be extended to accommodate covariates. For a proportional hazards model let $q_j = F_0(a_j)$, where F_0 is the baseline distribution function, and replace $(q_{j+1} - q_j)$ in the previous expression for p_i by $(q_{j+1}^{\psi_i} - q_j^{\psi_i})$, where $\psi_i = e^{\mathbf{x}_i^T \beta}$, for example. The log-likelihood is now a function of both $\mathbf{q}$ and β. For an accelerated life model, replace q_j by $(q_{i,j+1} - q_{ij})$ with $q_{ij} = F_0(\psi_i a_j)$.

4.4 Task Completion Times

Time and motion experts were once hailed as the saviours of British industry. Armed with stopwatch and clipboard they sallied forth to slay the dragons of inefficiency and skiving. On the ground, though, often found were much time and little motion. The 1960s Ealing comedy *I'm All Right Jack* encapsulated the situation beautifully. Desmond and Chapman (1993) modelled task-completion times in a large car plant: the views of the workforce on this exercise were not recorded in their paper.

The scenario for the data listed in Table 4.2 is a stroke rehabilitation centre in which the patients are asked to complete a certain physical task. The response variable is the time taken to complete the task, and the censoring here is assumed to be independent of the time. There are 48 patients (cases) and the covariates are age (years), sex (1 = male, 2 = female), and a categorical indicator representing one of three treatment regimes. The observations are listed

TABLE 4.2

Task Completion Times

Case	t	c	Age	Sex	tr	Case	t	c	Age	Sex	tr	Case	t	c	Age	Sex	tr
1	3.48	0	29	2	1	17	9.19	1	34	1	1	33	4.93	1	41	1	3
2	2.44	0	28	2	3	18	5.89	0	20	1	1	34	10.85	1	61	2	1
3	2.56	0	73	2	3	19	2.86	1	22	2	3	35	9.13	0	47	1	1
4	8.54	1	27	2	1	20	8.08	1	28	2	1	36	5.17	0	40	2	3
5	3.72	1	45	2	3	21	7.64	1	36	2	1	37	9.93	1	49	2	3
6	0.75	0	26	2	3	22	6.24	0	54	2	2	38	2.8	1	37	2	3
7	5.01	0	34	1	3	23	5.14	1	32	1	2	39	2.13	0	39	1	3
8	4.51	0	44	2	3	24	6.32	1	31	1	1	40	5.94	1	25	2	1
9	1.85	1	23	2	1	25	4.8	1	46	1	1	41	4.92	0	44	1	2
10	8.52	0	31	1	1	26	8.26	1	52	1	2	42	5.26	0	28	2	1
11	5.35	0	70	2	3	27	6.11	1	22	1	2	43	7.16	0	26	1	1
12	6.14	1	28	2	3	28	1.99	0	22	2	3	44	7.05	1	48	1	1
13	2.62	1	23	2	2	29	1.49	1	44	1	3	45	2.78	1	41	2	3
14	1.53	1	25	2	2	30	5.65	1	35	1	2	46	1.22	1	31	1	1
15	3.32	1	33	2	3	31	2.41	0	39	2	2	47	8.06	0	47	2	1
16	7.66	0	44	2	3	32	6.83	0	40	1	1	48	6.95	1	36	2	1

in Table 4.2 in which t is time (minutes), c is the censoring indicator, and tr is treatment. The data, though simulated, are representative of a real situation.

The data are stored in file `tasktimes.dat`, which has 1 header row plus 48 rows of data.

```
#task completion times  (Sec I.4.4)
dx1=read.table('tasktimes.dat',header=T); attach(dx1);
```

Next, the Kaplan–Meier estimate is obtained and a log-rank test performed to assess treatment differences.

```
#KM estimate
library(survival); sobj1=Surv(time,cens,type='right');
sfit1=survfit(sobj1~1+treat); summary(sfit1);
#log-rank test
survdiff(sobj1~1+treat,rho=0); #rho=0 gives log-rank test
```

Here is part of the output, with the KM estimate just for treatment 2.

```
              treat=2
time n.risk n.event survival std.err lower 95% CI upper 95% CI
1.53     9       1     0.889   0.105      0.7056       1.000
2.41     8       1     0.778   0.139      0.5485       1.000
2.62     7       1     0.667   0.157      0.4200       1.000
4.92     6       1     0.556   0.166      0.3097       0.997
5.14     5       1     0.444   0.166      0.2141       0.923
5.65     4       1     0.333   0.157      0.1323       0.840
6.11     3       1     0.222   0.139      0.0655       0.754
6.24     2       1     0.111   0.105      0.0175       0.705
8.26     1       1     0.000    NaN          NA          NA

Call:
survdiff(formula = sobj1 ~ 1 + treat, rho = 0)
          N Observed Expected (O-E)^2/E (O-E)^2/V
```

```
treat=1 20       12     16.80      1.369      3.99
treat=2  9        6      3.56      1.675      2.01
treat=3 19        9      6.65      0.834      1.19
 Chisq= 4.2  on 2 degrees of freedom, p= 0.12
```

Evidently, when the other covariates are ignored, differences between the treatment groups are not strong enough to show up significantly.

Let us now move on to consider possible effects of the covariates. The computation for fitting the PH model is performed by the R-function `coxph`. We apply it here as follows. Note that the objects from the previous computations, such as `sobj1`, are retained for use here; if one starts afresh, with a new R-window, they must all be re-created.

```
#Cox PH fit: survival package
sfit2=coxph(sobj1~age+sex+treat); summary(sfit2);
```

The output appears as follows:

```
Call:
coxph(formula = sobj1 ~ age + sex + treat)
  n= 48
          coef exp(coef) se(coef)      z Pr(>|z|)
age   -0.04704   0.95405  0.01735 -2.711 0.006710 **
sex   -0.16563   0.84736  0.36329 -0.456 0.648437
treat  0.81886   2.26791  0.21072  3.886 0.000102 ***
---
Signif. codes:  0 *** 0.001 ** 0.01 * 0.05 . 0.1   1

      exp(coef) exp(-coef) lower .95 upper .95
age      0.9541     1.0482    0.9222     0.987
sex      0.8474     1.1801    0.4158     1.727
treat    2.2679     0.4409    1.5006     3.428

Rsquare= 0.348   (max possible= 0.993 )
Likelihood ratio test= 20.54  on 3 df,   p=0.0001310
Wald test             = 18.58  on 3 df,   p=0.0003344
Score (logrank) test = 20.25  on 3 df,   p=0.0001508
```

Now treatment does achieve significance, as does age, but sex is nowhere near the magic 5% threshold. The baseline survivor function can now be obtained and plotted—see Figure 4.2.

```
#baseline survivor fn
sfn0=survfit(sfit2); plot(sfn0,xlab='time',ylab='baseline sfn');
title(main='Baseline survivor function -- Cox PH fit');
```

Last, residuals can be extracted and plotted as follows:

```
#residuals
res1=residuals(sfit2,type="martingale");
res2=residuals(sfit2,type="deviance");
par(mfrow=c(2,2)); nd=dim(dx1)[1];
indxplot(nd,res1,1,'martingale residuals','');
indxplot(nd,res2,1,'deviance residuals','');
plot(age,res1,xlab='age',ylab='',cex=0.5);
plot(age,res2,xlab='age',ylab='',cex=0.5);
```

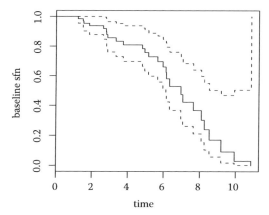

FIGURE 4.2
Task completions times: baseline survivor function.

The upper two frames of Figure 4.3 show index plots of the two types of residual, and in the lower two they are plotted against age; remember to paste `indxplot` into the R-window. You would not really expect these plots to show much untoward, since the data are simulated, although the martingale index plot does reveal one rather negative individual.

There are many other facilities in the R-package `survival`. Many happy hours can be spent selecting gems from the manual, like Just William left in charge of the sweetshop (Crompton, 1922).

4.5   Accelerated Life Models

The distribution-free development of the proportional-hazards model led to Cox's partial likelihood. That this depends on the sample times only through their ranks is not surprising when you think about it. Given an ordered sample

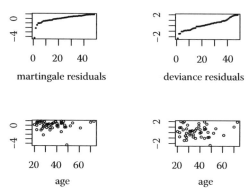

FIGURE 4.3
Task completions times: residual plots.

of observations, their distribution determines where they are bunched up and where they are spaced out. Discarding the distribution just leaves you with the ordering. Similar development of the accelerated life model proceeds via an unexpected route (at least, unexpected to me), namely, *linear rank tests*.

The accelerated life model was described briefly in Section 3.3. With slightly adapted notation, it may be written here as

$$Y_i = \psi(\mathbf{x}_i; \beta) + e_i,$$

where $Y_i = \log T_i$, the log-failure time, $\mathbf{x}_i$ is a covariate attached to T_i, β is a regression parameter, and the e_i are randomly sampled from some unspecified distribution.

Miller (1976) and Buckley and James (1979) considered the linear model, with $\psi(\mathbf{x}; \beta) = \mathbf{x}^T \beta$ and $\mathrm{E}(e_i) = 0$, and in which the observations can be right-censored. They proposed estimation methods that involve extracting a Kaplan–Meier estimate of the residual distribution for each new β in an iterative optimisation. Unfortunately, a change of β will often cause a re-ordering of the residuals and thus a discontinuity in the estimating function. The associated problems are discussed in the source papers.

Assuming that the form of model is correct, and that β is the "true" parameter, the residuals

$$e_i(\beta) = y_i - \psi(\mathbf{x}_i; \beta)$$

arising from a data set $\{(y_i, \mathbf{x}_i) : i = 1, \ldots, n\}$ should resemble a random sample. In particular, there should be no residual association between the $e_i(\beta)$ and the $\mathbf{x}_i$. This could be assessed in various ways, for example, using a rank correlation coefficient. However, here we summarise the methodology following Kalbfleisch and Prentice (2002, Chapter 7) and Lawless (2003, Chapter 8).

Order the residuals as $e_{(1)}(\beta) < \ldots < e_{(n)}(\beta)$ and denote the correspondingly ordered covariates by $\mathbf{x}_{(1)}, \ldots, \mathbf{x}_{(n)}$. If the $e_i(\beta)$ really do form a random sample all $n!$ orderings are equally likely, and then $(\mathbf{x}_{(1)}, \ldots, \mathbf{x}_{(n)})$ is just one such random permutation of the $\mathbf{x}_i$. So, to see if we have got the correct β, we check to see if $\mathbf{x}_{(i)}$ looks like a random permutation of the $\mathbf{x}_i$, that is, showing no obvious patterns. A *linear rank test statistic* for the hypothesised value of β has the form

$$\mathbf{v}(\beta) = \sum_{i=1}^{n} c_i \mathbf{x}_{(i)},$$

in which the c_i are given coefficients. Note that $\mathbf{v}(\beta)$ is a function of β because changing β can change the ordering of the residuals and thus swap around the $\mathbf{x}_{(i)}$. Moreover, it is a discontinuous function of β: if the values β_1 and β_2 give different orderings, there must be some point on a path between the two where a swap occurs. The c_i are known as *scores* and are given values chosen to highlight any suspected departure from the hypothesis. As is often the case, to construct a test statistic you have to know what kind of departure you are looking for. The scores c_i are often obtained via the *score statistic* based on a particular residual distribution.

For example, suppose that $\psi(x_i; \beta) = x_i\beta$, with scalar $x_i > 0$. Assume that the form of model is correct and that $\beta > \beta_0$, β_0 being the true value. Then

$$e_i(\beta) = e_i(\beta_0) - x_i(\beta - \beta_0),$$

which shows that the $e_i(\beta)$ will be negatively correlated with the x_i. In consequence, when the $e_i(\beta)$ are ordered, those with larger x_i values will tend to appear earlier on. In this case, taking the c_i to form an increasing sequence is likely to produce a value of $\mathbf{v}(\beta)$ smaller than expected.

Consider the null hypothesis $H_0 : \beta = \beta_0$. To perform a test based on $\mathbf{v}(\beta_0)$, we need to compare the observed $\mathbf{v}$ with its expected value under H_0. Now, the x_i themselves are not random, in the sense that most statistical analyses are conducted conditionally on the recorded covariate values. What is random is their reordering as the $\mathbf{x}_{(i)}$. So, the expected value of $\mathbf{v}$ is to be calculated as an average over the random orders, that is, with respect to the *permutation distribution* in which all $n!$ orderings are equally likely under H_0. Denoting this permutation averaging by E_p we have

$$E_p(\mathbf{x}_{(i)}) = \bar{\mathbf{x}} = n^{-1} \sum_{k=1}^{n} \mathbf{x}_k,$$

because each $\mathbf{x}_k$ appears in the ith position an equal number of times (actually, $(n-1)!$) in the $n!$ permutations. Likewise,

$$E_p(\mathbf{x}_{(i)}\mathbf{x}_{(j)}^T) = \begin{cases} n^{-1} \sum_{k=1}^{n} \mathbf{x}_k\mathbf{x}_k^T & (i = j) \\ \{n(n-1)\}^{-1} \sum_{k \neq l} \mathbf{x}_k\mathbf{x}_l^T & (i \neq j) \end{cases}$$

To see this, think about the number of times that $\mathbf{x}_k$ and $\mathbf{x}_l$ appear in the ith and jth positions. It is convenient to apply the constraint $\sum_{i=1}^{n} c_i = 0$; in the terminology of experimental design $(c_1, \ldots, c_n)$ is a *contrast vector*. The (permutation) mean and covariance matrix of $\mathbf{v}(\beta_0)$ under H_0 can now be evaluated as

$$E_p\{\mathbf{v}(\beta_0)\} = \mathbf{0}$$

and

$$\mathbf{C}_0 = \text{cov}_p\{\mathbf{v}(\beta_0)\}$$

$$= (n-1)^{-1} \left(\sum_{i=1}^{n} c_i^2 \right) \sum_{k=1}^{n} (\mathbf{x}_k - \bar{\mathbf{x}})(\mathbf{x}_k - \bar{\mathbf{x}})^T.$$

The details are left to you, the reader, as an exercise—enjoy. To test the departure of $\mathbf{v}(\beta_0)$ from the zero vector, it is natural to adopt the form $\mathbf{v}(\beta_0)^T \mathbf{C}_0^{-1} \mathbf{v}(\beta_0)$ as a test statistic. It has an asymptotic chi-square distribution with $\dim\{\mathbf{v}(\beta_0)\}$ degrees of freedom under H_0, subject to the proviso that $\mathbf{C}_0$ is non-singular.

For estimation $\mathbf{v}(\beta)$ can be employed as follows. A natural estimator for β is the value that minimises $Q(\beta) = \mathbf{v}(\beta)^T \mathbf{C}_\beta^{-1} \mathbf{v}(\beta)$, where $\mathbf{C}_\beta = \text{cov}_p\{\mathbf{v}(\beta)\}$, or $\mathbf{v}(\beta)^T \mathbf{A}\mathbf{v}(\beta)$ for any positive-definite $\mathbf{A}$. A corresponding (approximate) confidence region is the set of β such that $Q(\beta)$ does not exceed the appropriate chi-square quantile (with $\mathbf{A} \neq \mathbf{C}_{\beta}^{-1}$ the asymptotic chi-square distribution needs to be modified). However, this is easier said than done. Because $\mathbf{v}(\beta)$ is discontinuous in β, even non-monotonic, computation is not altogether straightforward.

Considerably more detail is given in the books cited above, including particular cases, the connection between linear rank statistics and efficient score statistics and weighted log-rank tests, the extension to accommodate censoring, asymptotic distribution theory, time-dependent covariates, and references to the literature—phew!

4.6 Exercises

4.6.1 Random Samples

1. Suppose that we have independent random samples from survivor functions $\bar{F}_1$ and $\bar{F}_2$, and that $F_2(t) = F_1(\alpha t)$, that is, an AL connection. Let t_{1q} be the upper q-quantile of F_1 and t_{2q} that of $F_2(t)$. How are t_{1q} and t_{2q} related? Does this suggest a plot to check the AL connection?

4.6.2 Partial Likelihood

1. Show that the partial likelihood function, with $\psi_x = \exp(\mathbf{x}^T \beta)$, is convex.

   Hint: for an arbitrary vector $\mathbf{z}$,

   $$\mathbf{z}^T \mathbf{V}(\beta)\mathbf{z} = \sum_{l=1}^{m} \mathbf{z}^T \mathbf{V}_l(\beta)\mathbf{z} = \sum_{l=1}^{m} \left\{ \mathbf{z}^T \mathbf{X}_l \mathbf{z} - (\mathbf{z}^T \mathbf{v}_l)^2 \right\}$$

   $$= \sum_{l=1}^{m} \left\{ \sum^{l} \pi_i z_i^2 - \left(\sum^{l} \pi_i z_i \right)^2 \right\},$$

   where $z_i = \mathbf{z}^T \mathbf{x}_i$ and $\pi_i = \psi_i / \sum^{l} \psi_i$. Let $\bar{z}_l = \sum^{l} \pi_i z_i$ and note that $\sum^{l} \pi_i = 1$. Then,

   $$\mathbf{z}^T \mathbf{V}(\beta)\mathbf{z} = \sum_{l=1}^{m} \sum^{l} \pi_i (z_i - \bar{z}_l)^2 \geq 0,$$

   so $\log P(\beta)$ has a negative-definite second derivative matrix as claimed.

2. The baseline hazard: follow through the *collecting terms* step.

4.6.3 Applications

1. Data sets are available for download from a variety of Web sites. For example, Collett (2003, Tables 8.3 and 8.4) gives some artificial data on cirrhosis patients, available on the CRC Press Web site. Such data

can be used for exploring the huge array of facilities in the R-package `survival`. Go to it.

4.6.4 Accelerated Life Models

1. Verify the formulae given for $E_p\{\mathbf{v}(\beta_0)\}$ and $\text{cov}_p\{\mathbf{v}(\beta_0)\}$.

4.7 Hints and Solutions

4.7.1 Random Samples

1. $F_2(t_{2q}) = F_1(\alpha t_{2q})$ and $F_1(t_{1q}) = q = F_2(t_{2q})$. Therefore, $\alpha t_{2q} = t_{1q}$. Plot: estimate a bunch of corresponding quantiles from the two samples and plot the points (a *qq plot*). Under AL one should obtain a near-straight line through the origin.

4.7.2 Accelerated Life Models

1. $E_p\{\mathbf{v}(\beta_0)\} = \sum_{i=1}^{n} c_i E_p(\mathbf{x}_{(i)}) = \sum_{i=1}^{n} c_i \bar{\mathbf{x}} = \mathbf{0}$.
   Again, since $\mathbf{v}(\beta_0)$ has zero mean,

$$\text{cov}_p\{\mathbf{v}(\beta_0)\} = E_p\{\mathbf{v}(\beta_0)\mathbf{v}(\beta_0)^T\} = \sum_{i,j} c_i c_j E_p\left(\mathbf{x}_{(i)}\mathbf{x}_{(j)}^T\right)$$

$$= \sum_{i=1}^{n} c_i^2 \left(n^{-1} \sum_{k=1}^{n} \mathbf{x}_k \mathbf{x}_k^T \right) + \sum_{i \neq j} c_i c_j \{n(n-1)\}^{-1} \sum_{k \neq l} \mathbf{x}_k \mathbf{x}_l^T.$$

But,

$$0 = \left(\sum_{i=1}^{n} c_i \right)^2 = \sum_{i=1}^{n} c_i^2 + \sum_{i \neq j} c_i c_j \text{ and } n^2 \bar{\mathbf{x}}\bar{\mathbf{x}}^T = \sum_{k=1}^{n} \mathbf{x}_k \mathbf{x}_k^T + \sum_{k \neq l} \mathbf{x}_k \mathbf{x}_l^T,$$

so

$$\text{cov}_p\{\mathbf{v}(\beta_0)\} = \left(\sum_{i=1}^{n} c_i^2 \right) \left(n^{-1} \sum_{k=1}^{n} \mathbf{x}_k \mathbf{x}_k^T \right) - \left(\sum_{i=1}^{n} c_i^2 \right) \{n(n-1)\}^{-1} \sum_{k \neq l} \mathbf{x}_k \mathbf{x}_l^T$$

$$= n^{-1} \left(\sum_{i=1}^{n} c_i^2 \right) \left\{ \sum_{k=1}^{n} \mathbf{x}_k \mathbf{x}_k^T - (n-1)^{-1} \left(n^2 \bar{\mathbf{x}}\bar{\mathbf{x}}^T - \sum_{k=1}^{n} \mathbf{x}_k \mathbf{x}_k^T \right) \right\}.$$

$$= (n-1)^{-1} \left(\sum_{i=1}^{n} c_i^2 \right) \left(\sum_{k=1}^{n} \mathbf{x}_k \mathbf{x}_k^T - n \bar{\mathbf{x}}\bar{\mathbf{x}}^T \right)$$

$$= (n-1)^{-1} \left(\sum_{i=1}^{n} c_i^2 \right) \sum_{k=1}^{n} (\mathbf{x}_k - \bar{\mathbf{x}})(\mathbf{x}_k - \bar{\mathbf{x}})^T.$$

5

Discrete Time

Parametric, non-parameric, and semi-parametric methods are covered in this chapter, the latter having come to the fore in recent years.

5.1 Random Samples: Parametric Methods

We assume for now that there are no explanatory variables: other situations will be considered below. Suppose that the sample comprises r_l observed failures and s_l right-censored ones at time τ_l ($l = 0, \ldots, m$); in keeping with the definitions of τ_0 and τ_m given in Section 2.3, we have $r_0 = 0, r_m \geq 0, s_0 \geq 0$ and $s_m = 0$. Obviously, with a finite sample the recorded values of r_l and s_l will all be zero for l beyond some finite point.

The likelihood contributions are $f(\tau_l; \theta)$ for an observed failure time τ_l and $\bar{F}(\tau_l; \theta)$ for one right-censored at τ_l; θ is a parameter (vector). The overall likelihood function is

$$L = \prod_{l=1}^{m} \{ f(\tau_l; \theta)^{r_l} \bar{F}(\tau_l; \theta)^{s_l} \},$$

where $\bar{F}(\tau_m; \theta)^{s_m} = 0^0$ is interpreted as 1; there is no explicit factor for $l = 0$ because $r_0 = 0$ and $\bar{F}(\tau_0; \theta) = 1$. This likelihood can be used for parametric inference in the usual way after substituting the chosen parametric forms for f and $\bar{F}$. Note that L does not involve s_0, reflecting the absence of information in lifetimes known only known to be positive.

5.1.1 Geometric Lifetimes

For the geometric distribution (Section 2.4) the likelihood function takes the form (with $\tau_l = l$)

$$L(\rho) = \prod_{l=1}^{m} \left[\{ \rho^{l-1}(1 - \rho) \}^{r_l} (\rho^l)^{s_l} \right] = \rho^a (1 - \rho)^b,$$

where $a = \sum\{ls_l + (l-1)r_l\}$ and $b = \sum r_l$. Thus, the maximum likelihood estimator is $\hat{\rho} = a/(a+b)$. The second derivative of the log-likelihood $l(\rho) = \log L(\rho)$ is

$$l''(\rho) = -a/\rho^2 - b/(1-\rho)^2$$

and so the asymptotic variance of $\hat{\rho}$ can be estimated by

$$\{-l''(\hat{\rho})\}^{-1} = ab/(a+b)^3.$$

To calculate the expected information $E\{-l''(\rho)\}$ we would need to know more about the censoring rule, which determines the distribution of the r_l and s_l. Suppose, for example, that the sample size $n = r_+ + s_+$ is fixed, and that everything is observed up to time m', so that $r_l = 0$ for $l > m'$ and $s_l = 0$ for $l \neq m'$. Then $(r_1, \ldots, r_{m'}, s_{m'})$ has a multinomial distribution with probabilities $\pi_l = \rho^{l-1}(1-\rho)$ for $l = 1, \ldots, m'$ and $\pi_{m'+1} = \rho^{m'+1}$. Hence,

$$E(a) = \sum_{l=1}^{m'} l\, E(s_l) + \sum_{l=1}^{m'} (l-1)\, E(r_l) = nm'\rho^{m'+1} + n\sum_{l=2}^{m'} (l-1)\rho^{l-1}(1-\rho)$$

$$= nm'\rho^{m'+1} - nm'\rho^{m'} + n\rho(1-\rho^{m'})/(1-\rho),$$

$$E(b) = \sum_{l=1}^{m'} E(r_l) = \sum_{l=1}^{m'} n\rho^{l-1}(1-\rho) = n(1-\rho^{m'}),$$

which gives $E\{-l''(\rho)\}$ explicitly in terms of n, m', and ρ.

5.1.2 Career Promotions

One of my cousins followed his father into the Civil Service, that much-envied job for life in former days. At one stage he wrote a long report on the state of the North Sea trawler fleet. This was during the period of demise of the British fishing industry, an entirely predictable result of handing our fishing grounds over to Europe. As in most organisations, and dare I say particularly in the Civil Service, it is a natural ambition to climb the ladder. In many cases promotions come round once a year (for those passed over shed a tear). So, the waiting time from starting on the bottom rung until first promotion is a discrete variable, the number of years spent toiling away in hope.

Table 5.1 presents some data of this type, concocted because such information is commonly not easy to access. The career progressions of 200 rookies were tracked over their first seven years. A summary of the waiting times until first promotion is given in the table. For example, in year 4 after appointment, 18 individuals were promoted and 4 departed to pastures new or were no longer in the running for other reasons. In their book for users of the Stata computing package, Rabe-Hesketh and Skrondal (2008) quote a study in which 301 male biochemists, starting as assistant professors at American universities, were tracked for 10 years.

TABLE 5.1

Career Promotions

No. years	1	2	3	4	5	6	7	Total
No. promotions	29	21	24	18	7	8	13	120
No. lost to view	18	10	9	4	3	2	34	80

The likelihood function for such frequency data has form

$$L = \prod_{j=1}^{m} (p_j^{r_j} q_j^{s_j}),$$

where m is the number of frequencies, here 7, $p_j = \mathrm{P}(W = j)$ for waiting time W, $q_j = \mathrm{P}(W > j)$, and r_j and s_j are the sample frequencies, observed and censored. We assume that the units (individuals) have independent outcomes (ignoring competition within departments, for instance), and that the censoring (departure of individuals from the fray) is independent (rejecting the unworthy charge that some might have flounced off due to failure to gain promotion).

Negative binomial variates can be generated as sums of geometric waiting times (Section 2.4). At a pinch, one might think of various hurdles to be jumped before promotion, for example, sterling effort, completing an important project successfully, and catching the eye of le Grand Fromage, each incurring a waiting time. Hurriedly passing over this somewhat dubious analogy, let us fit a negative binomial distribution to the data. The negative binomial probabilities to be applied here are given by

$$p_j = \binom{\kappa + j - 2}{\kappa - 1}\rho^{j-1}(1 - \rho)^{\kappa} \text{ for } j = 1, 2, \ldots,$$

with parameters $\kappa > 0$ and $0 < \rho < 1$. For computation we can use the recursions

$$p_{j+1} = p_j \rho(\kappa + j - 1)/j, \quad q_{j+1} = q_j - p_{j+1} \quad (j = 1, 2, \ldots),$$

with $p_1 = (1 - \rho)^{\kappa}$ and $q_1 = 1 - p_1$.

We use `fitmodel` as described in Section 3.2, here in conjunction with `rsampd1` (random samples, discrete time) specialised to the negative binomial distribution via `negbin1` (see the CRC Press Web site). The R-code is

```
#promotions data (Sec I.5.1)
pml=c(1,2,3,4,5,6,7,  29,21,24,18,7,8,13,  18,10,9,4,3,2,34);
dim(pml)=c(3,7); #data matrix
nn=301; md=7; #nn=no.promotees, md=no.promotion rounds
#fit neg binomial model
ir1=2; ir2=3; iopt=0; iwr=0; #see rsampd1 for these
adt=c(nn,md,ir1,ir2,iopt,iwr); #additional data passed through to rsampd1
np=2; par0=c(0.5,0.5); #np=no.parameters, par0=initial estimates
par1=fitmodel(rsampd1,np,par0,pml,adt); #par1=final estimates
iopt=1; adt=c(nn,md,ir1,ir2,iopt,2); rtn1=rsampd1(par1,pml,adt);
```

The resulting *mle* are $\hat{\rho} = 0.56$ and $\hat{\kappa} = 4.43$ with standard errors 0.09 and 1.44, respectively. Since $\hat{\rho}$ is well away from its boundary points (at 0 and 1), and its standard error is comparatively small, it seems safe to assume an approximate normal distribution for the estimator; this is implicit in quoting the usual plus-or-minus-two range for a 95% confidence interval (same as Appendix B). The last line of code given above has `opt` set to 1, which causes `rsampd1` to print out the estimated probabilities p_j and q_j and the log-likelihood contributions from the fitted model.

Although κ seems to be clear of the value 1, at which the waiting-time distribution would be geometric, we fit the latter for comparison. The fit may be accomplished simply by choosing the geometric distribution in `rsampd1` rather than the negative binomial. This yields $\hat{\rho} = 0.90$, with maximised log-likelihood 251.6448. That for the negative binomial fit is 234.388 and the likelihood-ratio chi-square value for comparing the two is 34.51, yielding p-value 5.4×10^{-21}. Of course, the geometric distribution cannot be ruled out with absolute certainty because it is possible that an event of probability 5.4×10^{-21} has occurred.

In the R-code given above, `ir1` and `ir2` identify the rows in the data matrix containing observed and censored times. So, to estimate the censoring distribution, we can just switch them. However, the result is unsatisfactory (try it)—maybe there is just too much censoring this way round and/or the censoring distribution is not negative binomial.

5.1.3 Probabilities of Observation versus Censoring

Consider the situation where a unit is liable to failure, at time T^f, or censoring (being lost to observation), at time T^c. Assume that T^f and T^c take values $0, 1, 2, \ldots$ and that they are independent. Let

$$p_j^f = P(T^f = j), \quad q_j^f = P(T^f > j), \quad p_j^c = P(T^c = j), \quad q_j^c = P(T^c > j),$$

and, for convenience, $p_{-1}^f = 0, q_{-1}^f = 1, p_{-1}^c = 0, q_{-1}^c = 1$. The probabilities of observed failure and censoring at time j are

$$P(T^f = j, T^c \geq j) = p_j^f q_{j-1}^c \text{ and } P(T^c = j, T^f > j) = p_j^c q_j^f;$$

in case $T^c = T^f$ we assume that failure is observed. The overall probabilities of observed failure and of censorship are obtained by summing these probabilities over j.

Suppose now that there are data in which the numbers observed and censored at time j are r_j and s_j. Then the likelihood function is

$$L = \prod_{j=0}^{\infty} \left(p_j^f q_{j-1}^c\right)^{r_j} \left(p_j^c q_j^f\right)^{s_j} = \prod_{j=0}^{\infty} \left(p_j^f\right)^{r_j} \left(q_j^f\right)^{s_j} \times \prod_{j=0}^{\infty} \left(p_j^c\right)^{s_j} \left(q_{j-1}^c\right)^{r_j}.$$

Arguing as for the continuous-time case (Section 3.2), we may focus upon the first term for inference about the T^f-distribution. Of course, we can likewise use the second product to estimate the T^c probabilities if we want to.

If the censoring process is not independent of the failure process, we have *dependent competing risks*, which is covered in Part III.

5.2 Random Samples: Non- and Semi-Parametric Estimation

We consider discrete lifetimes as described in Section 2.3. Thus, T takes values τ_l with probabilities $f(\tau_l)$ $(l = 0, 1, \ldots, m); 0 = \tau_0 < \tau_1 < \ldots < \tau_m, f(\tau_0) = 0$, $\bar{F}(\tau_m) = 0$, and m or τ_m or both may be infinite. For non-parametric estimation we use the hazard representation:

$$f(\tau_l) = h_l \prod_{s=1}^{l-1}(1 - h_s), \quad \bar{F}(\tau_l) = \prod_{s=1}^{l}(1 - h_s),$$

in which h_s is written for $h(\tau_s)$.

Suppose that there are r_l observed failures and s_l right-censored times at τ_l in the data. Then, with the usual assumptions of independence, the likelihood function can be expressed as

$$L = \prod_{l=1}^{m}\{f(\tau_l)^{r_l}\bar{F}(\tau_l)^{s_l}\} = \prod_{l=1}^{m}\left[\left\{h_l\prod_{s=1}^{l-1}(1-h_s)\right\}^{r_l} \times \left\{\prod_{s=1}^{l}(1-h_s)\right\}^{s_l}\right].$$

After collecting terms,

$$L = \prod_{l=1}^{m}\{h_l^{r_l}(1-h_l)^{q_l-r_l}\},$$

where

$$q_l = (r_l + s_l) + (r_{l+1} + s_{l+1}) + \cdots + (r_m + s_m);$$

q_l is the number of cases still *at risk* at time τ_l-, q_1 being the overall sample size. Here *at risk* means still at risk of being recorded as a failure, that is, cases that have not yet failed nor been lost to view by censoring and are therefore still under observation and liable to failure. So, q_l is the length of the queue for failure at time τ_l.

Treating the h_l as the parameters of the distribution, their maximum likelihood estimators are $\hat{h}_l = r_l/q_l$ $(l = 1, \ldots, m)$, obtained by solving $\partial \log L/\partial h_l = 0$. The $\hat{h}_l$ evidently result from equating h_l, the conditional probability of failure at time τ_l, to r_l/q_l, the proportion of observed failures among those liable to observed failure at that time. The corresponding estimators for $\bar{F}$ and f follow for $l = 1, \ldots, m$; the convention is applied that $\sum_{l=1}^{0}(1 - h_s)$ be taken as 1.

$$\hat{\bar{F}}(\tau_l) = \prod_{s=1}^{l}(1 - \hat{h}_s), \quad \hat{f}(\tau_l) = \hat{h}_l\prod_{s=1}^{l-1}(1 - \hat{h}_s).$$

Because $f(\tau_0) = 0$, $r_0 = 0$ and the formula $\hat{h}_l = r_l/q_l$ extends correctly to cover $l = 0$, and then $\hat{F}(\tau_0) = 1 - \hat{h}_0 = 1$. Likewise, because $\bar{F}(\tau_m) = 0$, $s_m = 0$ and $\hat{h}_m = r_m/q_m = 1$ (interpreting $0/0$ as 1, if necessary) and then $\hat{F}(\tau_m) = 0$ correctly. In principle, s_{m-1} could be transferred to r_m, since individuals observed to survive beyond time τ_{m-1} must then fail at time τ_m. However, this makes no difference to the estimates $\hat{h}_{m-1}$ and $\hat{h}_m$.

In finite samples it is possible for q_l to become zero first at $l = m' \le m$ and so $q_l = 0$ for $l = m', \ldots, m$; this is bound to occur if $m = \infty$. The estimates $\hat{h}_l = r_l/q_l$, as $0/0$, are then formally undefined for $l \ge m'$. This lack of an estimate for h_l ($l \ge m'$) can be put down to the fact that the conditioning event $\{T > \tau_{l-1}\}$, essential to its definition, has not been observed. However, if $s_{m'-1} = 0$, then $q_{m'-1} = r_{m'-1}$ and then $\hat{h}_{m'-1} = 1$, which gives $\hat{F}(\tau_{m'-1}) = 0$ and thence $\hat{F}(\tau_l) = 0$ for $l = m', \ldots, m$. If $s_{m'-1} > 0$, we know only that

$$\hat{F}(\tau_{m'-1}) \ge \hat{F}(\tau_l) \ge \hat{F}(\tau_{l+1}) \ge 0 \text{ for } l = m', \ldots, m-1,$$

with $\hat{F}(\tau_m) = 0$.

5.2.1 Career Promotions

We can apply `survfit` to these data as follows:

```
#promotions data (Sec I.5.2)
#Kaplan-Meier estimate
library(survival);
time=rep(c(1:7),2); cens=c(rep(1,7),rep(0,7));
freqs=c(29,21,24,18,7,8,13,  18,10,9,4,3,2,34);
sobj=Surv(time,cens,type='right',origin=0); #survival object
sfit1=survfit(sobj~1,weights=freqs); summary(sfit1);
plot(sfit1,conf.int=T,lty=3:3,
   xlab='time',ylab='survival prob (with 95% conf  limits)');
title(main='Kaplan-Meier plot', sub='');
```

The output is

time	n.risk	n.event	survival	std.err	lower 95% CI	upper 95% CI
1	200	29	0.855	0.0249	0.808	0.905
2	153	21	0.738	0.0320	0.677	0.803
3	122	24	0.593	0.0370	0.524	0.670
4	89	18	0.473	0.0388	0.402	0.555
5	67	7	0.423	0.0390	0.353	0.507
6	57	8	0.364	0.0388	0.295	0.448
7	47	13	0.263	0.0367	0.200	0.346

The estimated survivor function here may be compared with that under the fitted negative binomial model (Section 5.1), which is

```
0.97 0.91 0.82 0.70 0.58 0.46 0.36
```

The match is not good, so maybe the intial diffidence in adopting the negative binomial distribution for the waiting times is justified (to put it politely).

5.2.2 Large-Sample Theory

We now restrict attention to the case $m < \infty$. Then there is a finite number of parameters h_l and, under standard regularity conditions, the usual large-sample parametric theory will apply. Thus, the asymptotic joint distribution of the set of $(\hat{h}_l - h_l)$ will be multivariate normal with means zero and covariance matrix $E(\mathbf{H})^{-1}$, where $\mathbf{H}$ has (l, k) element

$$H_{lk} = -\partial^2 \log L / \partial h_l \partial h_k = \delta_{lk}\{r_l/h_l^2 + (q_l - r_l)/(1 - h_l)^2\};$$

Here δ_{lk} is the *Kronecker delta*, taking value 1 if $l = k$ and 0 if $l \neq k$. The diagonal form of $\mathbf{H}$, implying asymptotic independence of the $\hat{h}_l$, results from the orthogonal form of L (Cox and Reid, 1987). Using the observed, rather than the expected, information, we have the simpler form

$$\hat{H}_{lk} = \delta_{lk}\{q_l^2/r_l + q_l^2/(q_l - r_l)\} = \delta_{lk}q_l/\{\hat{h}_l(1 - \hat{h}_l)\},$$

and so $\mathrm{var}(\hat{h}_l)$ can be estimated by $\hat{h}_l(1 - \hat{h}_l)/q_l$. Applying the *delta method* (Appendix B) we have

$$\mathrm{var}\{\log \hat{\bar{F}}(t)\} = \mathrm{var}\left\{\sum_{\tau_s \leq t} \log(1 - \hat{h}_s)\right\} \approx \sum_{\tau_s \leq t}(1 - h_s)^{-2}\mathrm{var}(1 - \hat{h}_s)$$

$$\approx \sum_{\tau_s \leq t}(1 - \hat{h}_s)^{-2}\hat{h}_s(1 - \hat{h}_s)/q_s.$$

Thus, we obtain *Greenwood's formula*, so named after his 1926 contribution:

$$\mathrm{var}\{\hat{\bar{F}}(t)\} \approx \hat{\bar{F}}(t)^2 \sum_{\tau_s \leq t} \hat{h}_s/\{(1 - \hat{h}_s)q_s\}.$$

Actually, the formula for $\mathrm{var}\{\log \hat{\bar{F}}(t)\}$ is probably more useful than that for $\mathrm{var}\{\hat{\bar{F}}(t)\}$ because $\log \bar{F}(t)$ has less restricted range than $\bar{F}(t)$. The alternative transforms $\log\{-\log \bar{F}(t)\}$ and $\log[\bar{F}(t)/\{1 - \bar{F}(t)\}]$ are fully unrestricted, and some authors have recommended working in terms of these if variances are to be used in this context (e.g., Kalbfleisch and Prentice, 1980, Section 1.3).

In the absence of censoring, $s_l = 0$ for all l, and then $1 - \hat{h}_l = q_{l+1}/q_l$ and $\hat{\bar{F}}(\tau_l) = q_{l+1}/q_1$, the usual *empirical survivor function*, noting that $q_1 = q_0$ in this case. Also,

$$\hat{h}_s/\{(1 - \hat{h}_s)q_s\} = r_s/\{(q_s - r_s)q_s\} = (q_s - q_{s+1})/(q_{s+1}q_s) = q_{s+1}^{-1} - q_s^{-1}.$$

Then, writing $\hat{\bar{F}}_l$ for $\hat{\bar{F}}(\tau_l) = q_{l+1}/q_1$, the variance formula becomes, for $t = \tau_l$,

$$\mathrm{var}\{\hat{\bar{F}}(t)\} \approx \hat{\bar{F}}_l^2\left(q_{l+1}^{-1} - q_1^{-1}\right) = \hat{\bar{F}}_l(1 - \hat{\bar{F}}_l)/q_1,$$

the usual binomial variance estimate.

5.3 Explanatory Variables

Suppose now that there are explanatory variables, or covariates, so that the ith individual comes complete with a vector $\mathbf{x}_i$ of such at no extra charge.

5.3.1 Likelihood Function

Write $f(t; \mathbf{x}_i)$ and $\bar{F}(t; \mathbf{x}_i)$, respectively, for the density and survivor function of the failure time of the ith individual. The basic likelihood function of Section 5.1 is now filled out to

$$L = \prod_{l=1}^{m} \left\{ \prod_{i \in R_l} f(\tau_l; \mathbf{x}_i) \times \prod_{i \in S_l} \bar{F}(\tau_l; \mathbf{x}_i) \right\},$$

where R_l is the set of individuals observed to fail at time τ_l and S_l is the set of individuals right-censored at time τ_l. We take R_0 as empty, meaning that zero lifetimes are ruled out. Also, individuals in S_0 turned up to sign on at time $\tau_0 = 0$ but had disappeared before the first roll call at time τ_1. Again, R_m is the set of diehards who held out until time τ_m and S_m is empty. In principle, the set of individuals in S_{m-1} could be transferred to R_m since, although lost sight of just after time τ_{m-1}, they must fail at time τ_m. The hazard contributions will be denoted by

$$h_l(\mathbf{x}) = P(\textit{failure at time}\,\tau_l \mid \textit{survival past}\,\tau_{l-1}; \mathbf{x}) = f(\tau_l; \mathbf{x})/\bar{F}(\tau_{l-1}; \mathbf{x}),$$

and so, applying the product formulae for $\bar{F}$ and f (Section 2.3),

$$\bar{F}(\tau_l; \mathbf{x}) = \prod_{s=1}^{l}\{1 - h_s(\mathbf{x})\}, \quad f(\tau_l; \mathbf{x}) = h_l(\mathbf{x})\prod_{s=1}^{l-1}\{1 - h_s(\mathbf{x})\}.$$

Thus, the likelihood can be expressed in terms of the hazards as

$$L = \prod_{l=1}^{m} \left\{ \prod_{i \in R_l} \left[h_l(\mathbf{x}_i)\prod_{s=1}^{l-1}\{1 - h_s(\mathbf{x}_i)\} \right] \times \prod_{i \in S_l} \left[\prod_{s=1}^{l}\{1 - h_s(\mathbf{x}_i)\} \right] \right\}$$

$$= \prod_{l=1}^{m} \left\{ \prod_{i \in R_l} h_l(\mathbf{x}_i) \times \prod_{i \in R_l}\prod_{s=1}^{l-1} h_{si}^- \times \prod_{i \in S_l}\prod_{s=1}^{l} h_{si}^- \right\},$$

where we have written h_{si}^- for $1 - h_s(\mathbf{x}_i)$ and $\prod_{s=1}^{l-1} h_{si}^-$ is interpreted as 1 for $l = 1$. But, the terms involving h_{si}^- can be collected as

$$\prod_{i \in R_2} h_{1i}^- \times \prod_{i \in R_3}(h_{1i}^- h_{2i}^-) \times \cdots \times \prod_{i \in S_1} h_{1i}^- \times \prod_{i \in S_2}(h_{1i}^- h_{2i}^-) \times \cdots = \prod_{l=1}^{m}\prod_{i \in Q_l} h_{li}^-,$$

where

$$Q_l = S_l \cup R_{l+1} \cup S_{l+1} \cup \ldots \cup R_m \cup S_m = R(\tau_l) - R_l;$$

$R(\tau_l)$ is the *risk set* at time τ_l, that is, the set of individuals still at risk at time $\tau_l -$ of eventually being recorded as failures. Thus, the likelihood function becomes

$$L = \prod_{l=1}^{m} \left[\prod_{i \in R_l} h_l(\mathbf{x}_i) \times \prod_{i \in Q_l} \{1 - h_l(\mathbf{x}_i)\} \right]$$

$$= \prod_{l=1}^{m} \left[\prod_{i \in R_l} [h_l(\mathbf{x}_i)/\{1 - h_l(\mathbf{x}_i)\}] \times \prod_{i \in R(\tau_l)} \{1 - h_l(\mathbf{x}_i)\} \right].$$

5.3.2 Geometric Waiting Times

Recall the geometric model (Section 2.4). Suppose that ρ is a logit function of $\mathbf{x}$: $\log\{\rho_i/(1 - \rho_i)\} = \mathbf{x}_i^T \beta$. Then the discrete hazard satisfies

$$h_l(\mathbf{x}_i) = 1 - \rho_i = \left\{1 + \exp\left(-\mathbf{x}_i^T \beta\right)\right\}^{-1} \text{ and } h_l(\mathbf{x}_i)/\{1 - h_l(\mathbf{x}_i)\} = \exp\left(\mathbf{x}_i^T \beta\right),$$

and the likelihood for the regression parameter vector β is

$$L(\beta) = \prod_{l=1}^{\infty} \left[\prod_{i \in R_l} \exp\left(-\mathbf{x}_i^T \beta\right) \div \prod_{i \in R(\tau_l)} \left\{1 + \exp\left(-\mathbf{x}_i^T \beta\right)\right\}^{-1} \right].$$

5.3.3 The Driving Test

I passed my motorbike test at first attempt when I was 17 years old. When the applause dies down I should perhaps point out that the test in those days was somewhat less rigorous than it is now. The examiner first made sure that I could stay up on the bike for 25 yards without falling off, and then gave instructions for the "emergency stop." I was to ride around the block and stop quickly when he jumped out in front without warning from a concealed position. So, I came round the corner and could just make out, in the distance, a figure emerging from behind a tree waving a clipboard (no, not the tree). He was taking no chances! I duly stopped and was awarded a full driving licence.

The number of attempts that people make to pass the driving test would surely be of interest to various bodies: their own, clearly, but also motoring organisations, the Department of Transport, government statisticians, insurance companies, and others. However, there does not seem to be readily available data around. (A Google search throws up much individual discussion and polemic plus some driving school advertisements. Also catching the eye is a report of a Korean woman who took 960 attempts—no, me neither.) In the spirit of goodwill to all men I hereby bequeath some data to posterity, admittedly simulated, listed in Table 5.2.

TABLE 5.2

Numbers of Attempts at Passing the Driving Test

case	na	cens	age	sex	nl	case	na	cens	age	sex	nl	case	na	cens	age	sex	nl
1	2	1	21	2	4	17	1	1	22	1	16	33	1	1	20	1	12
2	3	1	19	2	0	18	1	0	23	1	6	34	8	1	19	2	0
3	1	1	20	1	4	19	1	0	19	2	14	35	1	1	23	2	9
4	1	0	20	1	4	20	1	1	22	1	14	36	9	1	23	2	1
5	2	1	22	1	0	21	5	1	22	2	0	37	2	1	25	2	8
6	1	1	17	1	7	22	1	1	18	1	8	38	1	0	21	1	12
7	12	0	22	2	0	23	1	0	24	1	9	39	1	1	17	2	1
8	10	1	17	1	0	24	1	1	26	2	0	40	4	1	26	2	0
9	1	1	24	2	8	25	1	0	18	2	6	41	1	0	23	2	8
10	1	0	20	2	3	26	3	1	24	1	2	42	1	0	22	2	6
11	1	1	23	2	25	27	1	1	20	2	13	43	1	1	22	2	0
12	1	0	18	2	3	28	1	1	18	2	5	44	5	1	19	2	0
13	1	0	25	1	6	29	1	1	20	2	5	45	3	1	22	2	7
14	1	0	22	2	8	30	1	1	18	2	11	46	1	1	20	1	13
15	1	1	21	2	11	31	1	0	23	1	14	47	2	1	21	2	4
16	1	1	21	1	14	32	7	1	22	2	0	48	1	1	21	2	5

In the table *na* is the number of attempts until success; *cens* is the censoring indicator (lost to follow-up); and the covariates are $x_1 = age$ (age at first attempt), $x_2 = sex$ (sex at first attempt), and $x_3 = nl$ (number of lessons taken before first attempt).

The geometric model could have been a contender here (like Marlon Brando). It is a repeated-trials process, though the structure implies that the probability of success for an individual might change as the attempts accumulate (up or down?) and the probabilities will almost certainly vary between individuals. Nevertheless let us start with a straightforward geometric model in which ρ is expressed in terms of the covariates via a logit model: $\rho_i = 1/(1 + e^{-x_i^T \beta})$. The R-code for loading the data from file `drivetest` and fitting the model is

```
#driving test data (Sec I.5.3)
#data input (case,nattempts,cens,age,sex,nlessons)
dx1=read.table('drivetest.dat',header=T); attach(dx1);
nd=48; md=6; dx2=as.numeric(unlist(dx1)); dim(dx2)=c(nd,md);
#fit geometric model
kt=2; kc=3; opt=0;
mx=3; kx=c(4,5,6); adt=c(nd,kt,kc,mx,opt,0,kx);
np=1+mx; par0=rep(0,np); par1=fitmodel(geom1,np,par0,dx2,adt); #full model
mx=0; kx=0; adt=c(nd,kt,kc,mx,opt,0,kx);
np=1+mx; par0=rep(0,np); par1=fitmodel(geom1,np,par0,dx2,adt); #null model
mx=1; kx=c(6); adt=c(nd,kt,kc,mx,opt,0,kx);
np=1+mx; par0=rep(0,np); par1=fitmodel(geom1,np,par0,dx2,adt); #x3=nlessons
```

The R-function `geom1` is listed on the CRC Press Web site. The resulting estimates of the regression coefficients are $(\hat{\beta}_0, \hat{\beta}_1, \hat{\beta}_2, \hat{\beta}_3) = (0.38, 0.07, -0.16, -0.20)$ with standard errors $se(\beta_1) = 0.10$, $se(\beta_2) = 0.55$, $se(\beta_3) = 0.05$. Only $\hat{\beta}_3$ exceeds its standard error here. Refitting the model without x_1 and x_2 yields log-likelihood -59.05816, to be compared with that of the full model,

-58.82742: this gives $\chi_2^2 = 0.46$ ($p = 0.79$), so it seems acceptable to omit age and sex from the covariate set, just leaving the number of lessons taken. Omitting all three covariates produces negative log-likelihood -69.02085 and $\chi_3^2 = 20.39$ ($p = 0.00014$), which is not acceptable, so x_3 should not be dropped.

Suggestions for hounding the data further are given below.

5.3.4 Proportional Hazards

Assume a PH model expressed as

$$\bar{F}(\tau_l, \mathbf{x}) = \bar{F}_0(\tau_l)^{\psi_x},$$

where $\psi_x = \psi(\mathbf{x}; \beta)$. Writing $h_s(\mathbf{x})$ for the discrete hazard $h(\tau_s; \mathbf{x})$, and h_{0s} for $h_0(\tau_s)$, we have

$$\bar{F}_0(\tau_l) = \prod_{s=1}^{l}(1 - h_{0s}), \quad \bar{F}(\tau_l, \mathbf{x}) = \prod_{s=1}^{l}\{1 - h_s(\mathbf{x})\},$$

which entails

$$h_s(\mathbf{x}) = 1 - (1 - h_{0s})^{\psi_x}.$$

Unfortunately, this form does not yield cancellation of the h_{0s} when a partial likelihood is set up along the lines outlined in Section 4.2, so we cannot follow that route to estimate β free of the baseline hazards. The alternative route, setting up the PH model directly as $h_s(\mathbf{x}_i) = \psi_i h_{0s}$, would lead to elimination of the h_{0s} as in Section 4.2 as before. However, again unfortunately, the resulting ratio, $\psi_{i_l}/\sum^{l}\psi_i$, is not the correct expression of the intended conditional probability, as will be explained below.

If we are willing to estimate β along with the baseline hazards, not free of them, then we can proceed as follows. Substitute the PH form for $h_s(\mathbf{x}_i)$ above into the likelihood function given above to obtain

$$L = \prod_{l=1}^{m}\left[\prod_{i \in R_l}\{1 - (1 - h_{0l})^{\psi_i}\} \times \prod_{i \in Q_l}(1 - h_{0l})^{\psi_i}\right].$$

This likelihood can now be maximised jointly over β and the h_{0l}. For standard asymptotic theory to apply there needs to be a finite, preferably small, number of parameters. For unrestricted h_{0l} this would entail finite m; otherwise, the h_{0l} would have to be restricted by being expressed as functions of a finite set of parameters.

5.3.5 Proportional Odds

An alternative model, which can be used to advantage with the partial likelihood approach, was proposed by Cox (1972). We begin with a more careful derivation of the conditional probability of Section 4.2; this is the probability that an individual i_l fails at time τ_l and the others in risk set $R(\tau_l)$ do not, given that there is one failure at time τ_l. Writing $h_l(\mathbf{x}_i)$ for $h(\tau_l; \mathbf{x}_i)$, this is

$$h_l(\mathbf{x}_{i_l}) \prod_{a \neq i_l} \{1 - h_l(\mathbf{x}_a)\} \div \prod_{i \in R(t_l)} \left[h_l(\mathbf{x}_i) \prod_{a \neq i} \{1 - h_l(\mathbf{x}_a)\} \right]$$

$$= \left[h_l(\mathbf{x}_{i_l}) / \{1 - h_l(\mathbf{x}_{i_l})\} \right] \div \prod_{i \in R(t_l)} [h_l(\mathbf{x}_i)\{1 - h_l(\mathbf{x}_i)\}] .$$

In the continuous case $h_l(\mathbf{x}_i)$ is replaced by $h(t_l; \mathbf{x}_i)dt$, and then, as $dt \to 0$, the original expression in Section 4.2 is recovered.

Cancellation of the baseline hazards in the present form will be achieved by assuming that

$$h_l(\mathbf{x}) / \{1 - h_l(\mathbf{x})\} = \psi(\mathbf{x}; \beta)\, h_{0l} / (1 - h_{0l}),$$

where $h_{0l} = h_0(\tau_l)$ is the baseline hazard contribution $P(T = \tau_l \mid T \geq \tau_l)$. The ratio $h_{0l}/(1 - h_{0l})$ is the *conditional odds* on the event $\{T = \tau_l\}$ under the baseline hazard. The assumption, then, is one of *proportional odds*, giving a PO model. With the usual choice, $\psi(\mathbf{x}; \beta) = \exp(\mathbf{x}^T \beta)$, this just boils down to a logit model:

$$\log [h_l(\mathbf{x}) / \{1 - h_l(\mathbf{x})\}] = \alpha_l + \mathbf{x}^T \beta,$$

where $\alpha_l = \log\{h_{0l}/(1 - h_{0l})\}$. Under PO the partial likelihood takes the same form as in Section 4.2. However, in replacing PH by PO we must bear in mind that the interpretation of the covariate effects changes because of the different way in which the baseline hazard is modified.

Tied failure times are very likely to occur in the discrete case. Formally, if individuals $i_1, \ldots, i_r$ all fail at time τ_l, the lth contribution to the partial likelihood is $\psi_{i_1} \ldots \psi_{i_r} / \sum^l \psi_{a_1} \ldots \psi_{a_r}$, where the summation is over all subsets $(a_1, \ldots, a_r)$ of individuals from $R(\tau_l)$; this form arises as described in Section 4.3. Gail et al. (1981) presented an algorithm for fast computation (see Section 5.6, Exercises). Otherwise, an approximate adjustment for (not too many) ties can be made as described for the continuous case. If there is a substantial proportion of ties in the data, and exact computation is unfeasible, one must settle for the kinds of alternative options described in Section 4.3.

The baseline hazards h_{0l} can be estimated as follows. Having obtained the estimates $\hat{\psi}_i = \psi(\mathbf{x}_i; \hat{\beta})$ from the partial likelihood, we can use the PO model to obtain

$$\hat{h}_l(\mathbf{x}_i) = \hat{\psi}_i h_{0l} / (1 - h_{0l} + \hat{\psi}_i h_{0l}) = \hat{\psi}_i / (\hat{\psi}_i + g_{0l}^{-1}),$$

where $g_{0l} = h_{0l}/(1 - h_{0l})$. Substitution into the likelihood function then yields

$$L = \prod_{l=1}^{m} \left\{ \prod_{i \in R_l} (\hat{\psi}_i g_{0l}) \times \prod_{i \in R(t_l)} (1 + \hat{\psi}_i g_{0l})^{-1} \right\},$$

which can be maximised over the g_{0l}. Note that L is of orthogonal form in the g_{0l}, thus simplifying the computation considerably.

5.3.6 The Driving Test

For comparison with the previous fit of this data, we feed it through the `coxph` mincer. The R-code is

```
#Cox PH fit
library(survival);
time=nattempts; sobj=Surv(time,cens,type='right');
sfit=coxph(sobj~age+sex+nlessons); summary(sfit);
```

The output is

```
Call:
coxph(formula = sobj ~ age + sex + nlessons)
  n= 48
             coef exp(coef) se(coef)      z Pr(>|z|)
age      -0.05841   0.94327  0.08010 -0.729 0.465927
sex       0.02380   1.02408  0.38269  0.062 0.950421
nlessons  0.11947   1.12690  0.03343  3.574 0.000352 ***
---
Signif. codes:  0 *** 0.001 ** 0.01 * 0.05 . 0.1   1
         exp(coef) exp(-coef) lower .95 upper .95
age         0.9433     1.0601    0.8062     1.104
sex         1.0241     0.9765    0.4837     2.168
nlessons    1.1269     0.8874    1.0554     1.203
Rsquare= 0.207   (max possible= 0.985 )
Likelihood ratio test= 11.16  on 3 df,   p=0.01089
Wald test            = 12.8   on 3 df,   p=0.005096
Score (logrank) test = 13.37  on 3 df,   p=0.003909
```

The qualitative result, that the number of previous lessons alone is siginificant, is the same as before.

5.3.7 The Baseline Odds

We go back to the full likelihood function, written as

$$L = \prod_{l=1}^{m} \left[\prod_{i \in R_l} [h_l(\mathbf{x}_i)/\{1 - h_l(\mathbf{x}_i)\}] \times \prod_{i \in R(\tau_l)} \{1 - h_l(\mathbf{x}_i)\} \right].$$

Now replace $h_l(\mathbf{x}_i)/\{1 - h_l(\mathbf{x}_i)\}$ by $\psi(\mathbf{x}_i; \hat{\beta})h_{0l}/(1 - h_{0l})$ and maximise the result over the h_{0l}. Explicitly, the function to be maximised is

$$\prod_{l=1}^{m} \left[\prod_{i \in R_l} \{\hat{\psi}_i h_{0l}/(1 - h_{0l})\} \times \prod_{i \in R(\tau_l)} \{1 + \hat{\psi}_i h_{0l}/(1 - h_{0l})\}^{-1} \right].$$

where we have written $\hat{\psi}_i$ for $\psi(\mathbf{x}_i; \hat{\beta})$.

5.4 Interval-Censored Data

The analysis outlined above, nominally for discrete failure times, actually applies without much damage to interval-censored, or grouped, continuous failure times of the kind recorded in life tables.

Suppose that the time scale is partitioned into intervals $(\tau_{l-1}, \tau_l]$ $(l = 1, \dots, m)$, with $\tau_0 = 0$ and $\tau_{m+1} = \infty$. The data comprise, for each l, a set R_l of r_l individuals with failures at unknown times during $(\tau_{l-1}, \tau_l]$, and a set S_l of s_l individuals with failure times right-censored at τ_l; the individuals in S_l are known to have survived the interval $(\tau_{l-1}, \tau_l]$ but were subsequently lost to view.

Assume for the moment that there are no covariates. The likelihood contribution for a time falling within the interval $(\tau_{l-1}, \tau_l]$ is $\bar{F}(\tau_{l-1}) - \bar{F}(\tau_l) = f(\tau_l)$, and that for one right-censored at τ_l is $\bar{F}(\tau_l)$. Thus, the overall likelihood function is

$$L = \prod_{l=1}^{m} \left[\prod_{i \in R_l} f(\tau_l) \times \prod_{i \in S_l} \bar{F}(\tau_l) \right].$$

This is formally the same as the form given in Section 4.1. There is a niggling unease that, in counting the likelihood contributions of times within an interval $(\tau_{l-1}, \tau_l]$ as $f(\tau_l)$, they are effectively being shifted out to the end of the interval at τ_l. But it is a question of interpretation: here $f(\tau_l)$ is the probability of T's falling within the interval, not of being equal to τ_l.

To accommodate covariates, as in Section 4.2, just replace $f(\tau_l)$ and $\bar{F}(\tau_l)$ by $f(\tau_l; \mathbf{x}_i)$ and $\bar{F}(\tau_l; \mathbf{x}_i)$.

5.4.1 Cancer Survival Data

The data (Section 1.2) provide an example of interval censoring, without covariates in this case. For the first three groups there are $m = 9$ intervals, with no censored times until the last; for the last group $m = 7$. The data can be assembled as displayed in the table in file `boag1.dat`, replacing + by 0. The computations can be performed by R-function `survfit`, as shown here. Note the `weights` option for grouped data.

```
#Boag data (Sec I.5.4)
library(survival);
bg1=read.table('boag1.dat',header=T); attach(bg1);
n1=dim(bg1)[1]; tv1=bg1[1:n1,2]; cv1=c(rep(1,n1-1),0); #times and censoring
par(mfrow=c(2,2));
labx=c('group a','group b','group c','group e'); #plot labels
wts=bg1[1:n1,3]; summ=boagfit(tv1,cv1,wts,1,labx[1]); summ;
wts=bg1[1:n1,4]; summ=boagfit(tv1,cv1,wts,1,labx[2]); summ;
wts=bg1[1:n1,5]; summ=boagfit(tv1,cv1,wts,1,labx[3]); summ;
wts=bg1[1:n1,6]; summ=boagfit(tv1,cv1,wts,1,labx[4]); summ;
```

The function `boagfit` (listed on the CRC Press Web site) is created just to avoid repeating the code for fits and plots four times. The printout from

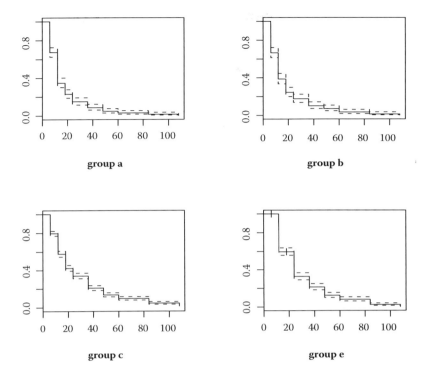

FIGURE 5.1
Survivor functions with 95% confidence bands.

`summary(survfit)` for `group a` is as follows:

```
time n.risk n.event survival std.err lower 95% CI upper 95% CI
   6    336     109  0.67560 0.02554     0.627348       0.7276
  12    227     110  0.34821 0.02599     0.300826       0.4031
  18    117      40  0.22917 0.02293     0.188359       0.2788
  24     77      27  0.14881 0.01942     0.115231       0.1922
  36     50      21  0.08631 0.01532     0.060949       0.1222
  48     29      13  0.04762 0.01162     0.029519       0.0768
  60     16       6  0.02976 0.00927     0.016163       0.0548
  84     10       5  0.01488 0.00661     0.006235       0.0355
 108      5       4  0.00298 0.00297     0.000420       0.0211
```

The plots of the estimated survivor functions, together with 95% confidence bands, are shown in Figure 5.1. They look fairly similar but we can perform a log-rank test (Section 4.3) using the R-function `survdiff`. Apparently, `survdiff` does not have the `weights` facility like `survfit`, so we use the function `boagexpand` to fill out the data matrix to have one row per case. The R-code is as follows:

```
bg2=boagexpand(n1,bg1); n2=bg2[1];
grp2=bg2[2:(n2+1)]; tv2=bg2[(n2+2):(2*n2+1)]; cv2=bg2[(2*n2+2):(3*n2+1)];
sobj2=Surv(tv2,cv2,type='right',origin=0);
survdiff(sobj2~grp2,rho=0);
```

The output is

```
         N Observed Expected (O-E)^2/E (O-E)^2/V
grp2=1 336      335      237     40.79      66.8
grp2=2 295      293      218     25.57      41.3
grp2=3 938      919      994      5.62      15.7
grp2=4 577      574      672     14.35      31.2
 Chisq= 127  on 3 degrees of freedom, p= 0
```

This makes it pretty clear that there are differences between the groups.

5.5 Frailty and Random Effects

The standard modification of a hazard profile in discrete time is to apply it to the odds ratio: replace $g_l = h_t/(1 - h_t)$ with Zg_t, where Z is a random effect with continuous distribution function K on $(0, \infty)$. (One could apply the factor Z to h_t directly, but then the maximum size of Z would need to be restricted to prevent Zh_t from exceeding 1, which is not allowed for proba- bilities such as discrete hazards.) Hence, h_s is replaced by $Zg_s(1 + Zg_s)^{-1}$ and $1 - h_s$ by $(1 + Zg_s)^{-1}$. The discrete density and survivor functions, conditioned on $Z = z$, then become

$$f(\tau_l \mid z) = P(T = \tau_l) = zg_l \prod_{s=1}^{l}(1 + zg_s)^{-1}$$

and

$$\bar{F}(\tau_l \mid z) = P(T > \tau_l) = \prod_{s=1}^{l}(1 + zg_s)^{-1}.$$

To obtain the unconditional functions we have to integrate over z:

$$f(\tau_l) = \int_0^\infty zg_l \prod_{s=1}^{l}(1 + zg_s)^{-1}dK(z) \text{ and } \bar{F}(\tau_l) = \int_0^\infty \prod_{s=1}^{l}(1 + zg_s)^{-1}dK(z).$$

The search for a conjugate form of K, for which the integrals can be explic- itly evaluated for all l, looks pretty hopeless. So, they need to be evaluated numerically, which is not a great recommendation for this type of model.

5.5.1 Geometric Distribution

The geometric distribution has constant hazard, so $g_s = g$ for all s. Then $\bar{F}(l \mid z) = (1 + zg)^{-l}$ ($l = 1, 2, \ldots$) and, taking Z to be exponentially distributed with mean ξ,

$$\bar{F}(l) = \int_0^\infty \xi^{-1}e^{-z/\xi}(1 + zg)^{-l}dz = (g\xi)^{-1}\int_0^\infty e^{-t/(g\xi)}(1 + t)^{-l}dt,$$

which is $(g\xi)^{-1}$ times the Laplace transform of the function $(1+t)^{-l}$ evaluated at $(g\xi)^{-1}$. Integration by parts yields, for $l > 1$,

$$\bar{F}(l) = \{g\xi(l-1)\}^{-1}\{1 - \bar{F}(l-1)\}.$$

The starting point for this iterative scheme is

$$\bar{F}(1) = (g\xi)^{-1} \int_0^\infty e^{-t/(g\xi)}(1+t)^{-1}dt = a\,e^a\,E_1(a),$$

where $a = (g\xi)^{-1}$ and $E_1(a) = \int_a^\infty t^{-1}e^{-t}dt$ is the *exponential integral* (Olver et al., 2010, Section 6.7). This is also an *incomplete gamma function* but, having shape parameter zero, is not returned properly by the R-function pgamma. The power series

$$E_1(a) = -\gamma - \log a - \sum_{k=1}^\infty (-1)^k a^k /(k!k)$$

is given by Olver et al. (2010, Section 6.6); here $\gamma = 0.577215649$ is Euler's constant. It is said to converge quite well for small-to-moderate values of a. Cody and Thacher (1969) give highly accurate approximations.

5.5.2 Random Effects

It is not obligatory to introduce random effects via frailty in proportional-hazards or proportional-odds models, though this is the construction that underlies the standard frailty method. There are other ways, sometimes more tractable, of introducing unobserved inter-unit variation into a model.

5.5.3 Beta-Geometric Distribution

Conditional on ρ, the discrete probability mass function of M, a random variable with geometric distribution, is

$$p_m(\rho) := \mathrm{P}(M = m \mid \rho) = \rho^{m-1}(1-\rho)\,(m = 1, 2, \ldots)$$

and the survivor function is

$$q_m(\rho) := \mathrm{P}(M > m \mid \rho) = \rho^m.$$

Take ρ to vary across units with a *beta distribution*: this has density

$$f(\rho) = B(\phi, \psi)^{-1}\rho^{\phi-1}(1-\rho)^{\psi-1}$$

TABLE 5.3

Numbers of Cycles to Pregnancy in Two Groups

Cycles	1	2	3	4	5	6	7	8	9	10	11	12	>12
Smokers	29	16	17	4	3	9	4	5	1	1	1	3	7
Non-smokers	198	107	55	38	18	22	7	9	5	3	6	6	12

Source: Weinberg, C.R. and Gladen, B.C., 1986, The Beta-Geometric Distribution Applied to Comparative Fecundability Studies, *Biometrics*, 42, 552. Reproduced by permission of the International Biometric Society.

with $\phi > 0$ and $\psi > 0$, and in which $B(\phi, \psi)$ denotes the *beta function*. The resulting unconditional functions are

$$p_m = B(\phi + m - 1, \psi + 1)/B(\phi, \psi)$$

$$= \psi(\phi + \psi + m - 1)^{-1} \prod_{j=1}^{m-1}\{(\phi + j - 1)/(\phi + \psi + j - 1)\} \ (m = 2, 3, \ldots),$$

with $p_1 = \psi/(\phi + \psi)$, and

$$q_m = B(\phi + m, \psi)/B(\phi, \psi) = p_m(\phi + m - 1)/\psi \ (m = 1, 2, \ldots).$$

Handy recurrence relations for computation are

$$p_{m+1} = p_m(\phi + m - 1)/(\phi + \psi + m) \ \text{ and } \ q_{m+1} = q_m - p_m.$$

The various formulae here are left to the Exercises in Section 5.6.

5.5.4 Cycles to Pregnancy

The title does not refer to the village midwife, who regularly cycles to pregnancy. The original data are from Weinberg and Gladen (1986), arranged in Table 5.3 as a 2×13 contingency table. A number of pregnant women were asked how many menstrual cycles it took to conceive since they first started trying for a baby. After some judicious selection, the data retained contained 100 smokers and 486 non-smokers.

A chi-square test for equality of distributions between smokers and non-smokers, after pooling the last five columns of the table, yields $\chi_8^2 = 22.28$, with p-value 0.005. So, there does seem to be a real difference between the two groups in the waiting-time distribution.

Adopting a geometric distribution for the number of cycles does not seem too far-fetched. For this one assumes that the successive attempts are independent and that the probability of success is the same for each. Part of the data selection was designed to exclude cases liable to any temporal development in this probability. Further, allowing the probability to vary across couples would seem to be prudent. Quarrelsome meddlers, like referees, might say that using the beta distribution for this purpose looks like the pure convenience of resorting to a conjugate distribution. But then, part of the fun in life is to gleefully embrace what the snooty disapprove of, as well as splitting the infinitive and ending a clause with a preposition. Nonchalantly flicking aside such detractors, let's embrace this approach without further ado, following the 1986 paper.

So, we apply the beta-geometric model to M, the number of cycles to conception for an individual couple: $1 - \rho$ is the probability of the happy event (or daunting prospect, depending on how you look at it) on any one cycle. The likelihood contribution from an individual couple is p_m if conception occurs at observed cycle m and q_m if conception has not occurred by cycle m (when the time is right-censored). The function `bgeoml` returns the log-likelihood for the beta-geometric distribution as employed here (see the Web site).

```
#cycles to pregnancy: parameter=(phi,psi)  (Sec I.5.5)
vv1=c(29,16,17,4,3,9,4,5,1,1,1,3,7); #smokers
vv2=c(198,107,55,38,18,22,7,9,5,3,6,6,12); #non-smokers
mc=13; iopt=0; adt=c(mc,iopt,0); #additional data for bgeoml
np=2; par0=c(1.5,1); #no.parameters, initial estimate
par1=fitmodel(bgeoml,np,par0,vv1,adt); #smokers
mean=par1[1]/(par1[1]+par1[2]); cat('\n mean: ',mean);
par1=fitmodel(bgeoml,np,par0,vv2,adt); #non-smokers
mean=par1[1]/(par1[1]+par1[2]); cat('\n mean: ',mean);
vv12=vv1+vv2; par1=fitmodel(bgeoml,np,par0,vv12,adt); #combined
mean=par1[1]/(par1[1]+par1[2]); cat('\n mean: ',mean);
```

We obtain *mle* $(\hat{\phi}, \hat{\psi}) = (1.92, 0.97)$ for the smokers and $(\hat{\phi}, \hat{\psi}) = (1.40, 1.04)$ for the non-smokers. The corresponding mean estimates for $1 - \rho$ are 0.34 and 0.43. The maximised log-likelihoods are -219.7234 and -892.3729, and that for a combined fit (smokers and non-smokers) is -1118.653. This gives a likelihood-ratio chi-square value $\chi_2^2 = 13.11$, with p-value 0.0014: so, a real difference between groups is supported by the fitted model.

5.5.5 The Driving Test

It is fairly certain that individuals will vary in their ability to pass the test. Suppose that, within the population of individuals determined by **x**, that is, those of the same age, sex, and number of previous lessons, the variation in ρ follows a beta distribution. Thus, the number of attempts now has a beta-geometric distribution and its parameters (ϕ, ψ) will presumably be related to **x** in some way. Probably the most natural approach would be to reparametrise in terms of the mean and variance (see Section 5.6, Exercises), and then relate the log-mean linearly to **x** while leaving the variance constant. This exciting project is left as an exercise.

5.6 Exercises

5.6.1 Random Samples

Insurance claim settlements. In the data matrix presented in Table 5.4 the (j, k)th element, r_{jk}, is the number of insurance claim settlements made in year k when the claim was first made in year j. For example, 13 claims were first made in

TABLE 5.4

Insurance Claim Settlements

Year	1	2	3	4	5	Total
1	592	213	57	14	7	984
2	–	702	208	61	13	1110
3	–	–	853	219	75	1346
4	–	–	–	982	315	1778
5	–	–	–	–	1168	2262

year 2 and finally settled in year 5: the corresponding likelihood contribution for these claims is $p(t)^{13}$, where the waiting time $t = 5 - 2$. Likewise, the contribution from the 1778 claims made in year 4, but still not settled within 2 years, is $q(1)^{481}$, where $481 = 1778 - 982 - 315$ and $q(1) = \mathrm{P}(t > 1) = 1 - p(0) - p(1)$.

Write out an expression for the likelihood function for the data and then specialise it to a geometric distribution, for which $f(t) = (1 - \rho)^t \rho$ for $t = 0, 1, \ldots$. Calculate the *mle* of ρ algebraically. As a check on the geometric distributional assumption, fit a model with parameter set $(p_0, \ldots, p_5)$, in which $p_r = \mathrm{P}(t = r)$ is not restricted to be of geometric form. Check by fitting the models using R-code.

5.6.2 Explanatory Variables

1. *The driving test.* Carry out the project described at the end of Section 5.5. The R-function bgeom can be used as a starting point.

2. (Gail et al. 1981) Let $\psi_1, \ldots, \psi_n$ be a given set of numbers, and let $b_{rn} = \sum \psi_{a_1} \ldots \psi_{a_r}$ be the sum of ψ-products over all selections of distinct indices $(a_1, \ldots, a_r)$ from $(1, \ldots, n)$, where $1 \le r \le n$; b_{rn} is the denominator in the partial likelihood contribution. Agree that $b_{rr} = \psi_1 \ldots \psi_r$, $b_{1n} = \psi_1 + \cdots + \psi_n$ and $b_{23} = \psi_1 \psi_2 + \psi_2 \psi_3 + \psi_3 \psi_1$. Verify the recursion $b_{rn} = b_{r,n-1} + a_n b_{r-1,n-1}$. Code it all up in R and check the results for some simple cases. Then send the code to me.

5.6.3 Gamma and Beta Distributions

1. The *gamma function* is defined, for $v > 0$, as

$$\Gamma(v) = \int_0^\infty x^{v-1} e^{-x} dx.$$

Integrate by parts to obtain $\Gamma(v) = (v - 1)\Gamma(v - 1)$ for $v > 1$. Agree that, when v is a positive integer, $\Gamma(v) = (v - 1)!$.

2. The *gamma distribution*, with shape parameter $v > 0$ and scale parameter $\xi > 0$, has density $f(x) = \Gamma(v)^{-1} \xi^{-v} x^{v-1} e^{-x/\xi}$ on $(0, \infty)$. Verify that $\int_0^\infty f(x) dx = 1$, that $\mathrm{E}(X) = v\xi$ and that $\mathrm{var}(X) = v\xi^2$.

3. The *beta function* is defined, for $a > 0$ and $b > 0$, as

$$B(a, b) = \int_0^1 x^{a-1}(1 - x)^{b-1}dx.$$

   Verify that $B(b, a) = B(a, b)$. For the identity $B(a, b) = \Gamma(a)\Gamma(b)/\Gamma(a+b)$ see, for example, Whittaker and Watson (1927, Section 12.41).

4. The *beta distribution*, with parameters $a > 0$ and $b > 0$, has density $f(y) = B(a, b)^{-1}y^{a-1}(1 - y)^{b-1}$ on $(0, 1)$. Verify that $\mu_Y = E(Y) = a/(a + b)$ and $\text{var}(Y) = \mu_Y(1 - \mu_Y)/(a + b + 1)$.

5.7 Hints and Solutions

5.7.1 Random Samples

1. The mle is $\hat{\rho} = 0.52$ with standard error 0.0049, and the maximised log-likelihood is -382.437. The result of fitting the unconstrained model is $\hat{p} = (0.57, 0.19, 0.068, 0.017, 0.0095)$, with maximised log-likelihood -98.5938. The log-likelihood ratio for comparing the fits takes value $2(382.437 - 98.5938) = 567.69$. Referring this to χ_4^2 the p-value is minute, so the geometric distribution is a poor model for these data.

2. In the recursion $b_{r,n-1}$ represents terms without a_n.

Part II

Multivariate Survival Analysis

Multivariate survival data come in all shapes and sizes. There are the routine cases, where there is a data matrix with each row containing a set of lifetime measurements for an individual unit. These might be repeated measures, where a hospital patient has had their times to perform a task recorded each day, or clustered data, where the ages of the animals in a litter to reach a given weight have been monitored. Often, each row will also carry a set of covariates for the individual. Then there are less familiar types of data that, nevertheless, are suitable for the same sorts of models and methods.

In this part of the book some standard cases and some non-standard are addressed. The latter include situations which, though unlikely on the face of it, can give rise to data falling within the domain of attraction of multivariate survival; by this is meant that their structure can be regarded as essentially the same. However, it is a large and growing subject so we can claim only a general introduction to the area here.

6

Multivariate Data and Distributions

We start off with a selection of data sets, together with the odd comment, before getting down to some more serious theory work on probability distributions in the second section. This short introductory chapter is rounded off with some exercises (see Section 6.3) for the terminally keen.

6.1 Some Small Data Sets

The data sets included here qualify mainly because they are small and uncomplicated. Nevertheless, they serve for illustration.

6.1.1 Repeated Response Times

There has been long-term concern about the effect on the brain of even small amounts of lead absorbed into the body, particularly among young children living in traffic-clogged cities. A series of experimental studies was conducted in the Biochemistry Department at Surrey University to study this.

In the data listed in Table 6.1 there are four groups, each of 10 rats, with Group 1 having the lowest dose of an analgesic drug and Group 4 the highest. The rats were exposed to a harmless sensory stimulus at times 0, 15, 30, and 60 minutes after administration of the drug; and their response times in seconds (t_1, t_2, t_3, t_4) were recorded on each occasion. The groups appear as the four blocks in the table. There are 160 observations of which 17 are right censored at 10 seconds.

The data have been used twice before to my certain knowledge. In Crowder (1985) they were (or *it was*, if you prefer) fitted with a multivariate Burr distribution, and some appreciation of the quantitative effect of the analgesic drug and its temporal profile was gained. Such knowledge served as a baseline for subsequent studies involving body lead levels. In Crowder (1998), a more general multivariate survival distribution was fitted, but there the focus was more on goodness of fit, based on residuals in particular.

TABLE 6.1

Repeated Response Times

Group 1				Group 2				Group 3				Group 4			
t_1	t_2	t_3	t_4	t_1	t_2	t_3	t_4	t_1	t_2	t_3	t_4	t_1	t_2	t_3	t_4
1.58	1.78	2.02	2.19	2.02	3.86	2.73	2.88	2.74	5.10	7.00	3.54	1.68	5.33	10.0	10.0
1.55	1.40	2.20	1.73	1.75	3.38	3.74	2.57	1.89	8.00	9.80	6.63	2.80	5.93	8.77	4.62
1.47	2.66	3.05	3.76	1.93	3.62	6.05	2.91	1.47	9.77	9.98	10.0	2.19	8.73	10.0	8.20
2.16	2.28	2.04	2.12	2.04	4.70	3.70	3.70	3.13	9.04	9.92	6.25	1.52	10.0	10.0	10.0
2.19	2.24	1.99	1.58	2.00	3.34	4.14	2.78	1.53	4.72	5.56	3.25	1.85	7.71	8.35	7.57
2.07	1.97	1.50	1.24	1.63	3.37	4.84	2.36	2.12	4.73	7.90	6.42	2.78	7.41	10.0	10.0
1.28	1.67	2.25	1.54	1.97	3.72	7.83	3.14	1.28	4.32	6.24	4.73	1.81	7.56	10.0	10.0
1.53	2.68	1.79	2.03	2.42	4.81	4.90	1.69	1.50	8.91	9.22	4.30	1.80	10.0	9.41	10.0
2.62	2.15	1.60	2.91	1.38	3.88	4.20	2.05	2.05	4.21	10.0	3.03	1.16	9.95	10.0	10.0
1.67	2.52	1.53	1.98	2.32	3.75	3.06	3.81	1.53	7.10	9.88	3.72	1.67	7.82	10.0	7.97

Source: Crowder, M.J., 1998, A Multivariate Model for Repeated Failure Time Measurements, *Scand. J. Statist.* 25, Table 1, 62. Reproduced by permission of Wiley-Blackwell Publishers.

6.1.2 Paired Response Times

The data here, like those in Table 6.1, arise from investigations at Surrey University concerning the effect on brain function of bodily absorbed lead. The data given in Table 6.2 arise from these studies: there are three groups of rats, with different body lead levels, and five dose levels of an analgesic drug in each group. The recorded times are seconds until the rat responds to a harmless tactile stimulus, the intensity of which is increased steadily from zero: there are two times per rat, one before administration of the drug and one after (t_1 and t_2). The times are given to the nearest 0.5 seconds and are right censored at 250.

The data made a previous highly acclaimed appearance in Crowder (1989), where they were fitted by a "multivariate distribution with Weibull connections." Some useful inferences arose from the analysis concerning the effect of lead and the drug on brain activity: details are given in the paper. The same data were again used for illustration in Crowder (1998), where more emphasis was placed on the goodness of fit of a particular type of mixture model.

6.1.3 Lengths and Strengths of Fibres

Table 6.3 gives a small subset of the data in Table 7.2 of Crowder et al. (1991). The fibres had been cut into lengths 5, 12, 30, and 75 mm, and each segment subjected to a steadily increasing load until breakage. The maximum load is 4.0; and if the fibre is still intact, at that point it is set free to go on its way without a stain on its character.

6.1.4 Household Energy Usage

One of the measures being promoted by governments, keen to polish their green credentials, is that of energy saving. (This seems to be a less contentious issue than energy production: instead of blotting the countryside with eyesore

TABLE 6.2

Paired Response Times: Pre-Drug (t_1) and Post-Drug (t_2)

Dose 0		Dose 0.1		Dose 0.2		Dose 0.4		Dose 0.8	
				Group 1					
t_1	t_2	t_1	t_2	t_1	t_2	t_1	t_2	t_1	t_2
47	29	43.5	107	40.5	103.5	45	241	48.5	250
36	35.5	48.5	109	56	203	40	218.5	60	250
51.5	48.5	37	89	57	127.5	52.5	250	39.5	250
39.5	77	55.5	93.5	58.5	114	57	211	39	245.5
31	32.5	27.5	62.5	33.5	95	40.5	117.5	28	224.5
32	49.5	41	71.5	27.5	106	26.5	250	36.5	215.5
35	47.5	36.5	92	40.5	130.5	58.5	179.5	35	249.5
50.5	55	35	79.5	38	131	35.5	193.5	43.5	250
		27.5	74	34.5	115	33	141.5		
		35	98						
				Group 2					
t_1	t_2	t_1	t_2	t_1	t_2	t_1	t_2	t_1	t_2
46	65	46	72	45.5	83	36	131	35.5	186
38	35.5	53	91	45.5	119	53	107	59	250
54	21.5	40.5	130.5	68.5	121.5	51	130.5	38.5	250
42.5	63	60.5	78	48.5	110	37	144	34.5	250
60.5	40.5	47	101.5	34	83.5	67.5	130	34.5	184.5
38.5	39	47	108	41	138	51	250	56.5	249
41	36.5	36	67	41.5	77	45.5	144	37.5	229
50	45.5	31.5	67.5			49.5	152	33.5	250
				Group 3					
t_1	t_2	t_1	t_2	t_1	t_2	t_1	t_2	t_1	t_2
70	59.5	67.5	103	66	98	49	126.5	68.5	250
51.5	42.5	50	98.5	42.5	103.5	57	111.5	29	250
38.5	40.5	58	68.5	55.5	85.5	62	99.5	41	204
55.5	64	46.5	70	52	93.5	47.5	145	40	148
59.5	79	59	70	69.5	109	39.5	170.5	65	250
60.5	64.5	71.5	53.5	44.5	114	44	157	62.5	250
29.5	40.5	30	99.5	45.5	89	27.5	120	35	214
42.5	41	29	69	29.5	87	75	150	45.5	197.5
30	35.5	25	70	29.5	93				

Source: Crowder, M.J., 1989, A Multivariate Distribution with Weibull Connections, *J. Roy. Statist. Soc. B 51*, Table 1, 104. Reproduced by permission of Wiley-Blackwell Publishers.

wind turbines they should put them inside Parliament, where the supply of hot air is far more reliable.) "Research has shown" (as they say in Sociology when stating the obvious) that electrical appliances in the home are often left switched on, either accidentally or deliberately, when not in use. To this end meters were installed in a number of households to monitor the times for which various appliances were switched on. The times were recorded over a run-in phase of four weeks and then over an experimental phase of four weeks. During the latter phase, devices were fitted that switch the appliance off automatically after a set time, the purpose being to avoid energy wasted

TABLE 6.3

Lengths and Strengths of Fibres

Fibre	5 mm	12 mm	30 mm	75 mm	Fibre	5 mm	12 mm	30 mm	75 mm
1	3.30	3.32	2.39	2.08	11	4.00	4.00	2.99	2.30
2	4.00	3.67	2.49	2.06	12	4.00	3.03	2.80	2.30
3	4.00	4.00	3.16	2.05	13	3.01	3.17	2.41	2.07
4	3.64	2.41	2.20	1.80	14	4.00	2.91	2.18	1.83
5	2.73	2.24	1.91	1.68	15	4.00	3.87	2.24	2.09
6	4.00	4.00	2.74	2.22	16	4.00	3.82	2.59	2.48
7	3.29	3.08	2.44	2.37	17	4.00	4.00	2.40	2.22
8	3.55	2.35	2.38	2.37	18	3.48	2.14	2.35	2.05
9	3.03	2.26	1.64	2.03	19	3.05	2.96	1.91	2.20
10	4.00	4.00	2.98	2.39	20	3.60	2.92	2.42	2.09

Source: Cox, D.R. and Oakes, D., 1984, *Analysis of Survival Data*, Chapman & Hall, London; Crowder, M.J., Kimer, A.C., Smith, R.L., and Sweeting, T.J., 1991, *Statistical Analysis of Reliability Data*, Chapman & Hall, London, Table 7.2. Reproduced by permission of Chapman & Hall/CRC Press.

when an appliance is not in active use. The data are commercially confidential but some typical values are shown in Table 6.4 for a lamp, and just for the last week in each phase. The data are of survival type in a broad sense: the times record how long the appliance survives until being switched off, and they are censored at 1440, the number of minutes in a day.

The location of the lamp, in the hall or sitting room or wherever, is not recorded in the data, and different locations might go some way toward explaining the large variation in usage between households. Moreover, the survey was conducted in the winter, and in some households the lamp seems to be left switched on most of the time. History does not record the energy usage by Florence Nightingale's lamp.

TABLE 6.4

Energy Usage Times for a Lamp

Household	Day (week 1)							Day (week 2)						
	1	2	3	4	5	6	7	1	2	3	4	5	6	7
1	196	165	109	75	37	12	101	158	161	43	90	145	39	75
2	34	44	7	33	38	34	42	5	14	13	13	6	4	3
3	40	59	25	70	45	80	54	48	4	47	3	22	11	6
4	344	61	299	112	22	188	128	87	61	73	21	151	128	69
5	106	49	57	37	103	90	57	22	40	54	87	73	46	25
6	95	82	174	101	104	69	125	53	4	38	75	21	54	41
7	168	137	116	107	62	111	105	48	85	44	58	71	149	49
8	169	163	148	115	154	156	75	24	59	54	74	68	31	81
9	1435	1439	1435	1438	635	45	1136	61	71	52	62	77	45	56
10	80	63	193	91	175	21	42	68	38	60	46	132	75	60
11	68	103	124	102	50	222	11	166	55	39	89	44	47	75
12	188	188	162	155	165	148	86	48	44	75	141	49	87	64
13	17	95	92	107	44	37	97	52	21	36	67	17	40	63
14	70	177	1437	900	1439	487	190	58	79	34	83	94	37	94

6.2 Multivariate Survival Distributions

6.2.1 Joint and Marginal Distributions

Let $\mathbf{T} = (T_1, \ldots, T_p)$ be a vector of failure times, a set of non-negative random variables. The *joint distribution function* is defined as $G(\mathbf{t}) = P(\mathbf{T} \leq \mathbf{t})$, where $\mathbf{t} = (t_1, \ldots, t_p)$ and $\mathbf{T} \leq \mathbf{t}$ means $T_j \leq t_j$ for each j; similarly, the *joint survivor function* is $\bar{G}(\mathbf{t}) = P(\mathbf{T} > \mathbf{t})$. If the T_j are jointly continuous, their joint density is $\partial^p G(\mathbf{t})/\partial t_1 \ldots \partial t_p$ or, equivalently, $(-1)^p \partial^p \bar{G}(\mathbf{t})/\partial t_1 \ldots \partial t_p$.

For individual components and subsets of components of $\mathbf{T}$ we will denote marginal distribution functions by

$$G_j(t) = P(T_j \leq t), \quad G_{jk}(s, t) = P(T_j \leq s, T_k \leq t), \text{ and so forth.}$$

Likewise for marginal survivor functions.

6.2.2 Conditional Distributions

Suppose that $\mathbf{T}$ is partitioned as $(\mathbf{T}_A, \mathbf{T}_B)$. Then, with $\mathbf{t} = (\mathbf{t}_A, \mathbf{t}_B)$,

$$P(\mathbf{T}_B > \mathbf{t}_B \mid \mathbf{T}_A > \mathbf{t}_A) = \bar{G}(\mathbf{t})/\bar{G}_A(\mathbf{t}_A)$$

could represent the probability that a unit passes a secondary battery of tests $(\mathbf{T}_B > \mathbf{t}_B)$ given that it passed an initial battery $(\mathbf{T}_A > \mathbf{t}_A)$. Successive examinations inflicted upon poor, long-suffering students are like this; other examples include quality control stages in manufacturing production and preliminary medical examinations to determine whether a patient is suitable for treatment. In survival this might mean that a unit has previously survived long enough $(\mathbf{T}_A > \mathbf{t}_A)$ to be included in the current data set, in which $\mathbf{T}_B$ is to be observed; this is an example of *left-censoring*. The similar quantity, $P(\mathbf{T}_B > \mathbf{t}_B \mid \mathbf{T}_A = \mathbf{t}_A)$, would be more suited for use as a predictive tool: $\mathbf{T}_A = \mathbf{t}_A$ has been observed, and on the basis of this information, we wish to assess the likelihood that $\mathbf{T}_B > \mathbf{t}_B$.

6.2.3 Dependence and Association

Independence of the T_j is defined equivalently by

$$G(\mathbf{t}) = \prod_{j=1}^{p} G_j(t_j) \text{ and } \bar{G}(\mathbf{t}) = \prod_{j=1}^{p} \bar{G}_j(t_j);$$

that these conditions are equivalent is the subject of one of the Exercises (see Section 6.3). When the T_j are not independent, the degree of dependence can be measured in various ways. For Gaussian-like distributions the correlations are the standard measures of dependence. But lifetime variables are usually markedly non-normal and observations are often right censored, which is no help for computing and interpreting sample correlations. In the absence of censoring, log-transformed lifetimes can often be close enough to Gaussian for practical purposes; alternatively, rank correlation coefficients can be applied to data. However, in general we look to dependence measures that are more widely applicable and interpretable. Conditional probabilities, like those mentioned above, are useful in this respect. In addition, two closely related overall measures are the ratios

$$R(\mathbf{t}) = G(\mathbf{t}) \Bigg/ \prod_{j=1}^{p} G_j(t_j) \ \text{ and } \ \bar{R}(\mathbf{t}) = \bar{G}(\mathbf{t}) \Bigg/ \prod_{j=1}^{p} \bar{G}_j(t_j);$$

if $R(\mathbf{t}) > 1$ or $\bar{R}(\mathbf{t}) > 1$ for all $\mathbf{t}$ there is *positive dependence*, and if $R(\mathbf{t}) < 1$ or $\bar{R}(\mathbf{t}) < 1$ for all $\mathbf{t}$ there is *negative dependence* (e.g., Lehmann, 1966; Shaked, 1982). For example, for a bivariate distribution

$$\bar{R}(\mathbf{t}) < 1 \ \text{ if and only if } \ P(T_2 > t_2 \mid T_1 > t_1) < P(T_2 > t_2),$$

that is , T_2 becomes less likely to exceed t_2 after it becomes known that $T_1 > t_1$.

Another way of looking at dependence is via regression, that is, considering the way in which the mean of $\mathbf{T}_B$, say, varies with $\mathbf{T}_A$. In survival analysis some of the more common survivor functions have explicit form, and then it is easier to work with quantiles than with means. Thus, taking $\mathbf{T}_B$ one-dimensional, for example, we solve

$$P(T_B > t_q \mid \mathbf{T}_A = \mathbf{t}_A) = q \ \text{ or } \ P(T_B > t_q \mid \mathbf{T}_A > \mathbf{t}_A) = q$$

for the upper tail $100q\%$ conditional quantile t_q of T_B. For example, if t_q increases with $\mathbf{t}_A$ we could conclude that T_B is positively associated with $\mathbf{T}_A$.

Oakes (1989) suggested a cross-ratio function for assessing dependence: in his notation this is

$$\theta^*(\mathbf{t}) = \bar{G}(\mathbf{t}) \, \partial^2 \bar{G}(\mathbf{t})/\partial t_1 \partial t_2 \div \{\partial \bar{G}(\mathbf{t})/\partial t_1\}\{\partial \bar{G}(\mathbf{t})/\partial t_2\},$$

a measure introduced by Clayton (1978). It is symmetric in t_1 and t_2, and can be expressed in terms of conditional hazards as

$$h(t_2 \mid T_1 = t_1) \div h(t_2 \mid T_1 > t_1),$$

which supports its claim to be a measure of dependence.

6.2.4   Hazard Functions and Failure Rates

A variety of these functions has appeared in the literature. A conditional hazard function for an individual component of $\mathbf{T}$ is defined as

$$h_j(t_j; \mathbf{t}) = \lim_{\delta \downarrow 0} \delta^{-1} P(T_j \leq t_j + \delta \mid \mathbf{T} > \mathbf{t}).$$

Thus, one might be monitoring a group of units and have a particular interest in the jth: with $\mathbf{t} = t\mathbf{1}_p = (t, \ldots, t)$, the probability in question is conditional on all units having survived to time t. This is the jth component of Johnson and Kotz's (1972) *vector multivariate hazard rate*, defined as $\mathbf{h}(\mathbf{t}) = -\nabla\{\log \bar{G}(\mathbf{t})\}$. Clayton (1978) used it for the bivariate case, and Clayton and Cuzick (1985) called $-\log \bar{G}(\mathbf{t})$ the *mortality potential*, since its partial derivatives are these hazard components.

A scalar *multivariate hazard function* can be defined as $h(\mathbf{t}) = g(\mathbf{t})/\bar{G}(\mathbf{t})$. In the bivariate case this is

$$h(t_1, t_2) = g(t_1, t_2)/\bar{G}(t_1, t_2) = \partial^2 \bar{G}(t_1, t_2)/\partial t_1 \partial t_2 \div \bar{G}(t_1, t_2).$$

The right-hand side here is not the second derivative of $\log \bar{G}(t_1, t_2)$, so this is not the direct extension of the univariate formula (Section 2.1).

6.2.5 Gumbel's Bivariate Exponential (Gumbel, 1960)

Gumbel suggested three different forms for bivariate exponential distributions. We will use the first of these for illustration at various points throughout this book on account of its simplicity. It has been pointed out elsewhere that it does not have any obvious justification as describing a realistic physical mechanism. This total lack of credibility is part of its appeal. It has joint survivor function

$$G(\mathbf{t}) = \exp(-\lambda_1 t_1 - \lambda_2 t_2 - v t_1 t_2),$$

in which $\lambda_1 > 0$ and $\lambda_2 > 0$; the dependence parameter v satisfies $0 \leq v \leq \lambda_1 \lambda_2$, $v = 0$ yielding independence of T_1 and T_2. The dependence ratio $\bar{R}(\mathbf{t})$ is equal to $\exp(-v t_1 t_2)$ here and, since this is always smaller than 1, the distribution exhibits only negative dependence.

The marginal survivor functions are $\bar{G}_j(t_j) = \exp(-\lambda_j t_j)$ $(j = 1, 2)$, and the conditionals are, for T_2 given T_1,

$$P(T_2 > t_2 \mid T_1 > t_1) = \exp(-\lambda_2 t_2 - v t_1 t_2)$$

and

$$P(T_2 > t_2 \mid T_1 = t_1) = (1 + v t_2/\lambda_1) \exp(-\lambda_2 t_2 - v t_1 t_2).$$

These probabilities are both decreasing in t_1, reflecting the negative dependence, and knowing that $T_1 = t_1$, rather than just $T_1 > t_1$, increases the likelihood that $T_2 > t_2$ by a factor $(1 + v t_2/\lambda_1)$.

The vector multivariate hazard here is

$$\mathbf{h}(\mathbf{t}) = -\nabla(-\lambda_1 t_1 - \lambda_2 t_2 - v t_1 t_2) = (\lambda_1 + v t_2, \ \lambda_2 + v t_1);$$

so, the hazard for each $\mathbf{T}$-component increases linearly with the value of the other.

For the conditional upper-tail T_2-quantiles we must solve for t_q

$$q = (1 + v t_q/\lambda_1) \exp(-\lambda_2 t_q - v t_1 t_q) \quad \text{or} \quad q = \exp(-\lambda_2 t_q - v t_1 t_q),$$

in which the first form is conditioned on $T_1 = t_1$ and the second on $T_1 > t_1$. The first is not soluble explicitly but the second yields $t_q = -\log q /(\lambda_2 + \nu t_1)$, entailing a given T_2-quantile decreasing with t_1.

6.3 Exercises

6.3.1 Joint and Marginal Distributions

1. Draw a diagram to convince yourself that

$$\bar{G}(t_1, t_2) + G_1(t_1) + G_2(t_2) - G(t_1, t_2) = 1.$$

Now go on to

$$\bar{G}(t_1, t_2, t_3) + G_1(t_1) + G_2(t_2) + G_3(t_3) - G_{12}(t_1, t_2) - G_{23}(t_2, t_3)$$
$$-G_{13}(t_1, t_3) + G(t_1, t_2, t_3) = 1.$$

Check that this works for $t_3 = 0$ and $t_3 = \infty$. What is the pattern?

2. Suppose that continuous $\mathbf{T} = (T_1, \ldots, T_p)$ has joint distribution function $G(\mathbf{t})$ and joint survivor function $\bar{G}(\mathbf{t})$. Verify that its density function can be obtained either as $\partial^p G(\mathbf{t})/\partial t_1 \ldots \partial t_p$ or as $(-1)^p \partial^p \bar{G}(\mathbf{t})/\partial t_1 \ldots \partial t_p$.

3. Derive $E(T_1 T_2) = \int \bar{G}(t_1, t_2) dt_1 dt_2$.

6.3.2 Dependence and Association

1. Independence of the T_j in $\mathbf{T} = (T_1, \ldots, T_p)$ is defined by $G(\mathbf{t}) = \prod_{j=1}^{p} G_j(t_j)$. Show that $G_{jk}(t_j, t_k) = G_j(t_j)G_k(t_k)$ follows, and likewise for any marginal distribution of dimension less than p. Show that independence can equivalently be specified by $\bar{G}(\mathbf{t}) = \prod_{j=1}^{p} \bar{G}_j(t_j)$. Thus, $R(\mathbf{t}) = 1 \Leftrightarrow \bar{R}(\mathbf{t}) = 1$.

2. Show that, for the bivariate case ($p = 2$), $R(t_1, t_2)$ and $\bar{R}(t_1, t_2)$ are roughly equivalent, in the sense that if one is greater than or less than 1 so is the other.

3. Convince yourself that the Oakes/Clayton cross-ratio satisfies

$$\theta^*(\mathbf{t}) = h(t_2 \mid T_1 = t_1) \div h(t_2 \mid T_1 > t_1).$$

4. Show that $\theta^*(\mathbf{t})$ is constant for the Clayton (1978) family of bivariate distributions, for which

$$\{\bar{G}(\mathbf{t})\} = (\bar{G}_1(t_1)^{-\nu} + \bar{G}_2(t_2)^{-\nu} - 1)^{-1/\nu}.$$

5. Suppose that $\mathbf{Y}$ has bivariate normal distribution $N_2(\mu, \Sigma)$, and let $\mathbf{Z}$ and $\mathbf{U}$ be as defined as in Section 1.4. Let $T_j = \exp(Y_j)$ ($j = 1, 2$), and

recall that $E(T_j) = e^{\mu_j + \sigma_j^2/2}$ and $var(T_j) = e^{2\mu_j + \sigma_j^2}(e^{\sigma_j^2} - 1)$. Verify that $E(e^{\alpha U_j}) = e^{\alpha^2/2}$ and then that

$$E(T_1 T_2) = e^{\mu_1 + \mu_2} e^{(\sigma_1^2 + 2\sigma_{12} + \sigma_2^2)/2}.$$

The covariance of T_1 and T_2 follows as

$$cov(T_1, T_2) = E(T_1 T_2) - E(T_1)E(T_2) = e^{(\mu_1 + \mu_2) + \frac{1}{2}(\sigma_1^2 + \sigma_2^2)} (e^{\sigma_{12}} - 1),$$

which yields correlation

$$(e^{\sigma_{12}} - 1) \div \sqrt{(e^{\sigma_1^2} - 1)(e^{\sigma_2^2} - 1)}.$$

6.4 Hints and Solutions

6.4.1 Joint and Marginal Distributions

2. Use Question 1.
3. Extend Exercise 9 of Section 2.7 to this bivariate case.

6.4.2 Dependence and Association

1. Set $t_l = \infty$ for $l \neq j, k$. Use Question 1 in Section 6.3.1.
2. Use Question 1 in Section 6.3.1 to derive

$$\bar{R}(t_1, t_2) = 1 + \{G(t_1, t_2) - G_1(t_1)G_2(t_2)\}/\{1 - G_1(t_1) - G_2(t_2)$$
$$+ G_1(t_1)G_2(t_2)\}$$
$$= 1 + \{R(t_1, t_2) - 1\}\{G_1(t_1)/\bar{G}_1(t_1)\} \{G_2(t_2)/\bar{G}_2(t_2)\}.$$

5. $$E(T_1 T_2) = \int_{-\infty}^{\infty} e^{y_1 + y_2} \, \phi_2(y_1, y_2; \mu, \Sigma) \, dy_1 dy_2$$

$$= (2\pi)^{-1} e^{\mu_1 + \mu_2} \int_{-\infty}^{\infty} \exp\{(\sigma_1 + \rho\sigma_2)u_1 + \sigma_2 a u_2\}$$
$$\times \exp\{-(u_1^2 + u_2^2)/2\} \, du_1 du_2$$
$$= e^{\mu_1 + \mu_2} \times E(e^{(\sigma_1 + \rho\sigma_2)U_1}) \times E(e^{\sigma_2 a U_2})$$
$$= e^{\mu_1 + \mu_2} \times e^{(\sigma_1 + \rho\sigma_2)^2/2} \times e^{(\sigma_2 a)^2/2} = e^{\mu_1 + \mu_2} e^{(\sigma_1^2 + 2\sigma_{12} + \sigma_2^2)/2}.$$

7

Some Models and Methods

7.1 The Multivariate Log-Normal Distribution

The normal distribution occupies a special place in statistics for a number of reasons. Here, we focus on the multivariate version in view of its familiar dependence structure in terms of correlations. To adapt it to survival analysis we need to transform the components to be positive: an exponential transformation produces the so-called *log-normal distribution*.

Let $\mathbf{Y}$ have distribution $N_p(\mu, \Sigma)$, p-variate normal with mean vector μ and covariance matrix Σ; μ has components μ_j, and Σ has diagonal elements σ_j^2 (the component variances) and off-diagonal elements σ_{jk} (the covariances). Then $\mathbf{T}$, defined component-wise by $T_j = \exp(Y_j)$, has p-variate log-normal distribution. The component means and variances are $E(T_j) = e^{\mu_j + \sigma_j^2/2}$ and $var(T_j) = (e^{\sigma_j^2} - 1)\mu_j^2$, and these and the the covariances and correlations are given as Exercises in Section 7.7. In any case, for survival analysis, quantiles are likely to be more useful than moments: component quantiles are obtained from Φ, the standard normal distribution function, but multivariate probabilities are not so easily computed (for example, Genz, 1993).

In studies of the effect of some treatment the recordings are often made pre and post. For example, the paired response times (Section 6.1) comprise records made once before and once after the administration of a drug. In such cases a bivariate distribution is the focus.

Suppose that $\mathbf{Y} = (Y_1, Y_2)^T$ has bivariate normal distribution $N_2(\mu, \Sigma)$, with mean $\mu = (\mu_1, \mu_2)^T$ and covariance matrix

$$\Sigma = \begin{pmatrix} \sigma_1^2 & \sigma_{12} \\ \sigma_{21} & \sigma_2^2 \end{pmatrix} = \begin{pmatrix} \sigma_1^2 & \sigma_1\sigma_2\rho \\ \sigma_1\sigma_2\rho & \sigma_2^2 \end{pmatrix},$$

where $\rho = \sigma_{12}/(\sigma_1\sigma_2)$ is the correlation coefficient. The bivariate normal density of $\mathbf{Y}$ is

$$\phi_2(\mathbf{y}; \mu, \Sigma) = \{\det(2\pi\Sigma)\}^{-1/2} \exp\left\{-\frac{1}{2}(\mathbf{y} - \mu)^T \Sigma^{-1}(\mathbf{y} - \mu)\right\},$$

and that of the standardised variables $Z_j = (Y_j - \mu_j)/\sigma_j$ is

$$\phi_2(\mathbf{z}; \mathbf{0}, \mathbf{R}) = (2\pi a)^{-1} \exp\left\{-\frac{1}{2}(z_1^2 - 2\rho z_1 z_2 + z_2^2)/a^2\right\},$$

in which $a^2 = 1 - \rho^2$ and $\mathbf{R} = \begin{pmatrix} 1 & \rho \\ \rho & 1 \end{pmatrix}$ is the correlation matrix. Independent components can be achieved with a further transformation: let

$$A = \begin{pmatrix} 1 & 0 \\ \rho & a \end{pmatrix} \quad \text{so that} \quad AA^T = \Sigma \text{ and } A^{-1} = a^{-1} \begin{pmatrix} a & 0 \\ -\rho & 1 \end{pmatrix}.$$

Expressing Σ as AA^T, where A is lower triangular, is an example of *Cholesky decomposition*. Then $\mathbf{U} = A^{-1}\mathbf{Z}$ has bivariate normal distribution with mean $\mathbf{0}$ (zero vector) and covariance matrix $\mathbf{I}$ (unit matrix), that is, the components, U_1 and U_2 are independent standard normal variates; their joint density is $\phi_2(\mathbf{u}; \mathbf{0}, \mathbf{I}) = (2\pi)^{-1} \exp(-\mathbf{u}\mathbf{u}^T/2)$. In the Exercises (see Section 7.7) this transformation is used to calculate the covariance and correlation of T_1 and T_2.

7.2 Applications

7.2.1 Household Energy Usage

We now switch on to the lamp data (Section 6.1). There could be serial correlation over days. In some cases a lamp might be used for illumination on dark days, and weather does go in spells. However, we will ignore this aspect for now. Plotting the data profiles, joining the points across time for each household, just produces a mess. This might be expected in view of the huge variations both within and between households. So, maybe the situation would be better treated to a log-normal, two-way ANOVA, with fixed effects for households and weeks, and with the days as seven replicates per cell. Here is the R-code: the log-times are formed as a long vector `y1`; then `x1` is generated to represent households and `x2` to represent weeks 1 and 2.

```
#lamp data (Sec II.7.2)
la1=read.table('lampc.dat',header=F); attach(la1);
nd=14; pdim=14; la2=as.numeric(unlist(la1[1:nd,2:(pdim+1)]));
#anova
y1=log(t(la2)); dim(y1)=c(nd*pdim);
x1=rep(1:nd,each=pdim); x1;
x2=rep(c(rep(1,pdim/2),rep(2,pdim/2)),nd); x2;
fit1=lm(y1 ~ x1+x2+x1*x2); summary(fit1);
fit2=lm(y1 ~ x1+x2); summary(fit2);
```

The first summary shows that the interaction term, `x1*x2`, can be dropped from the linear model. The second one shows that both `x1` and `x2` are highly significant (p-values 6×10^{-9} and 3×10^{-6}). Now, the primary issue here is whether the energy usage is significantly reduced when the device is fitted. We have statistical significance, but it is the magnitude of the reduction that is of real significance. The second summary gives the estimated coefficient of $x2$ as 0.647, with standard error 0.135. So, we can estimate the reduction factor as $\exp(-0.647) = 0.52$: an average reduction of 48% is pretty significant in real terms. Of course, the down-side of such energy-saving devices is having the power cut off when you don't want it to be cut off: so, during hours of concentrated typing of your magnus opus on the computer, make sure you save it at regular intervals.

7.2.2 Repeated Response Times

Let us apply the multivariate log-normal model to the data (Section 6.1). To begin with we ignore the right censoring: the response times given as 10 will be taken at face value. In addition to the usual bias induced by ignoring censoring, this will tend to diminish differences between groups, since there are more censored values in the last group. So, the data are y_{ij}, the log-time for animal i on occasion j ($i = 1, \ldots, n = 40$; $j = 1, \ldots, p = 4$). There are two covariates, *lead level* (group) and *time after exposure to stimulus* (0, 15, 30, and 60 minutes, yielding $t1, t2, t3, t4$). Some R-code for loading the data, summarising it, and then performing MANOVA is

```
#repeated response times (Sec II.7.2)
rt1=read.table('rtimes1.dat',header=T); attach(rt1); nd=40; p=4;
summary(rt1[1:nd,2:(p+1)]); #overall data summary
aggregate(rt1,by=list(group),mean); #group means
rt1a=matrix(0,nrow=nd,ncol=p+1); rt1a[1:nd,1]=group;
for (j in 1:p) { rt1a[1:nd,j+1]=log(rt1[1:nd,j+1]) }; #log-times
colnames(rt1a)=c('group','logt1','logt2','logt3','logt4');
aggregate(rt1a,by=list(group),mean); #group log-means
#manova
fity1=manova(rt1a[1:nd,2:(p+1)] ~ group); #one-way Manova
summary.aov(fity1); summary(fity1); #summary of results
```

The summary.aov(fity1) output shows one-way ANOVAS for each of the four responses, that is, those at 0, 15, 30 and 60 minutes after the stimulus. The F-ratios, all on 1 and 38 degrees of freedom, are 0.12, 161.23, 193.85, and 131.94; they're all highly significant except the first, which is presumably too early for the analgesic drug to have taken effect. The summary(fity1) shows the overall one-way MANOVA result: the F-ratio is 77.77 on 4 and 35 degrees of freedom. (Incidentally, if you are not terribly familiar with the material of this section, or even if you are, please buy Crowder and Hand [1990] and look at Chapters 4, 5, and 6.)

These results are more or less sufficient for what the investigators were mainly interested in. It seems that the analgesic drug, once it kicks in, produces slower reactions, as expected. How much slower is given in the tables of mean values, and these results serve as a baseline for later studies involving body lead levels. Moreover, the investigators can claim that they will not be "mis-lead" (sorry) by ignoring the censoring: these tests of group differences will be conservative. The data should not breathe a sigh of relief—it will be roughed up again in Chapter 8.

7.3 Bivariate Exponential (Freund, 1961)

In a two-component system the lifetimes, T_1 and T_2, are initially independent exponential variates. However, failure of one component alters the subsequent development of the other, the effect being to change the exponential rate parameter of the surviving component from λ_j to μ_j ($j = 1, 2$). The case $\mu_j > \lambda_j$ would be appropriate when the stress on the surviving component is increased after the failure, as when the load was previously shared. Conversely, $\mu_j < \lambda_j$ would reflect some relief of stress, as when the components previously competed for resources.

The joint survivor function can be derived as

$$\bar{G}(\mathbf{t}) = (\lambda_+ - \mu_2)^{-1}\{\lambda_1 e^{-(\lambda_+ - \mu_2)t_1 - \mu_2 t_2} + (\lambda_2 - \mu_2)e^{-\lambda_+ t_2}\}$$

for $t_1 \leq t_2$, assuming that $\lambda_+ = \lambda_1 + \lambda_2 \neq \mu_2$; the form for $t_1 \geq t_2$ is obtained from the above by interchanging λ_1 and λ_2, μ_1 and μ_2, t_1 and t_2. Setting $t_1 = 0$ in the form given for $\bar{G}(\mathbf{t})$, we obtain the marginal

$$\bar{G}_2(t_2) = (\lambda_+ - \mu_2)^{-1}\{\lambda_1 e^{-\mu_2 t_2} + (\lambda_2 - \mu_2)e^{-\lambda_+ t_2}\}.$$

This is a mixture of two exponentials, rather than a single one, so the joint distribution is not strictly speaking a bivariate exponential. Independence of T_1 and T_2 obtains if and only if $\mu_1 = \lambda_1$ and $\mu_2 = \lambda_2$.

7.3.1 Discrete-Time Version (Crowder, 1996)

Two sequences of Bernoulli trials are in progress, side by side in synchrony. After a failure in one sequence the other continues with success probability changed from π_j to ρ_j ($j = 1, 2$).

The marginal densities associated with this model are $g_j(t) = \pi_j^{t-1}(1 - \pi_j)$ ($j = 1, 2$; $t = 1, \ldots, m = \infty$), and the joint density of $\mathbf{T} = (T_1, T_2)$ is

$$g(\mathbf{t}) = \begin{cases} \pi_1^{t_1-1}(1 - \pi_1)\pi_2^{t_1}\rho_2^{t_2-t_1-1}(1 - \rho_2) & \text{for } t_1 < t_2 \\ \pi_2^{t_2-1}(1 - \pi_2)\pi_1^{t_2}\rho_1^{t_1-t_2-1}(1 - \rho_1) & \text{for } t_1 > t_2 \\ \pi_1^{t_1-1}(1 - \pi_1)\pi_2^{t_2-1}(1 - \pi_2) & \text{for } t_1 = t_2 \end{cases}$$

Note that $g(\mathbf{t}) \neq g_1(t_1)g_2(t_2)$ generally, so T_1 and T_2 are not independent (unless $\rho_1 = \pi_1$ and $\rho_2 = \pi_2$).

7.4 Bivariate Exponential (Marshall–Olkin, 1967)

Consider a two-component system subject to three types of fatal shock. The first type knocks out component 1, the second component 2, and the third both components. Suppose that the shocks arrive according to three independent Poisson processes of rates λ_1, λ_2 and λ_{12}, and let $N_j(t)$ be the number of shocks of type j occurring during time interval $(0, t)$. Then

$$\bar{G}(\mathbf{t}) = P(T_1 > t_1, T_2 > t_2) = P[N_1(t_1) = 0, N_2(t_2) = 0, N_3\{\max(t_1, t_2)\} = 0]$$

$$= e^{-\lambda_1 t_1} e^{-\lambda_2 t_2} e^{-\lambda_{12} \max(t_1, t_2)}.$$

An equivalent representation is in terms of the exponential waiting times W_j ($j = 1, 2, 3$) to the three types of shock: $T_1 = \min(W_1, W_3)$ and $T_2 = \min(W_2, W_3)$. Marshall and Olkin (1967a) gave two other derivations of this distribution, one via a different Poisson-shocks model, the other via the exponential *lack-of-memory* characterization.

The distribution has both a continuous and a singular component. The latter arises from the non-zero probability mass $P(T_1 = T_2) = \lambda_{12}/\lambda_+$, where $\lambda_+ = \lambda_1 + \lambda_2 + \lambda_{12}$, concentrated on the subspace $\{\mathbf{t} : t_1 = t_2\}$ of $(0, \infty)^2$. The precise forms of these components can be calculated as follows. We first find the continuous component of the distribution, that is, that which has a density. Now,

$$\partial^2 \bar{G}(\mathbf{t})/\partial t_1 \partial t_2 = \begin{cases} \lambda_2(\lambda_1 + \lambda_{12}) \exp(-\lambda_1 t_1 - \lambda_2 t_2 - \lambda_{12} t_1) = g_1(t) & \text{on } t_1 > t_2 \\ \lambda_1(\lambda_2 + \lambda_{12}) \exp(-\lambda_1 t_1 - \lambda_2 t_2 - \lambda_{12} t_2) = g_2(t) & \text{on } t_1 < t_2 \end{cases}$$

Integrating,

$$\bar{G}^C(\mathbf{t}) = \int_{t_1}^{\infty} dt_1 \int_{t_2}^{\infty} dt_2 \{\partial^2 \bar{G}(\mathbf{t})/\partial t_1 \partial t_2\}$$

$$= \int_{t_1}^{\infty} dt_1 \int_{t_2}^{t_1} dt_2 \{g_1(t)\} + \int_{t_1}^{\infty} dt_1 \int_{t_1}^{\infty} dt_2 \{g_2(t)\}$$

$$= e^{-\lambda_1 t_1 - \lambda_2 t_2 - \lambda_{12} \max(t_1, t_2)} - (\lambda_{12}/\lambda_+) e^{-\lambda_+ \max(t_1, t_2)} \quad \text{for } t_1 > t_2.$$

This formula, derived for the case $t_1 > t_2$, has been written in a symmetric form that also holds for $t_1 < t_2$. The total probability mass of the continuous component is $\bar{G}^C(\mathbf{0}) = 1 - \lambda_{12}/\lambda_+$, and the singular component is given by

$$\bar{G}^S(\mathbf{t}) = \bar{G}(\mathbf{t}) - \bar{G}^C(\mathbf{t}) = (\lambda_{12}/\lambda_+) e^{-\lambda_+ \max(t_1, t_2)}.$$

$\bar{G}^S(\mathbf{t})$ is the probability content of that part of the line $t_1 = t_2$ that lies north east of the point t.

Generalisations of the system, to multivariate and Weibull versions, have been developed by Marshall and Olkin (1967b), Lee and Thompson (1974), David (1974), Moeschberger (1974), and Proschan and Sullo (1974).

Proschan and Sullo (1974) calculated the likelihood functions for the bivariate Marshall–Olkin distribution under various observational settings including that of competing risks. Identifiability is established as a by-product. They also considered a model that combines the features of the Marshall–Olkin and Freund systems.

7.4.1 Discrete-Time Version (Crowder, 1996)

Consider two sequences of Bernoulli trials running in tandem. At any stage failure might strike either sequence, with probabilities $1 - \pi_j$ ($j = 1, 2$), or both together, with probability $1 - \pi_{12}$. The three failure mechanisms are assumed to operate independently. Unlike in the continuous case, ties can occur here even without dual strikes, that is, even if $\pi_{12} = 1$.

Let the numbers of failures of the three types over the first t trials be $N_1(t)$, $N_2(t)$ and $N_{12}(t)$. Then the joint survivor function of $\mathbf{T} = (T_1, T_2)$ can be derived as

$$\bar{G}(\mathbf{t}) = P(T_1 > t_1, T_2 > t_2) = P\{N_1(t_1) = 0, N_2(t_2) = 0, N_{12}(\max\{t_1, t_2\}) = 0\}$$
$$= \pi_1^{t_1} \pi_2^{t_2} \pi_{12}^{\max(t_1, t_2)}.$$

This has marginals $\bar{G}_j(t) = (\pi_j \pi_{12})^t$ ($j = 1, 2$), and dependence measure (Section 6.2) $\bar{R}(\mathbf{t}) = \pi_{12}^{-\min(t_1, t_2)}$. The marginal densities for the system are (for $j = 1, 2$ and $t = 1, \ldots, m = \infty$)

$$g_j(t) = (\pi_j \pi_{12})^{t-1}(1 - \pi_j \pi_{12}).$$

7.5 Some Other Bivariate Distributions

7.5.1 Block and Basu (1974)

A filleted version of the Marshall–Olkin distribution may be obtained by removing the singular part to produce an *absolutely continuous bivariate exponential distribution*. The joint survivor and density functions are

$$\bar{G}(\mathbf{t}) = e^{-\lambda_1 t_1 - \lambda_2 t_2 - \lambda_{12} \max(t_1, t_2)} - (\lambda_{12}/\lambda_+)e^{-\lambda_+ \max(t_1, t_2)} \text{ for } t_1 > t_2$$

and

$$g(t_1, t_2) = \begin{cases} \lambda_2(\lambda_1 + \lambda_{12}) \exp(-\lambda_1 t_1 - \lambda_2 t_2 - \lambda_{12} t_1) & \text{on } t_1 > t_2 \\ \lambda_1(\lambda_2 + \lambda_{12}) \exp(-\lambda_1 t_1 - \lambda_2 t_2 - \lambda_{12} t_2) & \text{on } t_1 < t_2 \end{cases}$$

7.5.2 Lawrance and Lewis (1983)

Let X_1, X_2, and U be independent, the first two having exponential distributions with mean 1 and the third a binary variable with $P(U = 0) = q$. Then, taking $T_1 = q\,X_1 + U\,X_2$ and $T_2 = U\,X_1 + q\,X_2$, we obtain joint density

$$g(t_1, t_2) = q^{-1}\exp\{-(t_1 + t_2)/q\} + I(q\,t_2 < t_1 < t_2/q)\,(1+q)^{-1}$$
$$\times \exp\{-(t_1 + t_2)/(1+q)\}.$$

The correlation is $\rho(T_1, T_2) = 3q(1-q)$, with maximum $3/4$ at $q = 1/2$.

Lawrance and Lewis (1983) gave several extensions and elaborations of the basic form described here. Raftery (1984, 1985) proposed similar schemes involving indicator variables.

7.5.3 Arnold and Brockett (1983)

The univariate Gompertz distribution has hazard function $\psi e^{\phi t}$, with $\psi > 0$ and $\phi > 0$. This represents an exponential rate of deterioration of a system. If we add the possibility of pure chance failure, represented by a constant hazard λ, we obtain the univariate Makeham survivor function

$$\exp\left\{-\int_0^t (\lambda + \psi e^{\phi s})ds\right\} = \exp\{-\lambda t + \psi\phi^{-1}(1 - e^{\phi t})\}.$$

Arnold and Brockett combined this with the Marshall–Olkin type of simultaneous failure (Section 7.4) to produce a bivariate survivor function of form

$$\bar{G}(t) = \exp\{-\lambda_1 t_1 - \lambda_2 t_2 - \lambda_{12}\max(t_1, t_2) + \psi_1\phi_1^{-1}(1 - e^{\phi_1 t_1})$$
$$+ \psi_2\phi_2^{-1}(1 - e^{\phi_2 t_2})\}.$$

7.5.4 Cowan (1987)

This has joint survivor function

$$\bar{G}(t_1, t_2) = \exp\left\{-\frac{1}{2}(t_1 + t_2 + \sqrt{(t_1 - t_2)^2 - 4\eta t_1 t_2})\right\},$$

with $0 \leq \eta \leq 1$. The distribution has unit exponential marginals.

7.5.5 Yet More Distributions

Many more survival distributions can be found in the literature; Hougaard (1987) reviewed the field up to that time. Here, as a somewhat random selection, we mention Ghurye (1987), Singpurwalla and Youngren (1993), and Kotz and Singpurwalla (1999); Petersen (1998) proposed an additive frailty model (Section 8.2); Walker and Stephens (1999) first presented a univariate *beta-log-normal family* of distributions and then showed how it may be generalised to the multivariate case.

7.6 Non- and Semi-Parametric Methods

Suppose that the marginal distributions are of primary interest, the dependence structure between components of $\mathbf{T} = (T_1, \ldots, T_p)$ being of secondary, or maybe no, interest. With this in mind we construct models for the marginals, say with parametric forms for the $\bar{G}_j(t; \mathbf{x}, \beta_j)$; as usual, $\mathbf{x}$ represents a covariate vector and the β_j are parameters to be estimated. Denote by $L_j(\beta_j)$ the likelihood function based on the T_j data. Then an estimating function can be formed as $\prod_{j=1}^{p} L_j(\beta_j)$ and maximised to produce estimates of the β_j. If the β_j are distinct, with no components in common, this amounts to p separate maximisations. If the T_j were independent $\prod_{j=1}^{p} L_j(\beta_j)$ would be the proper likelihood function. But we do not have to assume independence to use it as an estimating function: it will normally yield consistent estimators (assuming that the parametric specifications are correct), though the covariance matrix of the resulting $\hat{\beta}_j$ will be of "sandwich" form.

The *working independence* approach can be applied more widely. Thus, the L_j may be replaced by P_j, partial likelihoods (Section 4.2), in which case we make no distributional assumptions for the T_j. There might even be situations where a mixture of L_js and P_js is appropriate, reflecting different degrees of specification among the components of $\mathbf{T}$. The asymptotic theory goes through via the individual score functions, which are the derivatives of the $\log L_j$ and $\log P_j$ with respect to the β_js.

Suppose now that we have some, perhaps secondary, interest in the dependence structure. It may be that sufficiently informative residuals can be produced to examine the shape of dependence with a view to modelling it. Another approach is to insert the marginal estimates, that is, the $\bar{G}_j(t; \mathbf{x}, \hat{\beta}_j)$, into a copula; here the $\hat{\beta}$ are taken as known and only the copula parameters estimated. When there are no covariates, Kaplan–Meier estimates of the $\bar{G}_j(t)$ can be inserted. Yet another approach is to introduce a random effect, which will induce correlation: for example, the marginal hazards could be taken as $h_j(t; \mathbf{x}, \beta_j) = z h_0(t) \psi(\mathbf{x}, \beta_j)$ and the T_j components assumed to be independent given z, which has some assumed distribution on $(0, \infty)$. However, these approaches will make assumptions about the shape of the dependence, and the interpretation of the covariate effects will be modified because the β_j will be buried in the model formulae in a different way.

When there are no covariates one has a random sample of observations on $\mathbf{T}$ in which varying degrees of censoring can occur. For example, in the bivariate case there are four types of observation: (t_1, t_2), (t_1, t_2+), (t_1+, t_2), and (t_1+, t_2+), where $+$ denotes right-censoring. Fully non-parametric estimation of the joint survivor function $\bar{G}(\mathbf{t})$ is a tough problem with, as yet, no completely satisfactory solution. It's like those uncharted waters on the ancient mariner's map: *Here be Dragons*. Various suggestions have been made in the literature. For details and references see Hougaard (2000, Chapter 14), Kalbfleisch and Prentice (2002, Chapter 10), and Lawless (2003, Chapter 11).

Akritas and Van Heilegon (2003) have set out a method for estimation of the joint and marginal distributions from random bivariate samples when both components are subject to censoring. Their method is designed to improve upon Kaplan–Meier in the following way: Suppose that the components T_1 and T_2 are positively correlated, that T_1 is observed near the high end of its range, and that T_2 is right-censored near the lower end of its range. Kaplan–Meier makes no assumption about the likely value of T_2 beyond its censoring point. But, going on the high value of T_1, T_2 is more likely to be near the top end of its range, and this is usable information. The proposed method is based on estimating the conditional distributions, of $T_2 \mid T_1$ and $T_1 \mid T_2$, which is complicated by the censoring. Asymptotic properties of the estimators were derived, and the paper included a simulation study, comparing the proposal with previous methods, and an application to some data from McGilchrist and Aisbett (1991).

Liang et al. (1995) gave a review of regression methods for multivariate survival data up to that time. Liang et al. (1993, a slightly different *al.* in this earlier paper) considered marginal models. Marginal models are also considered by Li and Lagakos (2004), Lu and Wang (2005), and Peng et al. (2007); Cai (1999) presented some test statistics for assessing covariate effects based on marginal models of general form $h_j(t; \mathbf{x}, \beta_j) = h_{0j}(t) \exp(\mathbf{x}_j^T \beta)$ for T_j. Cai and Prentice (1997) had earlier considered estimation for this type of marginal model.

Dependence measures were covered by Fan et al. (2000), who proposed an estimator for a summary measure of dependence between bivariate failure times, and Pons (1986) and Lu (2007), who proposed tests of independence for bivariate survival data. Oakes (1986) considered semi-parametric inference in a bivariate association model. Segal et al. (1997) considered dependence estimation for marginal models.

Gray (2003) extended methodology for accelerated-life models (Section 4.5) to the case of clustered failure-time data. The title of the 2005 paper by Jones and Yoo is *Linear Signed-Rank Tests for Paired Survival Data Subject to a Common Censoring Time* and it does exactly what it says on the tin. Jung and Jeong (2003) proposed some *Rank Tests for Clustered Survival Data*.

Yin and Cai (2004) considered a multivariate model with additive hazards, perhaps arising from events of different types; their focus was on estimation of regression effects and the baseline hazard functions. Zeng and Cai (2010) investigated additive transformation models for the hazard function. Scholtens and Betensky (2006) tackled estimation of a bivariate distribution with one continuous and one discrete component. Quale and Van der Laan (2000) considered inference about the distribution of (T_1, T_2) when observation of the pair is conditional on $T_1 > C$ (or on $T_1 < C$), C being a third random time. For random samples of clustered survival data, for example, animal litters, Cai and Kim (2003) developed methodology for estimating quantiles; they used bootstrap and kernel smoothing to derive confidence intervals, in addition to presenting asymptotic theory, simulation, and an application.

7.7 Exercises

1. For each of the bivariate distributions listed in Sections 7.3, 7.4, and 7.5, derive the distributions of $\min(t_1, t_2)$ and $\max(t_1, t_2)$. In reliability, the former would relate to components in series and the latter to components in parallel (Section 2.6).

2. For each of the bivariate distributions, calculate the dependence measures described in Section 6.2.

3. Consider Freund's bivariate exponential distribution. Investigate the situation where only one of the component's rates is changed when the other fails. Do this for both continuous and discrete time versions.

8

Frailty, Random Effects, and Copulas

In applications the multivariate normal distribution has reigned supreme for decades and probably will continue to do so. It has friendly properties, including a readily interpretable dependence structure and a high degree of tractability (though not for multivariate censored data). However, in many situations it will be inappropriate to use the multivariate normal as a model for the data. Its univariate marginals, which are all normal, might not fit the application: this is either from data inspection or because of more fundamental reasons, for example, Pike (1966), Peto and Lee (1973), and Galambos (1978) argue for Weibull distributions in a certain contexts. Again, the multivariate normal dependence structure is tightly specified through its correlations and linear mean regressions with constant residual variance of one component on others. This might also be wrong for the situation at hand. Cook and Johnson (1981) made such points in their development of a particular class of non-normal multivariate distributions.

In this chapter some general methods for constructing non-normal multivariate distributions will be described. In particular, the distributions produced should be tractable for right-censored multivariate survival data.

8.1 Frailty: Construction

The univariate construction for this class of models was covered in Section 3.5. To develop it for the multivariate case we begin with the marginal survivor function of T_j: $\bar{G}_j(t) = \exp\{-H_j(t)\}$, where $H_j(t)$ is the integrated hazard function of T_j (Section 7.1). Suppose that (a) the H_j $(j = 1, \ldots, p)$ for an individual share a common random factor z, and that (b) z varies over the population of individuals with distribution function K on $(0, \infty)$. In some contexts z is known as the *frailty* of the individual (Vaupel et al., 1979). We now replace $H_j(t)$ with $zH_j(t)$, and make the key assumption that the T_j for an individual are conditionally independent given z. Then the joint survivor function conditional on z is

$$\bar{G}(\mathbf{t} \mid z) = \exp\left\{-z\sum_{j=1}^{p} H_j(t_j)\right\}.$$

The unconditional joint survivor function is now calculated by integrating over the z-distribution as

$$\bar{G}(\mathbf{t}) = \int_0^\infty e^{-zs} dK(z) = L_K(s),$$

say, where $s = H_1(t_1) + \cdots + H_p(t_p)$; L_K is the *Laplace transform*, and *moment generating function*, of the distribution K. The construction gives a multivariate distribution in which the component lifetimes are dependent, in general. However, the form of dependence is restricted by the symmetry of $\bar{G}$ in the H_j.

Suppose that $\mathbf{T} = (\mathbf{T}_A, \mathbf{T}_B)$ represents a partition of some significance. The marginal survivor function of $\mathbf{T}_A$ is obtained by setting $\mathbf{t}_B = \mathbf{0}$: $\bar{G}_A(\mathbf{t}_A) = L_K(s_A)$, in which s_A denotes what you think it denotes. So, all models of this type will have marginals of the same general form as the original.

Hougaard (2000) has five chapters on frailty, reflecting his substantial contributions to the literature; many references to his work, and that of others, are given in his book. Marshall and Olkin (1988) extended the approach to multivariate frailty. In the bivariate case, their construction, here based on survivor functions, is

$$\bar{G}(\mathbf{t} \mid \mathbf{z}) = \exp\{-z_1 H_1(t_1) - z_2 H_2(t_2)\} \quad \text{and} \quad \bar{G}(\mathbf{t}) = \int e^{-z_1 H_1(t_1) - z_2 H_2(t_2)} dK(\mathbf{z}),$$

where $K(\mathbf{z})$ is the bivariate distribution function of $\mathbf{Z} = (Z_1, Z_2)$. The marginal for T_1, obtained by taking $t_2 = 0$, is $\bar{G}_1(t_1) = \int e^{-z_1 H_1(t_1)} dK(z_1)$. For the version based on distribution functions replace $e^{-z_j H_j(t_j)}$ by $G_j(t_j)^{z_j}$ $(j = 1, 2)$; this does not give the same distribution as the $\bar{G}$ version since $\bar{G}_j(t_j)^{z_j} \neq 1 - G_j(t_j)^{z_j}$. The paper lists and derives a variety of properties and examples.

8.2 Some Frailty-Generated Distributions

We consider some particular cases, taking a Weibull form for the integrated hazard: $H_j(t) = (t/\xi_j)^{\phi_j}$ and then $s = \sum_{j=1}^p (t_j/\xi_j)^{\phi_j}$. For K the following choices prove to be fruitful.

8.2.1 Multivariate Burr (Takahasi, 1965; Clayton, 1978; Hougaard, 1984; Crowder, 1985)

This distribution has been derived, developed, discussed, and deployed by various authors. Taking K to be a gamma distribution with shape parameter ν and scale parameter ζ yields $\bar{G}(\mathbf{t}) = (1+\zeta s)^{-\nu}$, the survivor function of an MB (multivariate Burr) distribution. We can take $\zeta = 1$ by absorbing its scaling into the ξ_js. The marginal survivor function of component $\mathbf{T}_A$ in $(\mathbf{T}_A, \mathbf{T}_B)$ is $\bar{G}_A(\mathbf{t}_A) = (1+s_A)^{-\nu}$, of MB form. One conditional survivor function follows as

$$P(\mathbf{T}_B > \mathbf{t}_B \mid \mathbf{T}_A > \mathbf{t}_A) = (1+s)^{-\nu} \div (1+s_A)^{-\nu} = (1+s_B')^{-\nu},$$

where $s'_B = s_B/(1 + s_A)$; this is also of MB form, of dimension $\dim(\mathbf{t}_B)$, and with modified ξ-parameters $\xi'_j = \xi_j/(1 + s_A)$. The other conditional survivor function referred to in Section 6.2 is

$$P(\mathbf{T}_B > \mathbf{t}_B \mid \mathbf{T}_A = \mathbf{t}_A) = (1 + s'_B)^{-\nu-a},$$

where $a = \dim(\mathbf{t}_A)$; this is again of MB form. For data analysis, we note that the likelihood contribution for a case yielding observed $\mathbf{t}_A$ and right-censored $\mathbf{t}_B$ is

$$(-1)^a \, \partial^a \, \bar{G}(\mathbf{t}) / \partial \mathbf{t}_A = \prod_{j=1}^{a} (\nu + j - 1) \times \prod_A \left(\xi_j^{-\phi_j} \phi_j t_j^{\phi_j - 1} \right) \times (1 + s)^{-\nu-a},$$

where $\prod_A$ denotes product over the A-components. Some further properties are left to the Exercises (Section 8.7), and if you have not had enough of it, yet more appear in Crowder (1985). A test for independence based on a modified score statistic was proposed by Crowder and Kimber (1997); this involves non-standard asymptotic likelihood properties. Two applications were described in the paper, one each for censored and uncensored data.

8.2.2 Multivariate Weibull (Gumbel, 1960; Watson and Smith, 1985; Hougaard, 1986a,b)

Taking K to be a positive stable distribution with characteristic exponent $\nu \in (0, 1)$ yields $\bar{G}(\mathbf{t}) = \exp(-s^\nu)$, the survivor function of this particular MW (multivariate Weibull) distribution. With the partition $\mathbf{T} = (\mathbf{T}_A, \mathbf{T}_B)$ as before, the marginal survivor function of $\mathbf{T}_A$ is $\bar{G}_A(\mathbf{t}_A) = \exp(-s_A^\nu)$, also of MW form. The first conditional survivor function is

$$P(\mathbf{T}_B > \mathbf{t}_B \mid \mathbf{T}_A > \mathbf{t}_A) = \exp(-s^\nu) \div \exp\left(-s_A^\nu\right) = \exp\left(s_A^\nu - s^\nu\right),$$

which is not of MW form. However, let us extend the basic specification to $\bar{G}(\mathbf{t}) = \exp\{\kappa^\nu - (\kappa + s)^\nu\}$, introducing an extra parameter $\kappa > 0$. We then obtain marginal $\bar{G}_A(\mathbf{t}_A) = \exp\{\kappa^\nu - (\kappa + s_A)^\nu\}$ and conditional

$$P(\mathbf{T}_B > \mathbf{t}_B \mid \mathbf{T}_A > \mathbf{t}_A) = \exp\{\kappa^\nu - (\kappa + s)^\nu\} \div \exp(\kappa^\nu - (\kappa + s_A)^\nu)$$
$$= \exp\{(\kappa + s_A)^\nu - (\kappa + s)^\nu\},$$

both of which are members of this extended family of distributions. The other conditional survivor function, $P(\mathbf{T}_B > \mathbf{t}_B \mid \mathbf{T}_A = \mathbf{t}_A)$, is not held within the family. The introduction of κ was interpreted as a kind of pre-conditioning in Crowder (1989), where further details may be found. Some further properties are left to the Exercises (see Section 8.7).

Setting $\kappa = 0$ in $\bar{G}(\mathbf{t}) = \exp\{\kappa^\nu - (\kappa + s)^\nu\}$ yields the original form, though a goodness-of-fit test for the smaller model will be non-regular, $\kappa = 0$ being a boundary parameter value. Likewise, setting $\nu = 1$ makes κ disappear, also leading to non-regular problems. Tests based on modified score statistics and extreme values were devised and assessed for the univariate version of the distribution by Crowder (1990, 1996).

8.2.3 Distribution 3 (Crowder, 1998)

The random effect here is bivariate, with z_1 following a positive stable law with characteristic exponent $v \in (0, 1)$, and z_2^v independently distributed as gamma with shape parameter $1/\tau$. Some support for such a setup is given in the cited paper. Mixing of the conditional survivor function

$$\bar{G}(t \mid z_1, z_2) = \exp(-z_1 z_2 s)$$

over z_1 and z_2 yields

$$\bar{G}(t) = \left[1 - \tau v^{-1} \{\kappa^v - (\kappa + s)^v\} \right]^{-1/\tau},$$

in which $\kappa < (v/\tau)^{1/v}$ is introduced as a third structural parameter. It turns out that, in the bivariate case, the range of values of v can be extended beyond 1 provided that $v \leq 1 + (\tau + 1)\kappa^v$. Some typical density contours for the bivariate case are shown in Figure 8.1. The R-code for these plots is given below, in which the parameters are xi1=ξ_1, and so forth: the function cont1 is listed on the CRC Press Web site referred to in the Preface.

```
#bivariate contour plots for 'Distribution 3' (Sec II.8.2)
par(mfrow=c(2,2));
xi1=1; xi2=1; phi1=2; phi2=2; nu=2; tau=1.5; kap=1;
cont1(xi1,xi2,phi1,phi2,nu,tau,kap);
xi1=5; xi2=5; phi1=2; phi2=2; nu=2; tau=0.1; kap=1;
cont1(xi1,xi2,phi1,phi2,nu,tau,kap);
xi1=3; xi2=3; phi1=1.25; phi2=1.25; nu=1.5; tau=1; kap=0.7;
cont1(xi1,xi2,phi1,phi2,nu,tau,kap);
xi1=5; xi2=3; phi1=1.25; phi2=1.5; nu=2; tau=0.8; kap=1;
cont1(xi1,xi2,phi1,phi2,nu,tau,kap);
```

The distribution yields various special cases. Taking $v = 1$ gives the MB distribution with $\bar{G}(t) = (1 + \tau s)^{-1/\tau}$. Allowing $\tau \to \infty$ yields $\bar{G}(t) = \exp\left[v^{-1}\{\kappa^v - (\kappa + s)^v\} \right]$; now setting $\kappa = 0$ gives the MW distribution. Various other properties are left to the Exercises (Section 8.7) and more are given in the cited paper. Tests based on extreme values for a variety of parametric hypotheses were proposed for the univariate version of the distribution by Crowder (2000).

One aspect often of interest is the ratios of different components. This might be the case if, for example, T_1 is a baseline measurement and the rest are to be calibrated against it. In the bivariate case, with $R = T_2/T_1$, we have (in informal notation)

$$P(R > r) = \int_0^\infty P(T_2 > r t_1, \, T_1 = t_1) dt_1 = \int_0^\infty \left[-\partial \bar{G}(t_1, t_2)/\partial t_1 \right]_{t_2 = r t_1} dt_1.$$

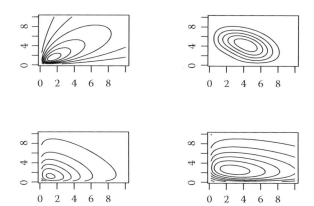

FIGURE 8.1
Some contour density plots for Distribution 3.

Suppose that $\phi_1 = \phi_2$ in the basic Weibull framework, representing the same scale for $\log T_1$ and $\log T_2$. Then, $s = (t_1/\xi_1)^\phi + (t_2/\xi_2)^\phi$, and the integral yields

$$P(R > r) = \{1 + (r/\xi)^\phi\}^{-1},$$

where $\xi = \xi_2/\xi_1$. Parameters (ν, κ, τ) have been eliminated by this transformation, leaving only the basic Weibull shape and scale. Also, only the ratio ξ_2/ξ_1 appears in the formula, so ξ_1 and ξ_2 are not separately estimable. The multivariate version, with $R_j = T_j/T_1$ and $\phi_j = \phi$ for all j, follows as

$$P(R_2 > r_2, \ldots, R_p > r_p) = \{1 + (r_2/\xi_{12})^\phi + \cdots + (r_p/\xi_{1p})^\phi\}^{-1},$$

in which $\xi_{1j} = \xi_j/\xi_1$.

8.2.4 Multivariate Beta-Geometric

A multivariate version of the beta-geometric distribution (Section 5.5) can be constructed as follows. Suppose that conditionally on random effect z, the discrete lifetimes $T_1, \ldots, T_p$ are independent geometric variates with $P(T_j > t) = (z\rho_j)^t$ for $t = 1, 2, \ldots$. If now z has distribution $beta(\phi, \psi)$ across units, the unconditional joint survivor function of the T_j is

$$P(\mathbf{T} > \mathbf{t}) = \int_0^1 \left\{ \prod_{j=1}^p (z\rho_j)^{t_j} \right\} B(\phi, \psi)^{-1} z^{\phi-1}(1-z)^{\psi-1}\, dz$$

$$= \left(\prod_{j=1}^p \rho_j^{t_j} \right) B(\phi + t_+, \psi)/B(\phi, \psi) = \left(\prod_{j=1}^p \rho_j^{t_j} \right) \prod_{l=1}^{t_+} \left(\frac{\phi + l - 1}{\phi + \psi + l - 1} \right),$$

where $t_+ = t_1 + \cdots + t_p$. For computation, where t_+ varies over units, denote $\prod_{l=1}^m (\frac{\phi+l-1}{\phi+\psi+l-1})$ by a_m and note that $a_m = a_{m-1}(\phi+m-1)/(\phi+\psi+m-1)$, starting at $a_1 = \phi/(\phi+\psi)$; the a_m can then be computed recursively and stored up to the maximum of the t_+ values. The marginal survivor functions are $\bar{G}_j(t) = \rho_j^t a_t$ $(j = 1, \ldots, p)$, and the dependence measure $\bar{R}(\mathbf{t}) = a_{t_+}/\prod_{j=1}^{p} a_{t_j}$.

8.2.5 Multivariate Gamma-Poisson

At first sight the Poisson is not a natural survival distribution. However, consider a series of opportunities generated by a Poisson process of rate λ. Suppose that, at each such opportunity, a particular event might occur (with probability p) or might not (with probability $1-p$). The discrete time (number of opportunities) to first event occurrence then has a Poisson distribution with mean $(\lambda p)^{-1}$.

For a multivariate version suppose that the waiting times $T_1, \ldots, T_p$ are conditionally independent given the random effect z, with T_j having a Poisson distribution of mean $(\lambda_j z)^{-1}$. To match the situation described in the previous paragraph we could now take z to have a beta density on $(0, 1)$, say proportional to $z^\alpha(1-z)$; integration over z will then produce a multivariate distribution for the T_j, as intended. Alternatively, suppose that z has distribution $gamma(\xi, \nu)$. Then the T_j have unconditional joint survivor function

$$P(\mathbf{T} = \mathbf{t}) = \int_0^\infty \left\{ \prod_{j=1}^{p} \{e^{-z\lambda_j}(z\lambda_j)^{t_j}/t_j!\} \Gamma(\nu)^{-1}\xi^{-\nu}z^{\nu-1}e^{-z/\xi} \, dz \right.$$

$$= \left\{ \prod_{j=1}^{p}(\lambda_j^{t_j}/t_j!) \right\} \xi^{t_+}(1 + \xi\lambda_+)^{-\nu-t_+} \Gamma(\nu)^{-1}\Gamma(\nu + t_+),$$

in which $t_+ = t_1 + \cdots + t_p$ and $\lambda_+ = \lambda_1 + \cdots + \lambda_p$.

8.2.6 Marshall–Olkin Families

Marshall and Olkin (1988) proposed the following construction. Given two univariate distribution functions, F_1 and F_2, and a bivariate one, $G(z_1, z_2)$, form

$$H(t_1, t_2) = \int F_1(t_1)^{z_1} F_2(t_2)^{z_2} dG(z_1, z_2).$$

The variables z_1 and z_2 may be thought of as random effects. Their entry into the construction is equivalent to that of the frailties above: instead of being displayed as multiplicative factors for the hazard function they are shown as producing *Lehmann alternatives* of F_1 and F_2. As a special case one may take $z_1 = z_2$, replacing $G(z_1, z_2)$ by a univariate $G(z)$. Survivor functions, $\bar{F}_j$, may be used in place of distribution functions, and the extension to higher dimensions is obvious. The cited paper explores a variety of possibilities and derives association measures and other properties.

8.3 Applications

A couple of examples are given here of applications of a frailty-induced model.

8.3.1 Paired Response Times

The data (Section 6.1) exhibit a prime case where ratios, of pre- and post-treatment observations, are appropriate. In fact, the original investigators insisted that a ratio-based analysis would be of most relevance to them, and the statistician should not bite the hand, and so forth. For group j and dose d_k the survivor function is modelled as

$$P(R > r) = \{1 + (r/\xi^{(jk)})^\phi\}^{-1} \text{ with } \log \xi^{(jk)} = \xi^{(j)} + \beta_1 d_k + \beta_2 d_k^2.$$

So, there is a group effect $\xi^{(j)}$ coupled with a quadratic curve in dose: the latter reflects the rise and subsequent fall in response time as the drug effect ramps up and then wears off, with $\beta_2 < 0$. The parameter set is $\theta = (\xi^{(1)}, \xi^{(2)}, \xi^{(3)}, \beta_1, \beta_2, \phi)$.

The data are first assembled as a 126×4 matrix, called `rt2b`, with columns Group Number, Dose, t_1, and t_2; the times were all scaled by 0.01 for good measure. The R-code for fitting the model is

```
#paired response times (Sec 8.3)
#model fit 1: ratios, all three groups
n1=44; n2=39; n3=43; nd=n1+n2+n3;
censr=2.5; mdl=1; np=6; opt=0; iwr=0;
adt=c(nd,censr,mdl,np,opt,iwr);
par0=c(1.15,1.15,1.15,3.46,0,0.69);
par1=fitmodel(distn3a,np,par0,rt2b,adt);
```

The function `distn3a` is listed on the Web site. The maximised log-likelihood value is -129.5849 and the *mle* is

$$(\hat{\xi}^{(1)}, \hat{\xi}^{(2)}, \hat{\xi}^{(3)}, \hat{\beta}_0, \hat{\beta}_1, \hat{\phi}) = (0.37, 0.11, 0.044, 3.76, -2.12, 1.69).$$

When t_2 is right-censored but t_1 is observed, we can take it that the ratio r exceeds the recorded value of t_2/t_1, and so the likelihood contribution is the survivor function for r. When t_1 is right-censored but t_2 is observed, we can take it that r is less than t_2/t_1, so the likelihood contribution is one minus the survivor function. However, when t_1 and t_2 are both right-censored, their ratio is not able to be assessed. As it happens, there are no doubly censored pairs in these data, but if there were we would have to either do something more complicated or just omit them from the analysis, assuming that the double-censoring was not informative. The eagle-eyed will spot that in function `distn3aa` (a satellite to `distn3a`), this omission is achieved by setting the likelihood contribution, `pr`, equal to 1. The inverse Hessian matrix is needed for standard errors of the *mle* and so forth, and `optim` returns an approximation.

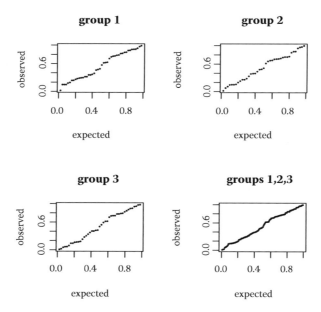

FIGURE 8.2
Uniform residual plots from ratio fits.

To extract uniform residuals from function `distn3a` call it with `opt` set to 1. They can then be plotted to produce Figure 8.2. The R-code is

```
#residual plots
opt=1; mdl=1; np=6; adt=c(nd,censr,mdl,np,opt,iwr);
par0=c(0.369,0.112,0.0437,3.76,-2.12,1.69);
rtn=distn3a(par1,rt2b,adt);
llkd=rtn[1]; pv=rtn[2:(nd+1)]; rv=rtn[(nd+2):(2*nd+1)];
par(mfrow=c(2,2));
u1=c(1:n1)/(n1+1); v1=unlist(sort(rv[1:n1]));
plot(u1,v1,type='p',cex=0.25,xlab='expected',ylab='observed');
title(main='group 1',sub=' ');
u2=c(1:n2)/(n2+1); v2=unlist(sort(rv[(n1+1):(n1+n2)])));
plot(u2,v2,type='p',cex=0.25,xlab='expected',ylab='observed');
title(main='group 2',sub=' ');
u3=c(1:n3)/(n3+1); v3=unlist(sort(rv[(n1+n2+1):nc]));
plot(u3,v3,type='p',cex=0.25,xlab='expected',ylab='observed');
title(main='group 3',sub=' ');
u4=c(1:nc)/(nc+1); v4=unlist(sort(rv[1:nc]));
plot(u4,v4,type='p',cex=0.25,xlab='expected',ylab='observed');
title(main='groups 1,2,3',sub=' ');
```

The residual plots do not give cause for concern—the points appear to follow the set route from (0, 0) to (1, 1) without too much deviation from the straight and narrow.

TABLE 8.1

Energy Usage Times for an Iron

Household	Day (week 1)							Day (week 2)						
	1	2	3	4	5	6	7	1	2	3	4	5	6	7
1	0	19	11	6	0	6	22	0	28	19	6	0	0	7
2	14	27	0	0	17	20	0	5	8	5	0	27	9	0
3	21	0	70	0	48	0	5	33	0	7	7	27	0	6
4	0	54	0	0	28	23	0	12	0	88	0	61	56	0
5	78	0	39	0	31	0	88	12	0	0	49	38	0	81
6	88	0	0	58	0	64	76	0	31	85	0	0	54	49
7	85	15	0	67	0	0	114	64	50	0	0	60	0	69
8	8	20	11	12	52	12	37	61	15	33	43	32	6	16
9	0	27	0	14	6	39	16	0	0	8	7	0	0	19
10	11	9	8	0	5	0	0	7	5	0	8	0	0	5
11	9	0	19	0	0	24	25	0	8	5	0	0	11	26
12	23	27	21	6	15	15	9	6	9	23	28	5	5	15
13	22	18	13	0	0	17	9	23	0	0	8	0	9	0
14	15	0	0	71	22	57	0	0	11	35	0	9	0	0

8.3.2 Household Energy Usage

Ironing has never been the most popular of chores. (There is the tale of the young man who a-courting did go to press his suit—after they're married she'll press his suit, with a little bit of luck!) The data in Table 8.1 are from the same source as the lamp data (Section 6.1) and have exactly the same structure. They are times (minutes) for which the iron has been left switched on each day. The ironing tends to be done rather sporadically over the week; also, the iron can be left switched on and unattended for periods while the factotum absconds to do something more pressing (sorry). (Incidentally, there is an old Somerset folk song concerning a maid "dashing away with the smoothing iron": I am not sure if this means that she was making off with stolen goods.)

Just as with the lamp data, usage is likely to vary considerably between households; for instance, I have heard that some jobs require a clean shirt and tie every day. So, a frailty model is called for and the MB is a convenient choice here. In addition, there is a significant proportion (0.36) of no-iron days: two households have no zero times but the others all have around half a dozen. Let us try a *zero-inflated* model that incorporates positive probability mass at zero and take this probability to be homogeneous over units. (The description zero-inflated is more commonly applied to discrete distributions, such as the Poisson, that already have some probability mass at zero.)

To construct the model we begin with the joint probability, conditional on the random effect (frailty) z, for vector $\mathbf{t} = (t_1, \ldots, t_p)$ and indicator vector $\mathbf{c}$, whose elements are $c_j = I(t_j = 0)$:

$$p(\mathbf{t}, \mathbf{c} \mid z) = p_0^{c_+} \times \prod_{\{j: \, t_j > 0\}} \{(1 - p_0)p_j(t_j \mid z)\},$$

where $p_0 = P(t_j = 0 \mid z)$, $c_+ = \sum_j c_j$, and $p_j(.)$ is the basic probability function adopted for t_j. Integrating over z as for the MB distribution, we obtain

$$p(\mathbf{t}, \mathbf{c}) = p_0^{c_+}(1 - p_0)^{p-c_+}(1 + s_1)^{-\nu}, \quad s_1 = \sum_{\{j: \, t_j > 0\}} H_j(t_j).$$

For the purpose of the illustration here we have taken the Weibull form $H_j(t) = (t/\xi_j)^\phi$, in which ξ_j takes one value for the first week and another for the second, and ϕ is taken to be the same throughout, both homogeneous over units. The R-code is

```
#iron data  (Sec 8.3)
iron1=read.table('iron.dat',header=F); nc=14; nd=7;
iron2=as.numeric(unlist(iron1)); dim(iron2)=c(nc,1+2*nd);
#fit zero-inflated MB model
opt=0; iwr=0; adt=c(nc,nd,opt,iwr);
np=5; par0=c(0,-1,0,3,3);
par1=fitmodel(ironc,np,par0,iron2,adt);
```

The results of fitting this model are as follows. Within the program the parameters are transformed to a log-scale, thus imposing a positivity constraint on them. It is the log-estimates that are printed out: these are $(-0.04, -0.94, 0.62, 3.07, 2.80)$. The *mle* for the untransformed parameters are then

$$(\hat{\nu}, \hat{p}_0, \hat{\phi}, \hat{\xi}_1, \hat{\xi}_2) = (0.96, 0.39, 1.87, 21.51, 16.46).$$

The value for $\hat{p}_0$ is close to the sample proportion of zeros, and that for $\hat{\nu}$ is close to 1, which corresponds to an exponential distribution for z over units. These noted values fall well within the implied standard error ranges. The function `ironc` can be found on the Web site noted in the Preface.

Various elaborations of the model are suggested. The parameters p_0 and ϕ could change between weeks 1 and 2; in addition, p_0 could vary over units, say with a beta distribution over which it is to be integrated like that of z. Serial correlation might be present: if no ironing is done on one day, does it become more likely that some will be done on the following day? Would it be better to adopt a model based on the assumption that ironing is there to be done every day but it is often held over to save getting the ironing board out? Please feel free to investigate these and other models at your leisure.

8.3.3 Cycles to Pregnancy

The data in Section 5.5 refer to the numbers of cycles to first pregnancy among two groups of couples. We extend this to the numbers of cycles to first and second pregnancies here, for a single group, and fit a bivariate beta-geometric model (Section 8.2). The frequencies in Table 8.2 are simulated: for example, eight couples waited three cycles for the first conception and four for the second; 7+ means seven or more cycles. A raw chi-square test for no association on the 7×7 contingency table yields chi-square value 54.80 with 36 degrees of freedom (p = 0.023).

TABLE 8.2

Cycles to Pregnancy

	1	2	3	4	5	6	7+
1	49	21	21	14	10	3	12
2	38	25	10	12	6	1	24
3	16	14	10	8	4	3	11
4	11	7	7	6	6	1	8
5	14	7	2	4	3	1	8
6	6	1	1	6	1	0	3
7+	18	14	10	3	7	6	27

The joint survivor function, as given in Section 8.3, is

$$\bar{G}(t_1, t_2) = P(T_1 > t_1, T_2 > t_2) = \left(\rho_1^{t_1}\rho_2^{t_2}\right)a_{t_1+t_2}$$

for $t_1, t_2 = 1, 2, \ldots$, where

$$a_m = \prod_{l=1}^{m}\left(\frac{\phi + l - 1}{\phi + \psi + l - 1}\right),$$

and the marginal survivor functions are $\bar{G}_j(t) = \rho_j^t a_t$ ($j = 1, 2$). The likelihood contributions for data like that shown are calculated as follows. Let $p(j, k) = P(T_1 = j, T_2 = k)$. Then, for $j = 1, 2, \ldots$,

$$p(1, j) = 1 - \bar{G}_1(1) - \bar{G}_2(j) + \bar{G}(1, j) - \sum_{k=1}^{j-1} p(1, k),$$

in which $\sum_{k=1}^{0}$ is interpreted as 0; a similar formula holds for $p(j, 1)$. For $j, k = 2, 3, \ldots$,

$$p(j, k) = \bar{G}(j - 1, k - 1) - \bar{G}(j - 1, k) - \bar{G}(j, k - 1) + \bar{G}(j, k).$$

Finally, the probabilities for the right-hand margin of the table are, for $j = 2, \ldots, 7$,

$$\bar{p}(j, 7) = P(T_1 = j, T_2 \geq 7) = \bar{G}(j - 1, 6) - \bar{G}(j, 6),$$

and those along the bottom margin are

$$\bar{p}(7, j) = P(T_1 \geq 7, T_2 = j) = \bar{G}(6, j - 1) - \bar{G}(6, j).$$

We now have lights, camera, and (Exercise 5, Section 8.7) action!

8.4 Copulas: Structure

Let (T_1, T_2) have bivariate distribution function $G(t_1, t_2)$, and let $G_1(t_1)$ and $G_2(t_2)$ be the associated marginal distribution functions. Suppose that we wish to retain the essential form of G but re-assign the marginals. Reasons for doing this will be discussed below. We can make the marginals uniform on $(0, 1)$ by applying the *probability integral transform* to each. Thus, taking $U_j = G_j(T_j)$ for $j = 1, 2$, we have

$$P(U_1 \leq u_1,\ U_2 \leq u_2) = P\{G_1(T_1) \leq u_1,\ G_2(T_2) \leq u_2\}$$

$$= P\{T_1 \leq G_1^{-1}(u_1),\ T_2 \leq G_2^{-1}(u_2)\}$$

$$= G\{G_1^{-1}(u_1),\ G_2^{-1}(u_2)\}.$$

(We have assumed here that the inverse functions, G_1^{-1} and G_2^{-1}, are well defined, for which it will suffice that G_1 and G_2 are continuous and strictly monotone increasing.) So, in the formula for $G(t_1, t_2)$, we just replace t_1 by $G_1^{-1}(u_1)$ and t_2 by $G_2^{-1}(u_2)$.

What we have done here is to tinker at the margins without meddling with the essential substance of the distribution: the core structure has been retained. For example, the dependence measure R is unchanged under the correspondence $u_j \leftrightarrow G_j(t_j)$:

$$R_U(u_1, u_2) = P(U_1 \leq u_1,\ U_2 \leq u_2)/\{P(U_1 \leq u_1)\,P(U_2 \leq u_2)\}$$

$$= P(T_1 \leq t_1,\ T_2 \leq t_2)/\{P(T_1 \leq t_1)\,P(T_2 \leq t_2)\} = R_T(t_1, t_2).$$

However, means, variances, correlations and the like will generally be changed by nonlinear transformations. We could go on and transform the marginals again to any other desired combination, for example, to Weibull by taking $Y_j = W^{-1}(U_j)$, say, where W is the univariate Weibull distribution function.

Under the correspondence $u_j \leftrightarrow G_j(t_j)$ we can write

$$G(t_1, t_2) = P(T_1 \leq t_1,\ T_2 \leq t_2) = P(U_1 \leq u_1,\ U_2 \leq u_2) = \mathcal{C}(u_1, u_2)$$

$$= \mathcal{C}\{G_1(t_1), G_2(t_2)\},$$

in which $\mathcal{C}$ is the *copula function*. The copula expresses the joint distribution function in terms of the marginals. It has a specified, and therefore restricted, dependence structure but with optional marginals. We could have based the development on survivor functions instead of distribution functions, and the idea is extended to more than two components in the obvious way. Formally, a copula is a multivariate distribution with standard uniform marginals (i.e., each uniform on $(0, 1)$). Moreover, although any multivariate distribution can generate a copula, via marginal transformation, there seems to be a tacit requirement that the $\mathcal{C}$-function should have simple form.

Regarding the usefulness of copulas, there is a variety of views. Lawless (2003) describes them as a "useful way to develop bivariate lifetime models," and draws attention to the fact that one can base them on "parametric or semi-parametric specifications" (p. 495) for the marginals. Hougaard (2000, p. 436) describes their use in a "combined approach": one specifies the marginals, including any dependence on covariates, and then inserts them into a suitable copula. On the other hand, the enthusiasm of some is containable: Mikosch

(2006) refers to the "emperor's new clothes," though some of his discussants beg to differ.

Estimation has received some attention in the literature. Genest and Rivest (1993) outlined a method based on the assumption of a copula of Archimedean type (see the following section): they described a semi-parametric estimator for random, bivariate, uncensored samples. Hanley and Parnes (1983) proposed a maximum-likelihood approach for non-parametric bivariate estimation from right-censored data, not assuming a copula of any given type; the algorithm is rather involved and they pointed out possible problems with insufficient data and unstable estimates.

8.5 Further Details

First, note that $C(u_1, 0) = 0$ and $C(0, u_2) = 0$ for all u_1 and u_2 in $(0, 1)$. Second, allowing $t_1 \to \infty$ in $G(t_1, t_2) = C\{G_1(t_1), G_2(t_2)\}$ yields $G_2(t_2) = C\{1, G_2(t_2)\}$, so $C(1, u) = u$, and likewise $C(u, 1) = u$, for $0 \le u \le 1$; this confirms uniformity of the marginals. Third, note that, for $0 \le u_1 \le v_1 \le 1$ and $0 \le u_2 \le v_2 \le 1$,

$$C(u_2, v_2) - C(u_2, v_1) - C(u_1, v_2) + C(u_1, v_1) \ge 0$$

(why is that?). These three properties can be used to formally define a copula. The trivial copula is $C(u_1, u_2) = u_1 u_2$: this serves as a limiting case of complete independence for comparing "real" copulas.

Figure 8.3 has four panels. In the first (top left) three bivariate normal density contours are shown: they are based on means $(0, 0)$, $(-3, 0)$, $(3, 0)$; each have unit standard deviations; and the correlation coefficients are 0, -0.5, and 0.5. The second panel (top right, man wearing goggles) shows the corresponding copulas; the two correlated ones have been reduced in size and shifted apart to separate them all visually in the plot—I will leave you to pick out which is which. In the third panel (bottom left) the marginals have been transformed to standard exponential distributions (i.e., with unit means), and in the fourth panel Weibull transformations with different shapes have been applied. (Maybe this will get onto the front cover.)

The R-code for drawing the pictures is

```
#R-code for Fig II.8.3 (Sec II.8.5)
npts=50; par(mfrow=c(2,2));
#panel 1: bivariate normals
xlo=-6; xhi=6; ylo=-4; yhi=4; c=4.61;
plot(c(0),c(0),xlim=c(xlo,xhi),ylim=c(ylo,yhi),xlab='',ylab='',type='n');
mu1=c(0,0); cmx1=c(1,0,0,1); dim(cmx1)=c(2,2);
xy1=ellipse1(mu1,cmx1,c,npts); arcs1(xy1,npts,'bivariate normals');
mu2=c(-3,0); cmx2=c(1,0.5,0.5,1); dim(cmx2)=c(2,2);
xy2=ellipse1(mu2,cmx2,c,npts); arcs1(xy2,npts,'');
mu3=c(3,0); cmx3=c(1,-0.5,-0.5,1); dim(cmx3)=c(2,2);
xy3=ellipse1(mu3,cmx3,c,npts); arcs1(xy3,npts,'');
#panel 2: bivariate uniforms
xlo=-0.1; xhi=1.1; ylo=-0.1; yhi=1.1; scale=0.5;
```

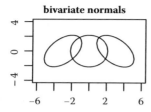

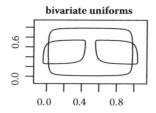

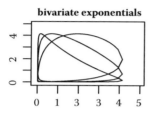

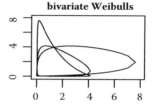

FIGURE 8.3
Copula density plots.

```
plot(c(0),c(0),xlim=c(xlo,xhi),ylim=c(ylo,yhi),xlab='',ylab='',type='n');
xu1=cop1(xy1,npts,0,0,1,1); arcs1(xu1,npts,'bivariate uniforms');
xu2=cop1(xy2,npts,-3,0,1,1); xv2=cp1(xu2,npts,0.2,0.5,scale);
arcs1(xv2,npts,'');
xu3=cop1(xy3,npts,3,0,1,1); xv3=cp1(xu3,npts,0.8,0.5,scale);
arcs1(xv3,npts,'');
#panel 3: bivariate exponentials
xlo=-0.1; xhi=5.0; ylo=-0.1; yhi=5.0; shape=1; scale=1;
plot(c(0),c(0),xlim=c(xlo,xhi),ylim=c(ylo,yhi),xlab='',ylab='',type='n');
xa1=cop2(xu1,npts,1,1,1,1); arcs1(xa1,npts,'bivariate exponentials');
xa2=cop2(xu2,npts,1,1,1,1); arcs1(xa2,npts,'');
xa3=cop2(xu3,npts,1,1,1,1); arcs1(xa3,npts,'');
#panel 4: bivariate Weibulls
xlo=-0.1; xhi=8; ylo=-0.1; yhi=8; shape=1; scale=1;
plot(c(0),c(0),xlim=c(xlo,xhi),ylim=c(ylo,yhi),xlab='',ylab='',type='n');
xa1=cop2(xu1,npts,1,1,1,1); arcs1(xa1,npts,'bivariate Weibulls');
title(main='bivariate Weibulls',sub=' ');
xa2=cop2(xu2,npts,1,0.7,1,1); arcs1(xa2,npts,'');
xa3=cop2(xu3,npts,1,1,1,0.7); arcs1(xa3,npts,'');
```

The article by Genest and MacKay (1986) is highly recommended: much of
this section and more than none of the Exercises are lifted shamelessly from
their account. They focussed upon so-called Archimedian copulas, which
have form

$$C(u_1, u_2) = \psi^{-1}\{\psi(u_1) + \psi(u_2)\},$$

in which the function ψ is defined on $(0, 1)$ with $\psi(1) = 0$, $\psi'(u) < 0$, and
$\psi''(u) > 0$; these conditions ensure that the inverse function, ψ^{-1}, has two

derivatives. Also, $C(u_1, u_2)$ is taken to be 0 whenever $\psi(u_1) + \psi(u_2) \leq \psi(0)$: if invoked, when $\psi(0) < \infty$, this constraint confines the region of positive density to a subset of the unit square $(0, 1)^2$ away from the origin.

Genest and MacKay (1986) also consider *Kendall's tau*, a measure of association. If (U_1, U_2) and (V_1, V_2) are independent observations, then τ is defined as the probability of concordance minus that of discordance:

$$\tau = P\{(U_1 - V_1)(U_2 - V_2) \geq 0\} - P\{(U_1 - V_1)(U_2 - V_2) < 0\}.$$

Various properties are listed for τ and a theorem is given to record that, for Archimedean copulas,

$$\tau = 1 + 4 \int_0^1 \{\psi(u)/\psi'(u)\} du.$$

Copulas seem to be growing in popularity, judging by the increasing numbers of articles devoted to theory and applications. We do not give anything like a comprehensive list here but refer to some recent work that will give an entry to the literature: see Yilmaz and Lawless (2011), Zhang et al. (2010), and Chen and Bandeen-Roche (2005).

8.6 Applications

8.6.1 Clayton Copula

This is of Archimedian genus, with $\psi(u) = (u^{-v} - 1)/v$: the full p-variate form is

$$C(\mathbf{u}) = \left\{ \sum_{j=1}^{p} u_j^{-v} - (p - 1) \right\}^{-1/v}.$$

The corresponding density function is the pth derivative of $C(\mathbf{u})$:

$$\partial^p C(\mathbf{u})/\partial u_1 \ldots \partial u_p = C(\mathbf{u})^{1+pv} \left\{ \prod_{j=1}^{p-1} (1 + jv) \right\} \left(\prod_{j=1}^{p} u_j^{-v-1} \right).$$

The copula can be given a frailty-based derivation as follows. We start with

$$\bar{G}(\mathbf{t} \mid z) = \exp \left\{ -z \sum_{j=1}^{p} H_j(t_j) \right\},$$

and then suppose that Z has a gamma distribution with shape parameter v and scale parameter 1. Integration over the frailty z yields unconditional joint survivor function

$$\bar{G}(\mathbf{t}) = (1+s)^{-\nu},$$

in which $s = \sum_{j=1}^{p} H_j(t_j)$; the multivariate Burr is a special case in which the H_j are Weibull based. The resulting marginals are $\bar{G}_j(t_j) = \{1 + H_j(t_j)\}^{-\nu}$, and so $H_j(t_j) = \bar{G}_j(t_j)^{-1/\nu} - 1$. Thus, we can write

$$\bar{G}(\mathbf{t}) = \left\{ \sum_{j=1}^{p} \bar{G}_j(t_j)^{-1/\nu} - (p-1) \right\}^{-\nu},$$

which is essentially the Clayton form. There is a slight difference here: in this derivation the marginals depend on ν, whereas in pure copula models the marginals are generally taken to be independent of any dependence parameters.

Let us apply this to the repeated response times (Section 6.1), for which $p = 4$, and for Weibull marginals we make the transformation

$$u_j = \exp\{-(t_j/\xi_j)^{\phi_j}\}, \quad t_j = \phi_j(-\log u_j)^{1/\phi_j}.$$

The data are as before (Section 7.2). For fitting the model, use

```
#repeated response times (Sec 8.6)
rt1=read.table('rtimes1.dat',header=T);
rt2=as.numeric(unlist(rt1)); nd=40; dim(rt2)=c(nd,5); pdim=4;
#Clayton copula -- group 1 only
n1=10; censr=10; opt=0; iwr=0; adt=c(n1,censr,pdim,opt,iwr);
np=9; par0=c(0.40,0.69,0.69,0.69,0.69,0,0,0,0);
par1=fitmodel(claycop,np,par0,rt2,adt);
```

The function `claycop` is listed on the CRC Press Web site noted in the Preface. As a special treat I leave the running of the program, and the inspection of results, to you. For further work see the Exercises (Section 8.7).

8.7 Exercises

8.7.1 Frailty-Generated Distributions

1. For the MB distribution (Section 8.2) verify the formulae quoted for the joint, marginal, and conditional survivor functions. Investigate further: derive the dependence measures, $\bar{R}(\mathbf{t})$ and $\theta^*(\mathbf{t})$, the various hazard functions, and quantiles of the marginal and conditional distributions. Calculate the survivor function of $U = \min(T_1, \ldots, T_p)$—notice anything? Suppose that $\mathbf{T}_i = (T_{i1}, \ldots, T_{ip})$ $(i = 1, \ldots, n)$ are independent and all with the same MB distribution, that is, they comprise a random sample from MB. Let $V_j = \min_i(T_{ij})$ be the minimum of the jth components of the $\mathbf{T}_i$. Calculate the multivariate survivor function of $\mathbf{V} = (V_1, \ldots, V_p)$.

2. Repeat Exercise 1 for the extended MW distribution.

3. Repeat Exercise 1 for the third distribution in Section 8.2.

4. Multivariate beta-geometric. Instead of replacing ρ_j by $z\rho_j$ (Section 8.2), try ρ_j^z with z having a gamma distribution.

5. Multivariate gamma-Poisson. Apply the first suggestion: derive the result of taking z to have a beta distribution. Extend this to more general beta forms.

8.7.2 Applications

1. Do what was suggested in Section 8.3: fit alternative models to the data sets.

2. Refer to the cycles-to-pregnancy data given in Section 8.3. Write some R-code, based on the formulae given, to fit the bivariate beta-geometric model.

8.7.3 Copulas

1. Derive the density function of the Archimedean copula as $-\psi''(c)\psi'(u_1)$ $\psi'(u_2)/\psi'(c)^3$, where $c = C(u_1, u_2)$.

2. Show that U_1 and U_2 in the Archimedean copula are independent if and only if $\psi(u) = -k \log u$ for some $k > 0$.

3. Derive the *Frechet Bounds*: $\max(0, u_1+u_2-1) \le C(u_1, u_2) \le \min(u_1, u_2)$. The upper bound is the bivariate distribution function of (U_1, U_2) when $U_2 = U_1$, the case of perfect positive correlation; the lower bound corresponds to $U_2 = 1 - U_1$, perfect negative correlation. (To verify this look at $P(U_1 \le u_1, U_2 \le u_2)$ in each case.) Transformation, $U_j = G_j(T_j)$, gives the bounds for general bivariate distributions.

4. $C(u_1, u_2)$ has a singular component, with probability mass $-\psi(0)/\psi'(0)$ concentrated on the one-dimensional subset defined by $\psi(u_1)+\psi(u_2) = \psi(0)$, if and only if $\psi(0)/\psi'(0) > 0$. See Theorem 1 of Genest and MacKay (1986): this exercise comprises a brisk walk to the library.

5. Verify that the Clayton (1978) copula, Archimedean with $\psi(u) = (u^{-\nu} - 1)/\nu$ and $\nu > 0$, produces positive dependence between U_1 and U_2. Also, as $\nu \to 0$, $\psi(u) \to -\log u$, which gives a valid copula, as do values of ν in $(-1, 0)$.

6. Some other famous copulas
   Farlie–Gumbel–Morgenstern: $C(u_1, u_2) = u_1 u_2 \{1 + \alpha(1 - u_1)(1 - u_2)\}$.
   Ali–Mikhail–Haq: $C(u_1, u_2) = u_1 u_2 / \{1 - \alpha(1 - u_1)(1 - u_2)\}$.
   Frank: $C(u_1, u_2) = \log_\alpha \{1 + (\alpha^{u_1} - 1)(\alpha^{u_2} - 1)/(\alpha - 1)\}$.
   Gumbel–Hougaard: $(-\log C(u_1, u_2))^\alpha = (-\log u_1)^\alpha + (-\log u_2)^\alpha$.

Pareto: $\bar{C}(u_1, u_2) = (u_1^{-1/\alpha} + u_2^{-1/\alpha} - 1)^{-\alpha}$ (Clayton form).

Durling–Pareto: $\bar{C}(u_1, u_2) = (1 + u_1 + u_2 + \beta u_1 u_2)^{-\alpha}$.

Gumbel Type I: $C(u_1, u_2) = 1 - e^{-u_1} - e^{-u_2} + e^{-(u_1 + u_2 + \alpha u_1 u_2)}$.

Gumbel–Barnett: $\bar{C}(u_1, u_2) = (1 - u_1)(1 - u_2)e^{-\alpha \log(1-u_1)\log(1-u_2)}$.
$\bar{C}(u_1, u_2)$ here denotes $P(U_1 > u_1, U_2 > u_2)$. Investigate these: if you find anything interesting let me know.

7. Push the analysis of Section 8.6 further: extract and plot residuals; try an enhanced model (such as a *Lehmann alternative* based on the survivor function, since one based on the distribution function does not extend the copula model); extend `claycop` to cope with censored values, like some of those in Groups 3 and 4 of the repeated response times; try a different copula.

8.8 Hints and Solutions

8.8.1 Copulas

1. Start with $\psi(c) = \psi(u_1) + \psi(u_2)$ and get differentiating.

2. The condition for independence is $\psi(u_1 u_2) = \psi(u_1) + \psi(u_2)$ and a double differentiation gives an easy differential equation for ψ.

3. The left-hand inequality follows from $C(u_1, u_2) = P(U_1 > u_1, U_2 > u_2) + u_1 + u_2 - 1$ (see Exercise 1 of Section 6.5). For the right-hand one note that ψ and ψ^{-1} are monotone decreasing, so as $u_1 \uparrow 1$, $\psi(u_1) \downarrow 0$ and $C(u_1, u_2) \uparrow C(u_1, 1) = u_1$.

5. $R(u_1, u_2) = (u_1^{-\nu} + u_2^{-\nu} - 1)^{-1/\nu} \div u_1 u_2 = \{1 - (1 - u_1^\nu)(1 - u_2^\nu)\}^{-1/\nu} > 1$ for $\max(u_1, u_2) < 1$. Also, as $\nu \to 0$, $\psi(u) \to -\log u$, which gives a valid copula, as do values of ν in $(-1, 0)$.

9

Repeated Measures

The data sets listed in Section 6.1 are of this type: they comprise several times to an event recorded for each of a number of individual units. In the first two data sets of Section 6.1 the units are laboratory animals, in the third they are fibres, and in the fourth they are households. Where there is a sense of monitoring a unit over a period of time, such as in the first set, the data are known as *repeated measures* or *longitudinal data*. In other cases, the data on a unit might be all collected at the same time, such as in the third set; a typical case in statistics textbooks is the weights of week-old (not weak old) animals in a litter. However, the statistical form of the data is similar, comprising sets of correlated readings; the term *clustered data* is often applied here.

The unavoidable aspect with multivariate data, survival or otherwise, is the correlation structure. Where does it come from? In frailty survival models it comes from the frailty: the components of the observed response are conditionally independent, given the random effect, but unconditionally dependent. In other cases, as detailed below, it can come from the temporal nature of the recordings—those further apart in time are often likely to be less correlated.

9.1 Pure Frailty Models: Applications

Standard MANOVA (Section 7.2) allows the covariance matrix to be estimated freely, along with the regression coefficients in the linear model for the means; it also assumes that it is homogeneous between groups. However, it is usually more satisfying to try to specify the covariance structure to reflect the situation at hand, that is, apply some statistical modelling. For example, if some animals are consistently quicker to respond than others, or if some households use more energy all round than others, a random effect term can be included in the model to reflect this. We assume here that the underlying quickness (of response) or profligacy (of energy usage) is unobserved; further, we do not normally treat such effects as parameters to be estimated.

9.1.1   Lengths and Strengths of Fibres

A case can be made that the strengths of materials should follow a Weibull
distribution (e.g., Galambos, 1978). Put that together with random variation
between units and you end up with a distribution like the MW or the MB,
as described in Section 8.2. The difference between them, or any other such
mixture, is the distribution of the frailties. There are usually no particular
scientific grounds for specifying the mixing distribution and, as has been
pointed out before, it is usually not well identified by the data. So we can be
pragmatic and select a convenient form in the hope that it will make little
difference. The MB is more amenable to computation than the MW, having
a less complicated form for the density, so we will fit it to the fibre strengths
listed in Section 6.1.

The p-variate survivor function for the MB distribution is $\bar{G}(\mathbf{t}) = (1 +
s)^{-\nu}$, where $s = \sum_{j=1}^{p} u_j$, with $u_j = (t_j/\xi_j)^{\phi_j}$. Formulae for the likelihood
contributions are given in Section 8.2. The R-code for loading the fibres data
from file `fibres1.dat`, and then for plotting the unit profiles, is

```
#fibres data (Sec 9.1)
fb1=read.table('fibres1.dat',header=T); attach(fb1);
nc=20; pdim=4; fb2=as.numeric(unlist(fb1[1:nc,2:(pdim+1)]));
dim(fb2)=c(nc,pdim); cens=4.00;
#plots
par(mfrow=c(2,2));
x1=c(1:4); y1=fb2[1,1:4];
plot(x1,y1,type='l',ylim=c(0.00,5.00),xlab='mm5-mm75',ylab='breaking strength');
for(ic in 2:nc) { y1=fb2[ic,1:4]; lines(x1,y1); };
```

From the plot, the left panel of Figure 9.1, it is evident that strength tends to
decrease with length.

We now fit the MB model, calling up function `mburr2` (listed on the CRC
Press Web site noted in the Preface).

```
#fit MB model
np=6; mdl=1; opt=0; cens=4; iwr=0;
adt=c(nc,pdim,np,mdl,opt,cens,iwr);
par0=rep(0,np); par1=fitmodel(mburr2,np,par0,fb2,adt);
```

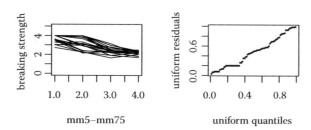

FIGURE 9.1
Breaking strengths and uniform residuals for the fibres data.

The maximised log-likelihood is 43.20558; and parameter estimates, after exponentiation of those used within the code, are (0.908, 9.89, 3.75, 3.22, 2.32, 2.07) with standard errors (0.37, 1.12, 0.26, 0.23, 0.15, 0.12).

We can extract and plot marginal uniform residuals. These can be computed from the marginal survivor function estimates as $r_{ij} = (1 + v_{ij})^{-\nu}$, where $v_{ij} = (t_{ij}/\hat{\xi}_j)^{\hat{\phi}}$ based on t_{ij} for subject i. The plot appears in the right-side panel of Figure 9.1. A more refined approach, described in Crowder (1989, Section 8.3) is to de-correlate the set of residuals for each individual by applying conditional transformations. (Of course, there will still be some residual correlation due to the sharing of the estimated parameters used to construct them.) Whether marginal residuals are plotted for each t-component separately or a combined plot of de-correlated residuals is made depends on the application. A combined plot of marginal residuals is given here, but this makes little difference to the interpretation—see the comments below.

```
#plot residuals -- all
par1=c(-0.0966,2.291,1.321,1.170,0.840,0.728);
nrs=nc*pdim; u1=c(1:nrs)/(nrs+1);
opt=1; adt=c(nc,pdim,np,mdl,opt,cens,0);
rtn=mburr2(par1,fb2,adt); rv=rtn[(nc+2):(nc+1+nrs)]; rv1=unlist(sort(rv));
plot(u1,rv1,type='p',cex=0.25,
     xlab='uniform quantiles',ylab='uniform residuals');
mtest=moran(rv1,nrs,np,1);
```

The residual plot should resemble a straight line from (0, 0) to (1, 1). Whether this one qualifies can be argued about over a pint. The *Moran test* produces chi-square value 157.32 on 80 degrees of freedom, with resulting p-value 5.6×10^{-7}, which is not very encouraging. Something clearly needs to be changed. The problem is not the correlation between marginal residuals: separate plots and tests for each of the four components (the fibre lengths) produces an equally dismal result. Maybe parameters ν and ϕ should be allowed to vary between fibre lengths instead of being constrained to be equal. Maybe the MB model itself is inappropriate. Such considerations can be argued over another pint—I will be available if you are paying. However, there are many things in life to the bottom of which one will never get, like a mug of bad beer.

9.1.2 Visual Acuity

The data in Table 9.1 have appeared previously in Crowder and Hand (1990, Examples 13.2 and 4.5) and Hand and Crowder (1996, Example 7.2). (We are doing our bit for the planet by recycling.) They arise from an experiment conducted at Surrey University in the late 1970s (happy days). A light is flashed, and the time lag (milliseconds) to an electrical response at the back of the cortex is recorded. Both left and right eyes were tested, through lenses of powers 6/6, 6/18, 6/36, 6/60; for example, 6/18 means that the lens magnification makes an object actually 18 feet away appear as if it were only 6 feet away. Seven students participated, each contributing the eight measures per row in the data matrix here; the figures have been scaled up by a factor 10.

The research question was whether the response time varies with lens strength, and various analyses have been given in the references quoted

TABLE 9.1

Visual Acuity

	Left Eye				Right Eye			
Student	6/6	6/18	6/36	6/60	6/6	6/18	6/36	6/60
1	116	119	116	124	120	117	114	122
2	110	110	114	115	106	112	110	110
3	117	118	120	120	120	120	120	124
4	112	116	115	113	115	116	116	119
5	113	114	114	118	114	117	116	112
6	119	115	94	116	100	99	94	97
7	110	110	105	118	105	105	115	115

Source: Crowder, M.J. and Hand, D.J., 1990, *Analysis of Repeated Measures*, Chapman & Hall, London, Examples 4.5, 13.2; Hand, D.J. and Crowder, M.J., 1996, *Practical Longitudinal Data Analysis*, Chapman & Hall/CRC Press, London, Example 7.2. Reproduced by permission of Chapman & Hall/CRC Press.

above. It was also noted previously that "seven student volunteers came forth and saw the light," for which awfulness I hereby apologise (though only a little).

The observations here are clustered, non-temporal, meaning that we do not look for serial correlation in successive measurements on a unit. A pure frailty model seems appropriate, to allow for possible inter-individual variation in response time. We will fit the MB distribution (Section 8.2) to the data. The data can be loaded from file `visac.dat`, in which they appear just as shown in Table 9.1, and the individual response profiles can be plotted as follows. (Note that the observations are scaled, divided by 100, in the code.)

```
#visual acuity (Sec 9.1)
va1=read.table('visac.dat',header=F); attach(va1);
nc=7; pdim=8; va2=as.numeric(unlist(va1[1:nc,2:(pdim+1)]));
va2=va2/100; dim(va2)=c(nc,pdim);
# plots
par(mfrow=c(2,2)); x1=rep(1:4); y1=va2[1,1:4]; y2=va2[1,5:8];
plot(x1,y1,type='l',ylim=c(0.9,1.25),xlab='left eye',ylab='response time');
for(ic in 2:nc) { y1=va2[ic,1:4]; lines(x1,y1); };
plot(x1,y2,type='l',ylim=c(0.9,1.25),xlab='right eye',ylab='response time');
for(ic in 2:nc) { y2=va2[ic,5:8]; lines(x1,y2); };
```

The plots, given in the top half of Figure 9.2, do not look promising: there appears to be no strong left–right effect, nor much of an overall effect of lens strength. Nevertheless, let us try an MB model in which the ξ_j are governed by parameters for left–right eye and lens strength, and with $\phi_j = \phi$ for all j. Formally, $\log \xi_j = \beta_{eye} + \gamma_{mag}$, where $\beta_{left} + \beta_{right} = 0$ for identifiability, and *mag* indicates lens magnification.

The R-code for fitting the MB model is as follows:

```
#fit MB model
np=7; par0=c(1,1.5,0,0,0,0,0);
mdl=2; opt=0; cens=max(va2)+1; iwr=0; adt=c(nc,pdim,np,mdl,opt,cens,iwr);
par1=fitmodel(mburr2,np,par0,va2,adt);
```

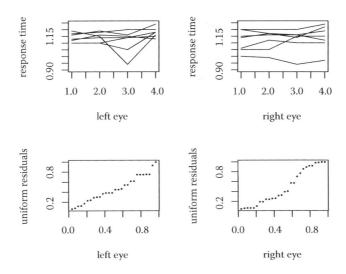

FIGURE 9.2
Visual acuity data: individual profiles and residuals from MB fit.

The maximised log-likelihood is -87.82607, and the *mle* are $\hat{\nu} = 0.789$, $\hat{\phi} = 3.435$, $\hat{\beta} = 0.003$, $\hat{\gamma}_1 = 0.16$, $\hat{\gamma}_2 = 0.16$, $\hat{\gamma}_3 = 0.15$, and $\hat{\gamma}_4 = 0.18$. These values confirm what was seen in the plots, and it is hardly worth extracting standard errors for them: (a) the standard errors are likely to be large because of the small sample size, and (b) even if not, the small differences between the γs are of little practical significance. Looking on the bright side, like the student volunteers, it may be that the experimenters were satisfied with the result: maybe it confirmed a hypothesis of no effect of lens strength, though it is difficult to prove a negative.

The residual plots appearing in the lower half of Figure 9.2 were produced as follows.

```
#plot residuals
parvec1=c(0.788,3.436,0.0028,0.156,0.156,0.154,0.182);
opt=1; iwr=2; adt=c(nc,pdim,np,mdl,opt,cens,iwr); rtn=mburr2(parvec1,va2,adt);
nr=nc*pdim; rv=rtn[(nc+2):(nc+1+nr)]; dim(rv)=c(nc,pdim);
p2=pdim/2; nr2=nc*p2; u1=c(1:nr2)/(nr2+1);
u2a=rv[1:nc,1:p2]; u2b=rv[1:nc,(p2+1):pdim];
v2a=unlist(sort(u2a)); v2b=unlist(sort(u2b));
plot(u1,v2a,type='p',cex=0.25,xlab='left eye',ylab='uniform residuals');
plot(u1,v2b,type='p',cex=0.25,xlab='right eye',ylab='uniform residuals');
```

9.2 Models with Serial Correlation: Application

Repeated measures and longitudinal data is a huge subject, and we will confine the treatment here to one or two ill-chosen examples. Massive tomes on the topic include Crowder and Hand (1990), Hand and Crowder (1996),

Diggle et al. (2002) and Fitzmaurice et al. (2009); however, none of the later books match the sheer, unadulterated brilliance of the first (according to my mum). The modelling of covariance structures, in the particular way applied here, is discussed at length in the first two books. Longitudinal statistics and event history analysis are nicely drawn together in the papers by Tsiatis and Davidian (2004) and Farewell and Henderson (2010); these topics have traditionally been treated separately, but the acquisition of more complete data nowadays prompts such amalgamation of methodology.

9.2.1 Repeated Response Times

The data (Section 6.1) involve four response times for each of 40 animals in four groups of 10. It was subjected to MANOVA in Section 7.2: the linear model for T_{ij}, the response time for animal i on occasion j, in such an analysis is

$$\log T_{ij} = \mu_j + \gamma_{g(i)} + e_{ij}$$

in which μ_j is the overall mean response on occasion j ($j = 1, \ldots, p = 4$), and $\gamma_{g(i)}$ is the mean for group $g(i)$ to which animal i belongs ($i = 1, \ldots, n = 40$); the $\gamma_{g(i)}$ sum to zero for identifiability. The residual vectors $e_i = (e_{i1}, \ldots, e_{ip})$ are independently distributed as $N(\mathbf{0}, \mathbf{E})$, in which $\mathbf{E}$ is unconstrained, no better than it ought to be, just positive definite, like all self-respecting covariance matrices.

Let us now see whether we can tie the covariance structure down a little more thoughtfully. Animals do tend to differ in their reaction times, some being naturally slower than others, and responses far apart in time are likely to be less highly correlated than those close in time. To express this, consider an enhanced model,

$$\log T_{ij} = \gamma_{g(i)} + b_i + e_{ij},$$

in which the b_i are unobserved random effects independently distributed as $N(0, \sigma_b^2)$: slower animals will have larger b_i values. For simplicity we have dropped the μ_j term, so we are not encumbered with identifiability constraints. Also, for the temporal attenuation, constrain $\mathbf{E}$ to have Markov form: $E_{jk} = \sigma_e^2 \rho^{|s_j - s_k|}$, where $\sigma_e^2 = \text{var}(e_{jj})$, $0 < \rho < 1$, and the s_j are the times at which the responses were recorded (0, 0.25, 0.5, and 1 hours). The resulting covariance matrix for the $\log T_{ij}$ is Σ, with elements

$$\Sigma_{jk} = \text{cov}(\log T_{ij}, \log T_{ik}) = \text{cov}(b_i + e_{ij}, b_i + e_{ik}) = \sigma_b^2 + E_{jk};$$

it has been assumed here that the b_i are uncorrelated with the e_{ij}.

The R-code for data input is as referred to in Section 7.2 and that for fitting the model is

```
#repeated response times (Sec 9.2)
rt1=read.table('rtimes1.dat',header=T); attach(rt1); nd=40; pdim=4;
rt1a=matrix(0,nrow=nd,ncol=p+1); rt1a[1:nd,1]=group;
for (j in 1:p) { rt1a[1:nd,j+1]=log(rt1[1:nd,j+1]) }; #log-times
```

```
#linear model with structured covar
ng=4; mdl=1; opt=0; adt=c(nd,pdim,ng,mdl,opt,0);
np=3; par0=c(1,1,0); #lmcov1(par0,rt1a,adt);
par1=fitmodel(lmcov1,np,par0,rt1a,adt); #equal group means
mdl=0; adt=c(nd,pdim,ng,mdl,opt,0);
par1=fitmodel(lmcov1,np,par0,rt1a,adt); #unconstrained group means
```

and the function `lmcov1` is listed on the CRC Press Web site noted in the Preface. There are two fits, the first with the full model and the second with the group means constrained to be equal—you can see how this is done in the code. The log-likelihoods are -3.2385 (unconstrained group means) and -96.5639 (equal group means). This yields chi-square value 199.60 on 12 degrees of freedom for testing the hypothesis of equal group means, a huge value reflecting the previous MANOVA result given in Section 7.2. Incidentally, the estimate for ρ is driven down to zero, somewhat surprisingly, so it seems that serial correlation is not evident in the data. Further, we can compare the normal log-likelihood from fitting the group means with an unstructured co-variance (as in MANOVA): this gives -9.6652. The resulting chi-square value is 12.85 on 8 degrees of freedom, giving p-value 0.12. On this test, at least, independence of the t-components seems tenable.

In all of the above, censoring has been ignored. To pursue this issue, with log-normal models, is unattractive because of the difficulty of computing the multivariate normal distribution function. For the bivariate case, Drezner and Wesolowsky (1989) give a viable algorithm, but going beyond that seems to be much harder (for example, Genz, 1993). There are plenty of packages listed on the CRAN Web site for R. One sees `lmec`, for mixed-effects linear models with censoring, which looks promising but, according to the online manual, the program is for left-censored observations, where some values fall below the threshold sensitivity of the monitoring instrument. With characteristic generosity I leave the search for suitable software as an exercise for the readers, asking that they be similarly generous and get back to me when they have found some.

9.3 Matched Pairs

Suppose that the data comprise pairs of times, $\{(t_{i1}, t_{i2} : i = 1, \ldots, n\}$, and that t_{ij} has a covariate $\mathbf{x}_{ij}$ attached to it. For example, the two times might arise from human twins: one is randomly assigned to training regime A, the other to B, and their subsequent task-completion times are observed. In this case, the covariate indicates which regime the individual underwent, together with any other explanatory variables of study interest.

Parametric models for such data have been described in preceding sections. We look here at a semi-parametric approach for which a proportional-hazards model is adopted:

$$h_j(t) = h_0(t)\psi_{ij}$$

in terms of a baseline hazard h_0 and a positive function ψ_{ij}, conventionally of form $\psi_{ij} = \exp(\mathbf{x}_{ij}^T \beta)$ ($j = 1, 2$). The argument now is the same as that in Section 4.2. Given that one of t_1 and t_2 is the first to fail at time t, the probability that it is t_{ij} is $\psi_{ij}/(\psi_{i1} + \psi_{i2})$. The resulting partial likelihood function, taken over n pairs whose first failure time is observed, is

$$P(\beta) = \prod_{i=1}^{n} \psi_{ij_i}/(\psi_{i1} + \psi_{i2}),$$

where $j_i = 1$ if $t_{i1} < t_{i2}$ and $j_i = 2$ if $t_{i2} < t_{i1}$. The form of $P(\beta)$ is just that of the likelihood for binary logistic regression, where the response variable takes value 1 or 2, that is, it is just j_i. So, the standard analysis (for example, Cox, 1970) applies.

9.4 Discrete Time: Applications

The literature is not overburdened with consideration of discrete-variable repeated measures. Molenbergh et al. (2007) provides some guidance for repeated-counts data, with an emphasis on random effects and over-dispersion. Shih (1997) proposed a family of discrete multivariate survival distributions: the marginal distributions may be specified as required, and the associations are defined by constant odds ratios.

9.4.1 Proportional Odds

Recall the PO model described in Section 5.5 for the univariate case: the univariate hazard function at time τ_s is h_s, and the odds ratio, $g_s = h_s/(1 - h_s)$, is replaced by zg_s, where z is a random effect with continuous distribution function K on $(0, \infty)$. Correspondingly, h_s is replaced by $zg_s(1 + zg_s)^{-1}$ and $1 - h_s$ by $(1 + zg_s)^{-1}$. The unconditional survivor function is evaluated as

$$P(T > \tau_l) = \int \bar{F}(\tau_l \mid z) dK(z) = \int \left\{ \prod_{s=1}^{l} (1 + zg_s)^{-1} \right\} dK(z).$$

For the multivariate version we assume that the T_j are conditionally independent, given z, and that the jth odds ratio component, g_{js}, is replaced by zg_{js}. The unconditional joint survivor function is then given by

$$\bar{G}(\mathbf{t}) = P(T_1 > \tau_{l_1}, \ldots, T_p > \tau_{l_p}) = \int \left\{ \prod_{j=1}^{l_p} \prod_{s=1}^{t_j} (1 + zg_{js})^{-1} \right\} dK(z).$$

On the bright side, numerical evaluation of this integral is hardly any worse than that for the univariate case.

9.4.2 Beta-Geometric Model

Sometimes special constructions are available. Recall from Section 5.5 the application of a beta random effect to a geometric waiting-time distribution. There, the discrete survivor function, conditional on ρ, was $\bar{F}(l \mid \rho) = \rho^l$ $(l = 1, 2, \ldots)$, with $0 < \rho < 1$, and the distribution of ρ across units was taken to be beta with density $f(\rho) = B(\phi, \psi)^{-1} \rho^{\phi-1} (1-\rho)^{\psi-1}$. This can be extended to the multivariate case quite straightforwardly. Suppose that the components T_j are conditionally independent, T_j having geometric$(a_j \rho)$ distribution for given a_j. Then their joint survivor function is

$$P(T_1 > l_1, \ldots, T_p > l_p) = \int_0^1 \left\{ \prod_{j=1}^p (a_j \rho)^{l_j} \right\} f(\rho) d\rho$$

$$= \prod_{j=1}^p a_j^{l_j} B(\phi + l_+, \psi) / B(\phi, \psi)$$

$$= \prod_{j=1}^p a_j^{l_j} \prod_{j=1}^{l_+} \{(\phi + j - 1)/(\phi + \psi + j - 1)\},$$

where $l_+ = l_1 + \cdots + l_p$.

9.4.3 Bird Recapture

North and Cormack (1981) reported some data, listed in Table 9.2, of interest to ornithologists (also known as twitchers). Apparently, *birding*, or *birdwatching*, is a predominantly male pursuit and has been linked by psychologists to the hunting instinct. Ignoring ribald comments from the floor, the data accumulated can be of core importance in the framing of environmental and conservation policy. The figures cover four years of data collection. They show the numbers of herons ringed as nestlings (*nringed*) and then the number later recaptured in the same year ($t = 0$) and after one, two, and three years ($t = 1, 2, 3$). For example, of the 282 birds ringed in the third year of the study, 21 were recaptured in the same year and 5 one year later. We assume here that the recaptures of individual birds are independent events. However, it is possible that there might be some unrecorded year effect, in which case the

TABLE 9.2

Recapture Counts of Ringed Birds

Year	Nringed	$t = 0$	$t = 1$	$t = 2$	$t = 3$
1	282	24	14	4	1
2	236	32	5	3	—
3	282	21	5	—	—
4	450	32	—	—	—

Source: North, P. and Cormack, R.M., 1981, On Seber's Method for Estimating Age-Specific Bird Survival Rates from Ringing Recoveries, *Biometrics*, 37, 103–122. Reproduced by permission of the International Biometric Society.

recapture times for a given year of ringing would be dependent, giving rise to multivariate data.

To construct a statistical model let us begin with simple assumptions that each bird has some constant probability of being recaptured in any one year, that this probability is the same for all birds, that the birds act independently, and that the counts represent first recaptures only. In consequence, the counts in each row should conform to a geometric distribution. Clearly, they do not. One might swallow a slightly dubious assertion that successive ratios of counts are equal across row 1, for instance, as they should be for a geometric distribution. But what happened to the rest of the original 282 birds—there are far too many left over. We can patch up this model a bit by supposing that there is a proportion, say p_0, of birds who will evade recapture forever— ones that just disappear from view. Maybe some have flown off to that great nest in the sky prematurely. So, the model for the year T of recapture for an individual bird is

$$P(T > l) = p_0 + (1 - p_0)\rho^{l+1}, \quad P(T = l) = (1 - p_0)(1 - \rho)\rho^l \quad (l = 0, 1, \ldots).$$

In a health context, where T is the time to succumb to a disease, p_0 would be the probability of being immune.

The R-code for loading the data and fitting the model is

```
#bird recapture data (Sec 9.4)
bd1=c(282,24,14,4,1,236,32,5,3,-9,282,21,5,-9,-9,450,32,-9,-9,-9);
dim(bd1)=c(5,4); bd2=t(bd1); bd2;
nc=4; mc=5; iwr=0; adt=c(nc,mc,iwr);
np=2; par0=c(-1,-1);
par1=fitmodel(bird1,np,par0,bd2,adt);
```

and the standard errors are computed as usual. Within the function `birds1`, which is listed on the CRC Press Web site noted in the Preface, the parameters are expressed in logit form: $p_0 = (1+e^{-\theta_1})^{-1}$ and $\rho = (1+e^{-\theta_2})^{-1}$. This ensures the constraints that both lie between 0 and 1. The resulting *mle* are $\hat{p}_0 = 0.883$ and $\hat{\rho} = 0.268$ with standard errors 0.029 and 0.090. The latter are obtained by transforming those from the inverse Hessian: the R-code for this is

```
p0=1/(1+exp(-par1[1])); rho=1/(1+exp(-par1[2])); cat('\n p0,rho: ',p0,rho);
se=c(0.09066,0.2036);
q1=p0*(1-p0); q2=rho*(1-rho); s1=sqrt(q1)*se[1]; s2=sqrt(q2)*se[2];
cat('\n std errors (untransformed): ',s1,s2);
#std errors
p0=1/(1+exp(-par1[1])); rho=1/(1+exp(-par1[2])); cat('\n p0,rho: ',p0,rho);
se=c(0.09066,0.2036);
q1=p0*(1-p0); q2=rho*(1-rho); s1=sqrt(q1)*se[1]; s2=sqrt(q2)*se[2];
cat('\n std errors (untransformed): ',s1,s2);
```

There is more to do. Expected frequencies, based on the *mle*, can be computed and compared with those observed. The model can be refitted, allowing p_0 or ρ or both to differ between ringing years. A frailty version can be tried, with either p_0 or ρ or both varying across individual birds with specified distributions. These investigations are generously left to be done as Exercises (Section 9.6).

TABLE 9.3

Antenatal Knowledge in Two Groups of Mothers-to-Be

MTB	Group	Pre-Course				Post-Course				MTB	Group	Pre-Course				Post-Course			
1	1	4	15	28	4	4	17	29	4	13	2	0	5	12	2	0	11	20	1
2	1	4	13	26	4	4	15	29	4	14	2	4	12	20	2	3	14	26	4
3	1	4	12	19	1	4	14	26	5	15	2	4	11	17	3	4	12	24	5
4	1	3	11	16	3	3	17	24	5	16	2	5	9	22	2	5	13	27	3
5	1	5	12	27	3	5	18	29	5	17	2	4	12	22	5	5	16	26	5
6	1	4	11	17	2	4	15	26	4	18	2	1	4	9	2	2	9	17	5
7	1	4	12	22	3	5	17	28	5	19	2	5	16	28	4	5	18	28	4
8	1	4	14	16	5	4	14	25	5	20	2	4	14	23	2	4	16	27	3
9	1	5	14	20	3	5	16	29	5	21	2	5	16	26	3	5	16	29	4
10	1	4	13	25	2	4	14	25	4										
11	1	3	10	21	3	3	14	26	5										
12	1	5	15	25	2	5	16	25	4										

Source: Hand, D.J. and Crowder, M.J., 1996, *Practical Longitudinal Data Analysis*, Chapman & Hall/CRC Press, London, Table B19. Reproduced by permission of Chapman & Hall/CRC Press.

9.4.4 Antenatal Knowledge

Table 9.3 arises from an antenatal study in which mothers-to-be (MTB) were quizzed on their knowledge about certain aspects of baby birth and care. It is tempting to link this application with one in Section 5.5, but we shall resist. The data here are lifted bodily from Hand and Crowder (1996, Table B19). There are two groups of MTB, 9 of whom attended a training course (Group 2) and 12 that did not (Group 1, control). Each filled in a questionnaire, once before the course and once after, yielding four responses each time. These response variables assess the level of knowledge on four different scales: they are the proxy survival times. (Think of each test as a sequence of questions and the score as how far you get through it before coming a cropper.) The numbers of questions on the four tests are 5, 20, 30, and 5, respectively.

To assess the value of the training course we assume that assignment to it, or to control, was random: one might suspect that, in the prenatal environment run by well-meaning people, some direction might have taken place, but this is unrecorded in the data. We can just note that the overall success rate (proportion of correct answers) in phase 1 (pre-course) for Group 1 is 0.69 and that for Group 2 is 0.62. Also, we will assume that, after the first phase, some MTB did not learn from the experience or go away and bone up on the subject independently. Even so, one cannot help noticing that the passage of time alone seems to have improved the performance of both groups: indeed, Figure 9.3 shows that this effect was universal. Nevertheless, the figures suggest that some MTB might ask for their money back. Anyway, the pregnant question is, did the treatment group improve more than the control group? A rough calculation produces an average success rate ratio (post/pre) for Group 1 as 1.23 and for Group 2 as 1.24: so, there is hope yet, but not much.

Let us start with a model in which every MTB in Group 1, and those in Group 2 before training, has probability p_1 of answering each question

group 1   **group 2**

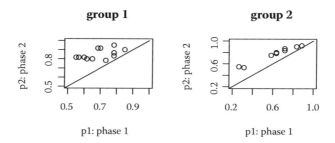

FIGURE 9.3
Success rates p_1 (pre-course) and p_2 (post-course) for Groups 1 and 2.

correctly and independently; for MTB in Group 2 after training let the probability be p_2. The group sizes are $n_1 = 12$ and $n_2 = 9$, and let m_j denote the number of questions in test j, and r_{ij} the number of correct responses by MTB i on test j. Then the likelihood function for MTB i is

$$L_i = b_{i1}b_{i2}p_1^{s_{i1}}(1-p_1)^{m_+ - s_{i1}}p_2^{s_{i2}}(1-p_2)^{m_+ - s_{i2}},$$

where $m_+ = \sum_{j=1}^4 m_j$, $b_{i1} = \prod_{j=1}^4 \binom{m_j}{r_{ij}}$, $b_{i2} = \prod_{j=5}^8 \binom{m_j}{r_{ij}}$, $s_{i1} = \sum_{j=1}^4 r_{ij}$, and $s_{i2} = \sum_{j=5}^8 r_{ij}$; also, $p_2 = p_1$ for MTB in Group 1. You will notice that this is just a glorified binomial setup, arising from a rather simplistic model, but it serves to dig some footings. Maximisation of the likelihoods for the two groups produces *mle* $\hat{p}_1 = 0.77$ for Group 1 and $\hat{p}_1 = 0.62$ and $\hat{p}_2 = 0.77$ for Group 2; in fact, the Group 1 probabilities go from 0.62 (pre-course) to 0.85 (post-course), averaging out at 0.77. This looks a bit odd: maybe the groups have been labelled the wrong way round or, as might be suspected, some of the selection and treatment of individuals has not been quite as advised. It will be safer, then, not to assume that $p_1 = p_2$ for Group 1 in what follows.

It is to be expected that some MTB will be more clued up than others at the outset. For example, the success rates (proportions of correct answers out of 60) for the 12 MTB in Group 1, phase 1, have sample mean 0.69 and sample variance 0.0094. But the theoretical, binomial-based variance is 0.0036, so we have extra-binomial variation by a factor 2.6. Hence, rather than assume the same values of p_1 and p_2 for all, these parameters should be allowed to vary randomly over the population of MTB, that is, random effects (frailties) should be introduced. For a preliminary assessment we compute the success ratios (post/pre) for each MTB: the average of these ratios in Group 1 is 1.23, and that in Group 2 is precisely the same, 1.23 (yes, *precisely* the same according to the accuracy of my computer). Are we downhearted? Well, yes, but let's soldier on anyway.

For each group we have two probabilities, p_1 and p_2, so we will need a *bivariate beta distribution*. The standard version has density

$$f(p_1, p_2) = \kappa^{-1}p_1^{a-1}p_2^{b-1}(1 - p_1 - p_2)^{c-1}$$

on the simplex $S_2 = \{p_1 > 0, \ p_2 > 0, \ p_1 + p_2 < 1\}$, with parameters $a > 0$, $b > 0$ and $c > 0$, and $\kappa = \Gamma(a)\Gamma(b)\Gamma(c)/\Gamma(a+b+c)$. The details for what follows are covered in the Exercises (Section 9.6). To match the $beta(\phi, \psi)$ marginal distribution of p_1 we take $a = \phi$ and $b+c = \psi$. Then, for the marginal distribution of p_2 to be $beta(\nu, \tau)$ we take $b = \nu$ and $\tau = a + c = \phi + \psi - \nu$. Thus, we have not got complete freedom to choose the marginals—there is one constraint on the four parameters concerned, namely, $\phi + \psi = \nu + \tau$. The likelihood contribution for MTB i in Group g is then

$$\int_{S_2} L_{ig} f(p_1, p_2) dp_1 dp_2.$$

The likelihood formulae, which are given in the Exercises (Section 9.6), look a bit unwieldy but are easy to code up. However, blindly rushing in and fitting the model to the data can lead to hours of frustration with optimisation failure and much code checking and rechecking—don't ask me how I know. The belated insight is that in the bivariate beta distribution p_1 and p_2 are negatively correlated, whereas in the data they are positively correlated. One could replace p_2 by $1 - p_2$ to obtain a modified version of the bivariate beta. Alternatively, one can reverse the responses in phase 2, that is, replace r_{ij} by $m_j - r_{ij}$. The second option has been chosen here since this preserves the standard bivariate beta form.

Some R-code for reading in the data and fitting the model is as follows. Function bbv1 is listed on the Web site noted in the Preface.

```
#antenatal data (Sec 9.4)
ante=read.table('antenatal.dat',header=F);
nc=21; mc=10; p=8; antf=as.numeric(unlist(ante)); dim(antf)=c(nc,mc);
#reverse scores for phase2 (for negative correln)
for(ic in 1:nc) {
  antf[ic,7]=5-antf[ic,7]; antf[ic,8]=20-antf[ic,8];
  antf[ic,9]=30-antf[ic,9]; antf[ic,10]=5-antf[ic,10]; };
#fit model
n1=12; ant1=antf[1:n1,3:10];
mc=4; mv=c(5,20,30,5); iwr=0; adt=c(n1,mc,mv,iwr);
par0=log(c(0.7,0.7,0.5)); fitmodel(bbv1,np,par0,ant1,adt);
n2=9; ant2=antf[(n1+1):(n1+n2),3:10];
adt=c(n2,mc,mv,iwr);
par0=log(c(0.7,0.7,0.5)); fitmodel(bbv1,np,par0,ant1,adt);
```

The (untransformed) *mle* are $(\hat{\phi}, \hat{\psi}, \hat{\nu}) = (0.72, 0.60, 0.60)$ for Group 1 and $(\hat{\phi}, \hat{\psi}, \hat{\nu}) = (0.73, 0.59, 0.59)$ for Group 2. The corresponding means for p_1 and p_2 in Group 1 are both 0.54, and the standard deviations are both 0.33; for Group 2 the means are both 0.55 and the standard deviations both 0.33. This result reflects the previous calculation. Reporting this negative finding back to the midwives will probably not help to confute the widespread mistrust of statistics in general and of statisticians in particular.

9.5   Milestones: Applications

A common example of repeated measures is the set of times to reach certain milestones for a unit. In this section some examples are discussed.

9.5.1   Educational Development

Among the plethora of statutory edicts in education are defined stages of development of which the Early Years Foundation Stage is the first, applying to children from birth to five years old. The Early Learning Goals comprise six areas of Learning and Development. (In the official literature all these nouns are given capital letters, presumably to emphasise the Importance of the Work on the Creation and Issue of Guidelines and Directives by the Department of Education and Skills; at least, that was what the department was called last week.) The six areas progress from things like *sitting quietly* and *maintaining attention* up to *engaging in representational play*—use your imagination.

Of course, the records on individual children are sensitive and confidential, so the figures in Table 9.4 are concocted. They represent times, in months, for a dozen children to reach the defined levels of development. In practice, such times are interval censored because of the gap periods between assessments. For the purpose of illustration, however, we will take the figures as read. Also, in practice, if a child is seen to be late in reaching the normal stages, intervention takes place in the form of extra help.

The data can be stored in a file in the usual format with one row per child, and then read in and plotted to produce Figure 9.4 as follows. The censored values, 60+, are conveniently represented as 61 in the data file.

```
#educational development (Sec 9.5)
kd1=read.table('littlehorrors.dat',header=F); attach(kd1);
nc=12; pdim=6; levs=c(0,1,2,3,4,5,6);
tj=c(0,kd1[1,2:(pdim+1)]);
plot(tj,levs,xlim=c(0,60),type='l',xlab='months',ylab='levels');
points(tj,levs,cex=0.5);
for(ic in 2:nc) { tj=c(0,kd1[ic,2:(pdim+1)]);
  points(tj,levs,cex=0.5); lines(tj,levs); };
```

Model fitting and analysis are left to the Exercises (Section 9.6).

TABLE 9.4

Milestone Months

Child	Times						Child	Times					
1	8	16	26	31	37	47	7	8	16	26	31	36	46
2	10	20	33	40	47	60	8	9	18	29	35	41	52
3	8	16	26	31	37	48	9	13	26	43	52	60+	60+
4	12	24	39	47	55	60+	10	8	16	26	31	36	46
5	10	19	31	38	45	58	11	12	24	40	49	57	60+
6	11	22	36	44	52	60+	12	8	15	25	31	36	46

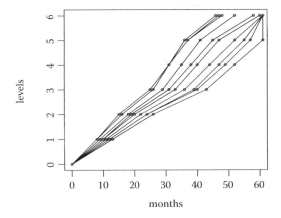

FIGURE 9.4
Educational development for 12 children over 60 months

9.5.2 Pill Dissolution Rates

The dissolution of the monasteries by Henry VIII in the 16th century was a rather more portentous event than the dissolution of a few pills in the 20th century in a laboratory in Surrey. Nevertheless, the latter event was also dutifully recorded for posterity, and the results appear in Table 9.5. In addition to the biochemical action of medicines, an important aspect is their palatability and availability: the headache cannot be put on hold while the pill undergoes a leisurely dilution process over several days. Each row of the table results from monitoring a single pill: the times (seconds) at which certain fractions of it remain undissolved are noted. The four groups represent different storage conditions, and the figures appear to be of somewhat uneven accuracy though no reason was ever forthcoming for this.

The data appeared previously in Crowder (1996), where a simple dissolution model was developed and some corresponding statistical analysis performed. Details may be found in the paper.

9.5.3 Timber Slip

Wood has ever been part of our lives. Archaeologists reveal evidence of its use by our distant ancestors, and even today Scotsmen are wont to throw telegraph poles around at the least provocation. In the construction industry wood is still a major resource. Being a natural material it exhibits a degree of uncontrolled variability. An aspect of importance to designers and architects, using timber in the support of structures, is the slippage properties of wood. In the Materials Department at Surrey University experiments were conducted to quantify slippage. Planks were clamped between two concrete surfaces, and a steadily increasing load was applied. The slippage was monitored every two seconds, and the loads at which it attained certain prespecified

TABLE 9.5

Dissolution Times of Pills

Fractions Undissolved						Fractions Undissolved					
0.90	0.70	0.50	0.30	0.25	0.10	0.90	0.70	0.50	0.30	0.25	0.10
Group 1						**Group 3**					
13	16	19	23	24	28	11.7	16.9	22.0	27.6	29.2	34.5
14	18	22	26	28	32	13.5	18.7	24.9	30.0	31.8	37.5
19	24	28	33	33	39	12.0	17.1	22.7	28.4	30.2	35.8
13	17	21	25	26	29	12.1	16.7	21.1	26.8	29.0	33.8
14	16	19	23	25	27						
13	16	19	23	24	26						
Group 2						**Group 4**					
13	17	21.5	26	27.8	33.5	11.0	14.5	19.0	24.5	26.0	32.0
11.5	15.5	20.5	24.5	26.7	31.3	14.0	19.0	24.0	30.0	31.5	39.0
10.4	14.4	18.4	23.5	25.1	29.6	11.0	14.0	17.5	22.0	23.5	29.5
11.1	15.1	19.0	23.6	24.7	28.7						

Source: Crowder, M.J., 1996, Some Tests Based on Extreme Values for a Parametric Survival Model, *J. Roy. Statist. Soc.*, 58, 417–424. Reproduced by permission of Wiley-Blackwell Publishers.

levels (millimetres) were recorded. Thus, the data are times (=loads) to reach milestones: here the milestones form a sequence of slip amounts.

The data in Table 9.6 give the loads for eight planks at slippages shown in the top row; they are borrowed from Table A15 of Hand and Crowder (1996), and a similar data set for 34 planks is given there in Table B3.

The data here are unusual in that n (sample size) is small and p (dimensionality) is large. So, there will be decent information on which to base an intra-unit model, describing slippage versus load, but not much on inter-unit variation. Modelling and fitting are left to the Exercises (Section 9.6).

9.5.4 Loan Default

A money lender (bank, mortgage company, etc.) will normally have a variety of portfolios of loans. For simplicity consider a particular portfolio for which

TABLE 9.6

Timber Slip

Slip	0.10	0.20	0.30	0.40	0.50	0.60	0.70	0.80	0.90	1.00	1.20	1.40	1.60	1.80
1	2.38	4.34	6.64	8.05	9.78	10.97	12.05	12.98	13.94	14.74	16.13	17.98	19.52	19.97
2	2.69	4.75	7.04	9.20	10.94	12.23	13.19	14.08	14.66	15.37	16.89	17.78	18.41	18.97
3	2.85	4.89	6.61	8.09	9.72	11.03	12.14	13.18	14.12	15.09	16.68	17.94	18.22	19.40
4	2.46	4.28	5.88	7.43	8.32	9.92	11.10	12.23	13.24	14.19	16.07	17.43	18.36	18.93
5	2.97	4.68	6.66	8.11	9.64	11.06	12.25	13.35	14.54	15.53	17.38	18.76	19.81	20.62
6	3.96	6.46	8.14	9.35	10.72	11.84	12.85	13.83	14.85	15.79	17.39	18.44	19.46	20.05
7	3.17	5.33	7.14	8.29	9.86	11.07	12.13	13.15	14.09	15.11	16.69	17.69	18.71	19.54
8	3.36	5.45	7.08	8.32	9.91	11.06	12.21	13.16	14.05	14.96	16.24	17.34	18.23	18.87

Source: Hand, D.J. and Crowder, M.J., 1996, *Practical Longitudinal Data Analysis*, Chapman & Hall/CRC Press, London, Tables A15, B3. Reproduced by permission of Chapman & Hall/CRC Press.

both the loan amount and the repayment schedule are the same for each of the borrowers. For example, it might be that a fixed amount is to be repaid each month over 36 months. It is not uncommon for borrowers to default on their loans, that is, cease to repay before the full term is up. In that case the loss to the lender is nominally the amount still outstanding, though some part of this is often recovered by various means. The aspect pertinent to survival analysis is the stage at which default occurs, that is, how long the borrower manages to continue repayments.

Assume that the repayment stages are $1, 2, \ldots, m$, which is the set of discrete survival times in this context. Let $q_1, \ldots, q_m$ be the probabilities of default at these times; the probability of full repayment is then $1 - q_+$, where $q_+ = \sum_{j=1}^{m} q_j$. The corresponding discrete hazards are $h_1 = q_1$ and, for $j = 2, \ldots, m$,

$$h_j = q_j/(1 - q_1 - \cdots - q_{j-1}).$$

Suppose that the portfolio comprises n independently acting individual loans, and let r_j be the number defaulting at stage j. The joint probability distribution of $\mathbf{r} = (r_1, \ldots, r_m)$ is multinomial with probability mass function

$$p(\mathbf{r}) = k_m(\mathbf{r}) \left(\prod_{j=1}^{m} q_j^{r_j} \right) (1 - q_+)^{n - r_+},$$

where $r_+ = \sum_{j=1}^{m} r_j$ and $k_m(\mathbf{r}) = n!/\{(\prod_{j=1}^{m} r_j!)(n - r_+)!\}$. The survivor function of the underlying lifetime T is $P(T > l) = 1 - q_1 - \cdots - q_l$. This can also be derived equating $P(T > l)^n$ to $P(r_1 = \cdots = r_l = 0)$, both representing the probability of no defaults up to stage l (Section 9.6, Exercises). This latter method can be applied to more sophisticated models as described in Crowder and Hand (2005). The event $\{T > m\}$, which has probability $1 - q_+$, represents completed repayment of the full amount.

In the stratospheric world of finance certain models for loan default have been proposed for the case of a single repayment, the case $m = 1$. There are two major versions, one associated with Credit-Metrics and the KMV Corporation and the other with CreditRisk+ and Credit Suisse Financial Products. These models start with the binomial distribution of r_1 and allow $n \to \infty$. The CM version has $r = O(n)$ and ends up with a distribution, dubbed the *normal inverse*, for the proportion of defaulters; the CR version has $r = O(1)$ and ends up with a Poisson distribution for r in the usual way. There are some additional but elementary aspects incorporated in the models. The framework is straightforwardly generalised to the multinomial case, and various refinements can be applied. The resulting models can be induced to yield a survivor function for individual default times in ways similar to that given above for the basic multinomial. The result, then, is a variety of multivariate joint distributions for $\mathbf{r} = (r_1, \ldots, r_m)$, from most of which can be extracted a univariate distribution for T. See Crowder and Hand (2005) for details.

The process can be viewed the other way round, that is, the focus might be on the successive times at which given numbers or proportions of defaults occur. For example, one can record the stages at which default proportions reach 1%, then 5%, and so forth. With a portfolio of contemporaneous loans, all starting at roughly the same time, such a record can provide information on the viability of the loan package and perhaps help to formulate contingency planning. Let T_1, T_2, ... be the milestone times at which the default proportions reach prespecified levels $p_1 < p_2 < ....$ Then,

$$P(T_1 = l) = P(s_{l-1} < np_1) - P(s_l < np_1),$$

where s_l denotes $r_1 + \cdots + r_l$. In the basic multinomial case, without refinements, the second probability on the right-hand side can be computed from the binomial distribution of s_l, which has parameters n and $q_1 + \cdots + q_l$; likewise the first probability. Continuing,

$$P(T_1 = l_1, \ T_2 = l_2) = P(s_{l_1-1} < np_1, \ s_{l_1} \geq np_1, \ s_{l_2-1} < np_2, \ s_{l_2} \geq np_2),$$

which can be computed from the joint (trinomial) distribution of s_{l_1} and s_{l_2}; a rather inelegant, but easily computable, double summation over the relevant pairs (s_{l_1}, s_{l_2}) would seem to be called for. The process can be continued on to T_3, and so forth, and carried over to the more sophisticated models, though I am not aware that this aspect has been tackled anywhere. However, the present purpose is achieved: we have obtained a multivariate distribution of milestone lifetimes for inclusion in this chapter. (Eye thang ewe one and all—no autographs.)

9.6 Exercises

9.6.1 Bird Recapture Data

1. Try the various extensions mentioned at the end of the application described in Section 9.3. For this you will need to modify the R-codes, including the function `birds1`.

9.6.2 Some Background for the Bivariate Beta Distribution

1. The bivariate beta distribution has density

$$f(y_1, y_2) = \kappa^{-1} y_1^{a-1} y_2^{b-1} (1 - y_1 - y_2)^{c-1}$$

on the simplex $S_2 = \{y_1 > 0, \ y_2 > 0, \ y_1 + y_2 < 1\}$, with parameters $a > 0, b > 0$, and $c > 0$ and constant $\kappa = \Gamma(a)\Gamma(b)\Gamma(c)/\Gamma(a+b+c)$. (The experts will recognise this as a two-variate Dirichlet distribution.) Verify that the density integrates to unity. Hint: In $\int_0^1 \int_0^1 f(y_1, y_2) dy_1$, dy_2 substitute $y_1 = z_1$, $y_2 = z_2(1-z_1)$, so the region of integration becomes $\{0 < z_1 < 1, \ 0 < z_2 < 1\}$ and the Jacobian of the transformation is $\partial(y_1, y_2)/\partial(z_1, z_2) = (1 - z_1)$. Then the integral becomes

$$\int_0^1 dz_1 \int_0^1 dz_2 \{z_1^{a-1} z_2^{b-1} (1 - z_1)^{b+c-1} (1 - z_2)^{c-1}\},$$

which yields (using Exercise 1) $B(a, b+c)B(b, c) = \Gamma(a)\Gamma(b)\Gamma(c)/\Gamma(a+b+c)$. Also verify that

$$\mu_1 = E(Y_1) = a/(a + b + c),$$
$$\sigma_1^2 = \text{var}(Y_1) = \mu_1(1 - \mu_1)/(a + b + c),$$
$$\sigma_{12} = \text{cov}(Y_1, Y_2) = -\mu_1\mu_2/(a + b + c).$$

Finally, verify that the marginal distribution of Y_1 is beta with parameters a and $b + c$.

2. The integral to be evaluated for the likelihood function has general form

$$\int_{S_2} y_1^{a-1} y_2^{b-1} (1 - y_1 - y_2)^{c-1} (1 - y_1)^l (1 - y_2)^m dy_1 dy_2,$$

where S_2 is the simplex $\{y_1 > 0, y_2 > 0, y_1 + y_2 < 1\}$ and l and m are positive integers. Make the substitution $y_1 = z_1$, $y_2 = z_2(1 - z_1)$, as in the preceding exercise, to obtain

$$\int_0^1 dz_1 \int_0^1 dz_2 \left[z_1^{a-1} z_2^{b-1} (1 - z_1)^{b+c+l-1} (1 - z_2)^{c-1} \{1 - z_2(1 - z_1)\}^m\right],$$

and then, by expanding the last factor, show that this can be expressed as

$$\sum_{k=0}^m \binom{m}{k}(-1)^k B(a, b + c + l + k) B(b, c + k).$$

9.6.3 Antenatal Data

1. For the simple binomial model, derive the overall log-likelihood for group 1 ($n_1 = 12$) as

$$l(p_1, p_2) = constant + s_{+1} \log p_1 + (n_1 m_+ - s_{+1}) \log(1 - p_1)$$
$$+ s_{+2} \log p_2 + (n_1 m_+ - s_{+2}) \log(1 - p_2).$$

where $s_{+j} = \sum_{i=1}^{n_1} s_{ij}$ ($j = 1, 2$). Differentiate with respect to p_1 and p_2 to obtain the score vector. The derivatives here do not involve the other parameter: the likelihood is *orthogonal* (Appendix D). Set the score vector to zero and insert the sample values to compute the *mle*. Differentiate again to obtain the information matrix (Hessian), invert it, and extract standard errors. Do this for group 2 ($n_2 = 9$) also.

2. The bivariate beta density to be applied is

$$f(p_1, p_2) = \kappa^{-1} p_1^{\phi-1} p_2^{\nu-1} (1 - p_1 - p_2)^{\psi-\nu-1},$$

in which $\kappa = \Gamma(\phi)\Gamma(\nu)\Gamma(\psi - \nu)/\Gamma(\phi + \psi)$. Verify that this gives the required marginals: $beta(\phi, \psi)$ for p_1 and $beta(\nu, \tau)$ for p_2, with $\phi + \psi = \nu + \tau$. The likelihood contribution from MTB i in Group 2 is

$$\int_{S_2} L_{i2} f(p_1, p_2) dp_1 dp_2.$$

Show that this equals

$$b_{i1} b_{i2} \kappa^{-1} \int_{S_2} p_1^{\phi+s_{i1}-1}(1 - p_1)^{m_+ - s_{i1}} p_2^{\nu+s_{i2}-1}(1 - p_2)^{m_+ - s_{i2}}$$
$$\times (1 - p_1 - p_2)^{\psi-\nu} dp_1 dp_2,$$

which is evaluated (using the preceding exercise) as

$$b_{i1} b_{i2} \kappa^{-1} \sum_{k=0}^{m_+ - s_{i2}} \binom{m_+ - s_{i2}}{k}(-1)^k B(\phi + s_{i1}, \psi + s_{i2} + m_+ - s_{i1} + 1 + k)$$
$$\times B(\nu + s_{i2}, \psi - \nu + 1 + k).$$

9.6.4 Binomial Waiting Times

1. Verify that the likelihood contribution for MTB i in Group 1 can be expressed as

$$\int_0^1 L_{i1} f(p_1) dp_1 = b_{i1} b_{i2} B(\phi, \psi)^{-1} B(\phi + s_{i1} + s_{i2}, \psi + 2m_+ - s_{i1} - s_{i2})$$
$$= b_{i1} b_{i2} \prod_{k=1}^{s_{i1}+s_{i2}} (\phi + s_{i1} + s_{i2} - k)$$
$$\times \prod_{k=1}^{2m_+ - s_{i1} - s_{i2}} (\psi + 2m_+ - s_{i1} - s_{i2} - k)$$
$$\div \prod_{k=1}^{2m_+} (\phi + \psi + 2m_+ - k).$$

9.6.5 Milestones Data

1. Fit models to the data as indicated and draw conclusions.
2. *Loan default.* Denote by $\sum^l$ the sum over the set $(r_{l+1}, \ldots, r_m)$ whose members are positive integers with sum not exceeding n. Then, under

the basic multinomial model,

$$
\begin{aligned}
\mathrm{P}(T > l)^n &= \sum^{l} \left\{ k_m(0, \ldots, 0, r_{l+1}, \ldots, r_m) \left(\prod_{j=l+1}^{m} q_j^{r_j} \right) (1 - q_+)^{n - r_{l+1} - \cdots - r_m} \right\} \\
&= \sum^{l} \left\{ k_m(0, \ldots, 0, r_{l+1}, \ldots, r_m) \left(\prod_{j=l+1}^{m} \{ q_j / (1 - q_+) \}^{r_j} \right) \right\} \times (1 - q_+)^n \\
&= \left[1 + \sum_{j=l+1}^{m} \{ q_j / (1 - q_+) \} \right]^n \times (1 - q_+)^n = (1 - q_1 - \cdots - q_l)^n,
\end{aligned}
$$

as claimed.

10

Recurrent Events

We give here a small introduction to a large subject: it is covered in far greater breadth and depth by Cook and Lawless (2007). Nevertheless, an outline of the basic aspects can be presented and appreciated here. A special issue of *Lifetime Data Analysis* (2010) has been devoted to the topic; this gives an interesting perspective on current research in the area, and one can pursue particular areas via the references listed in the papers.

You might guess from the name that the event in question is non-terminal for a unit. Moreover, the times of its recurrence are random, generated by some stochastic mechanism about which we wish to learn. For example, some people have been known to catch a cold every now and then, while others (to my intense irritation) never seem to catch one. It is said (mainly by wives) that women have minor colds while men have major flu. Also, the propensity to come down with a cold varies over the year, with a hazard function that increases in the winter. The weather is then a significant covariate for the process.

Recurrent events differ from repeated measures in at least one way: in the former the times of events determine the observation process; in the latter the observation occasions tend to be determined externally. With recurrent events n (the number of units observed) can be small with p (average number of events per unit) large, or n large with p small. An example of the former would be the 10-year record of burglaries in a small number of London boroughs. We will mainly be looking at the latter case.

10.1 Some Recurrence Data

We will base some of the discussion in this chapter around the data listed in Table 10.1. The data arise from a clinical trial of two treatment regimes for a certain medical condition. There are 29 patients, and the case variables in the table are patient number, treatment regime, sex (1 = female, 2 = male), age (years), and the number of occurrences of the condition. The subsequent columns contain severity scores (on an increasing scale 0 to 5) followed by the days on which the episodes occurred. For each patient the start of treatment is labelled as day 0, when an episode is in progress, and the last score is 0, on the final day of observation. Unfortunately, the fog of time has closed

TABLE 10.1

Recurrence Episodes for 29 Patients

Case	Variables			Score										Days								
1	2	1	51	3	2	1	0	–	–	–	–	–	–	0	4	8	–	–	–	–	–	–
2	2	2	26	4	2	1	1	0	–	–	–	–	–	0	2	4	6	–	–	–	–	–
3	2	2	34	7	2	2	1	1	1	1	0	–	–	0	2	4	7	10	12	17	–	–
4	1	2	24	5	3	1	0	1	0	–	–	–	–	0	4	6	10	14	–	–	–	–
5	2	2	37	3	2	1	0	–	–	–	–	–	–	0	3	7	–	–	–	–	–	–
6	1	2	40	3	3	3	0	–	–	–	–	–	–	0	3	7	–	–	–	–	–	–
7	2	2	20	4	2	2	1	0	–	–	–	–	–	0	3	6	13	–	–	–	–	–
8	1	2	14	5	2	2	2	1	0	–	–	–	–	0	1	5	7	9	–	–	–	–
9	2	1	26	5	3	2	2	1	0	–	–	–	–	0	4	8	10	18	–	–	–	–
10	2	2	37	9	3	2	2	1	1	1	1	1	0	0	3	7	10	14	17	21	24	28
11	1	2	44	7	3	2	2	1	1	1	0	–	–	0	2	3	9	15	17	21	–	–
12	1	2	21	5	2	2	1	1	0	–	–	–	–	0	1	5	8	12	–	–	–	–
13	2	2	24	4	3	2	1	0	–	–	–	–	–	0	3	6	10	–	–	–	–	–
14	1	1	15	3	3	1	0	–	–	–	–	–	–	0	1	5	–	–	–	–	–	–
15	1	2	60	3	2	1	0	–	–	–	–	–	–	0	2	4	–	–	–	–	–	–
16	2	1	31	6	2	2	2	2	1	0	–	–	–	0	4	6	10	14	16	–	–	–
17	1	1	23	3	3	1	0	–	–	–	–	–	–	0	3	7	–	–	–	–	–	–
18	1	1	21	2	1	0	–	–	–	–	–	–	–	0	3	–	–	–	–	–	–	–
19	1	2	46	7	3	2	2	2	1	1	0	–	–	0	4	6	9	11	13	17	–	–
20	2	1	17	3	2	2	0	–	–	–	–	–	–	0	2	6	–	–	–	–	–	–
21	1	2	13	3	3	2	0	–	–	–	–	–	–	0	3	7	–	–	–	–	–	–
22	1	2	17	3	2	1	0	–	–	–	–	–	–	0	3	5	–	–	–	–	–	–
23	2	1	39	5	2	2	2	1	0	–	–	–	–	0	2	5	8	12	–	–	–	–
24	2	1	16	5	3	2	1	1	0	–	–	–	–	0	4	7	14	18	–	–	–	–
25	2	2	17	7	3	3	1	1	1	1	0	–	–	0	4	7	11	13	17	21	–	–
26	1	1	29	4	2	1	1	0	–	–	–	–	–	0	3	10	14	–	–	–	–	–
27	2	1	44	6	2	2	0	0	1	0	–	–	–	0	2	6	10	13	15	–	–	–
28	1	2	30	5	4	2	2	1	0	–	–	–	–	0	3	7	10	17	–	–	–	–
29	2	2	15	9	2	2	1	1	1	1	1	0	–	0	3	5	7	9	11	14	16	21

in around the details, and we will assume that the last day marks termination unconnected with condition.

Figure 10.1 follows Figure 6.1 of Cook and Lawless (2007). It might be called an *abacus plot*, showing the individual time processes: there is one row per unit (patient) with the occurrence times marked along it. Each unit has first mark at time 0, and last mark at the censoring time. The R-code for reading in the data (which has −9s for the dashed positions in the table) and constructing the plot is as follows:

```
#recevnts data (Sec 10.1)
rv1=read.table('revnts1.dat',header=F);
nc=dim(rv1)[1]; mc=dim(rv1)[2];
rv2=as.numeric(unlist(rv1)); dim(rv2)=c(nc,mc); #rv2;
#abacus plot
jcols=c(15:23); abacus1(rv2[1:nc,jcols],nc,9);
```

The R-function `abacus1` is listed on the CRC Press Web site noted in the Preface.

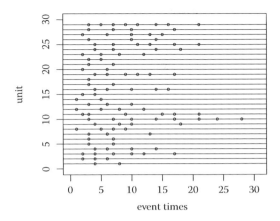

FIGURE 10.1
Abacus plot of individual time processes.

10.2 The Event Rate

Perhaps the first question concerns the rate of occurrence of events as time passes. To put it another way, how do the inter-event gap times progress over time? In the case of the 29 patients data, does treatment successfully make the gap times tend to increase over time? Imagine moving a vertical line steadily across the abacus plot, left to right, passing over the marks as it goes. The question is, will they come thick and fast early on, and then thin out later, indicating increasing gap times? Of course, we must allow for the fact that, as we traverse the plot, some units bow out (are right-censored), and so the marks would naturally thin out even without any change in underlying density.

Let $N_i(t)$ be the number of events recorded for unit i over time interval $(0, t]$; the interval contains the upper end point (square bracket) but not the lower one (round bracket). Also, suppose that unit i is observed only over time interval $[0, b_i]$. Then, the total number of events, across all n units, over a small time interval $(t-dt, t]$ is $\sum_{i=1}^{n} d N_i(t)$, where $d N_i(t) = N_i(t) - N_i(t-dt)$, and the number of units contributing to this sum is $n(t) = \sum_{i=1}^{n} I(b_i \le t)$. Hence, the sample *cumulative mean function* at time t, say $m(t)$, is obtained by summing these elemental contributions, the $n(s)^{-1} \sum_{i=1}^{n} d N_i(s)$, over $0 \le s \le t$. But $d N_i(s)$ is only non-zero at the event times for unit i, so $\sum_{i=1}^{n} d N_i(s)$ is only non-zero at the event times across all the units: call these $s_1, s_2, \ldots$. Then,

$$m(t) = \sum_{S(t)} n(s)^{-1} d N_i(s),$$

where $S(t) = \{s \le t; s \in (s_1, s_2, \ldots)\}$.

In continuous time, when, strictly speaking, no two s_i coincide, $\sum_{i=1}^{n} d N_i(s)$ would be 1 at each $s = s_i$ and 0 everywhere else. So, the mean function would

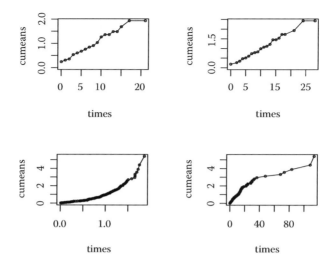

FIGURE 10.2
Cumulative mean plots.

be just a sequence of small blips along the time scale. The cumulative mean function is a smoother operator.

The top row of Figure 10.2 shows cumulative mean plots for the two treatment regimes. The plots look pretty straight to me, suggesting a fairly constant event rate with no strong trend in the inter-event gaps as time goes on. The R-code is as follows:

```
#cumeans (Sec 10.2)
par(mfrow=c(2,2)); trt=rv2[1:nc,2]; ntv=rv2[1:nc,5];
nn=c(0,0); nn[2]=sum(trt-1); nn[1]=nc-nn[2];
for(j in 1:2) { nj=nn[j]; ntvj=subset(ntv,trt==j);
   tx1=subset(rv2[1:nc,jcols],trt==j); dim(tx1=c(nj,9));
   cumean(tx1,nj,ntvj,2,1,1); };
```

To see what the plot should look like when the event rate is increasing or decreasing, an example of each type was simulated, and these are shown in the bottom half of Figure 10.2. The simulated processes do show trends, one curving upward and the other downward: the rate functions in the R-code were concocted to achieve just this.

```
#simulated Poisson processes
nc1=25; mt1=5; tmax=100; lam1=1; lam2=5;
ntv1=rep(mt1,nc1); tw1=matrix(0,nc1,mt1); tw2=tw1;
for(ic in 1:nc1) {
    tw1[ic,1:mt1]=simpoiss(mt1,tmax,hfunc1,lam1,0);
    tw2[ic,1:mt1]=simpoiss(mt1,tmax,hfunc2,lam2,0); };
rtn1=cumean(tw1,nc1,ntv1,1,1,1);
rtn2=cumean(tw2,nc1,ntv1,1,1,1);
```

The functions cumean (cumulative mean) and simpoiss (simulated Poisson process) are listed on the CRC Press Web site noted in the Preface. The method employed by `simpoiss` is described in the next section. Briefly mentioned there also are ways of taking account of the possible influence of explanatory variables. The data of Section 10.1 have covariates sex and age, in addition to treatment regime. In fact, it is even more involved: the events are graded according to severity. In this short account we will not go there—in any case, for such more elaborate analyses one generally needs a lot more data.

10.3 Basic Recurrence Processes

10.3.1 Poisson Processes

Consider a process in continuous time that produces an event every now and then according to some stochastic mechanism. For the *Poisson process* the probability of an event's occurring during small time interval $(t - dt, t]$ is independent of what happened up to time $t - dt$ and is proportional to dt. More formally, the defining properties for a Poisson process of constant *rate* ρ are

$$P\{d\,N(t) = 1 \mid \mathcal{H}(t-)\} = \rho dt + o(dt) \text{ and } P\{d\,N(t) > 1 \mid \mathcal{H}(t-)\} = o(dt),$$

where $\mathcal{H}(t-)$ denotes the history of the process up to but not including time t. Implications include:

1. the past history is forgotten because the probabilities do not depend on $\mathcal{H}(t-)$;
2. the probability of two or more events occurring at the same time is zero;
3. the probability of no event during the small time interval is $1 - \rho dt + o(dt)$.

Also consequent is that the probability of r events occurring over time interval $(0, t]$ is

$$P\{N(t) = r\} = e^{-\rho t}(\rho t)^r / r!,$$

and the distribution of waiting times between events is exponential with mean $1/\rho$: $P(W > w) = e^{-\rho w}$ (Exercises, Section 10.8). Since $E\{N(t)\} = \rho t$, ρ is indeed the average *rate* of occurrence of events.

The process can be generalised in various ways. One such is to allow ρ to vary with time, as $\rho(t)$ say. In this wider context the Poisson process with constant ρ is described as *homogeneous* and that with time-varying $\rho(t)$ is *non-homogeneous* or *time dependent*. The distribution of $N(t)$ is then Poisson with mean $\mu(t) = \int_0^t \rho(s)ds$ (Exercises, Section 10.8). The waiting time to the first event has survivor function $P(W > w) = \exp\{-\mu(w)\}$. However, unlike the homogeneous case, the successive waiting times are not independent. To see this, suppose that you know how long the first wait, w_1, was. This then tells you how far along the $\rho(t)$ scale you are for starting on the second wait, and so how long that wait is likely to be (assuming that you know the $\rho(t)$ function). The information is only useless if $\rho(t)$ is the same at all points along the time scale, that is, the process is homogeneous.

10.3.2 Renewal Processes

This is like a homogeneous Poisson process, with independent and identically distributed inter-event waiting times. However, the waiting-time distribution can be more general, not necessarily exponential. For the moment we consider only continuous W-distributions: in particular, this rules out multiple events occurring at the same time since $P(W = 0) = 0$. Also, we will assume that an event occurs at time 0, so that the first waiting time has the same distribution as the subsequent ones.

The connection between waiting times and cumulative numbers of events is

$$P\{N(t) \geq r\} = P(W_1 + \cdots + W_r \leq t),$$

and then $P\{N(t) = r\}$ can be calculated as $P\{N(t) \geq r\} - P\{N(t) \geq r + 1\}$. In general, the distribution of sums of W_js is pretty intractable. An approach via generating functions can sometimes be useful, the moment generating function of a sum of independent variates being the product of the individual ones. There again, one can always resort to simulation.

10.3.3 Recurrence of Medical Condition

For homogeneous Poisson and renewal processes, the successive inter-event gap times are *iid* (independent and identically distributed). In the top half of Figure 10.3 the average gap lengths (1st gap, 2nd gap...) are plotted; in the bottom half successive gaps are related in a scatter plot. The R-code is

```
#plots of successive gap times (Sec 10.3)
par(mfcol=c(2,2)); ml=8; trt=rv2[1:nc,2]; ntv=rv2[1:nc,5];
nn=c(0,0); nn[2]=sum(trt-1); nn[1]=nc-nn[2]; jcols=c(16:23);
for(j in 1:2) { nj=nn[j]; ntvj=subset(ntv,trt==j);
  tx1=subset(rv2[1:nc,jcols],trt==j); dim(tx1=c(nj,9));
  mtitle='regime 1'; if(j==2) mtitle='regime 2';
  gaps1(tx1,nj,ml,1,mtitle); };
#plots of successive gap times (Sec II.10.3)
par(mfcol=c(2,2)); ml=8; trt=rv2[1:nc,2]; ntv=rv2[1:nc,5];
nn=c(0,0); nn[2]=sum(trt-1); nn[1]=nc-nn[2]; jcols=c(16:23);
for(j in 1:2) { nj=nn[j]; ntvj=subset(ntv,trt==j);
  tx1=subset(rv2[1:nc,jcols],trt==j); dim(tx1=c(nj,9));
  mtitle='regime 1'; if(j==2) mtitle='regime 2';
  gaps1(tx1,nj,ml,1,mtitle); };
```

Function `gaps1` is listed on the CRC Press Web site (see Preface). The gaps do seem to be decreasing over time, but there does not seem to be strong evidence against an *iid* hypothesis here.

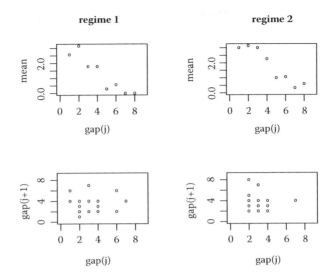

FIGURE 10.3
Successive inter-event times.

10.3.4 Simulation

Simulation can be used in various ways in the present context. One can construct plots under known conditions to compare with what the sample plot looks like. This was done for the *cumean* plots in the previous section. In a similar vein, one can pick out some particular aspect of the situation, simulate many copies to estimate its distribution, and then see where the sample value fits in. If it does not, for example, if the sample value falls way out in the tail of the simulated distribution, then the assumptions underlying the simulation are at odds with the sample.

The method used here is based on the *probability integral transform*: if the random variable W has (continuous, strictly monotone decreasing) survivor function $\bar{G}$, then $\bar{G}(W)$ is uniformly distributed on $(0, 1)$.

For the non-homogeneous Poisson process suppose that the hazard or rate function is $h(t)$, and let $H(t) = \int_0^t h(s)ds$ be the integrated hazard. Then, assuming that the first event is at time 0 and the jth occurs at time t_{j-1}, the waiting time until the $(j+1)$th event is $W_j = t_j - t_{j-1}$ with survivor function

$$\bar{G}(w) = \exp\{-\int_{t_{j-1}}^{t_j} h(s)ds\} = \exp\{H(t_{j-1}) - H(t_j)\}.$$

Let $U = \bar{G}(W_j)$: then $H(t_j) = H(t_{j-1}) - \log U$. So, t_j can be generated by sampling U from the uniform distribution on $(0, 1)$ and then taking $t_j = H^{-1}\{H(t_{j-1}) - \log U\}$. Equivalently, $-\log U$ can be sampled from an exponential distribution of mean 1.

For a renewal process, with inter-event waiting-time survivor function $\bar{G}$, just sample independent uniform U_js and take $W_j = \bar{G}^{-1}(U_j)$ for $j = 1, 2, \ldots$.

10.4 More Elaborate Models

The basic Poisson and renewal process models are liable to a considerable amount of elaboration. The rate functions can be made to depend on covariates, to depend on the previous process history, and to incorporate random effects. Such modelling, together with the associated methodology, is covered in detail by Cook and Lawless (2007). Multiple events can also be accommodated: this will be familiar to Londoners—you wait ages for a bus and then three come along all at once.

Take machine breakdown events as an example. Repairs that restore the unit to good-as-new create a renewal process of breakdowns. Repairs that fix the problem without affecting the underlying ageing (breakdown-causing) process of the unit create a Poisson process, the ageing being reflected in the $\rho(t)$ function. Breakdowns that affect the future performance of the machine, perhaps through partial repair, create a non-Poisson, non-renewal process.

10.4.1 Poisson Process

In the notation of Section 10.2, $dN(t) = 1$ if an event occurs at time t and $dN(t) = 0$ if not. In terms of the indicator function, $dN(t) = I(event\ at\ time\ t)$ and, symbolically,

$$\mathrm{P}\{dN(t) = 1\} = \rho(t)dt \ \text{ or } \ \mathrm{E}\{dN(t)\} = d\mu(t).$$

The equivalence of the two equations is based on the fact that for a binary variate, its expectation is equal to the probability of its taking value 1. The function $\mu(t)$ need not now be differentiable, with derivative $\rho(t)$, or even continuous; it can have jumps in the range 0 to 1.

Covariates can be accommodated: let $x(t)$ denote the covariate history up to time t. It is assumed here that the covariates are *external*; *internal* covariates are trickier to handle (Section 3.6). Applying a *relative risk* formulation,

$$\rho(t \mid \mathbf{x}) = \rho_0(t)\psi(\mathbf{x}; \beta),$$

where, commonly, $\psi(\mathbf{x}; \beta) = \exp(\mathbf{g}^T\beta)$, $\mathbf{g}$ being some vector of functions of $\mathbf{x}(t)$. When the covariates are constant over time, this is a *proportional-hazards* model and $N(t)$ is Poisson-distributed with mean $\mu_0(t)\psi(\mathbf{x}; \beta)$. For time-varying covariates, the mean is $\int_0^t \rho_0(s)\psi(\mathbf{x}; \beta)ds$; $\psi(\mathbf{x}; \beta)$ here depends on s through $\mathbf{x}(s)$ and so cannot be taken outside the integral.

10.4.2 Intensity Functions

The preceding framework can be generalised as follows; Replace $\rho(t)$ by $\lambda(t)$ and $\mu(t)$ by $\Lambda(t)$, respectively; λ and Λ are the standard notations for the *intensity* and *cumulative intensity* functions. These are now allowed to depend on the process history $\mathcal{H}(t-)$ up to time $t-$:

$$P\{d\,N(t) = 1 \mid \mathcal{H}(t-)\} = \lambda(t)dt \ \text{ or } \ E\{d\,N(t) \mid \mathcal{H}(t-)\} = d\,\Lambda(t).$$

The history, $\mathcal{H}(t-)$, can be extended to include covariates. As an added bonus, the second form of the equation, for $E\{d\,N(t) \mid \mathcal{H}(t-)\}$, can accommodate multiple events, where $d\,N(t)$ can be greater than 1.

The setup is given some theoretical underpinning in Part IV. There the intensity functions are slightly augmented to encompass an observation indicator $Y(t)$: $Y(t) = 1$ if the unit is under observation at time t and $Y(t) = 0$ if not. This gives a flexible framework in which $Y(t)$ can switch on and off to reflect intermittent participation of a unit in the recording of data. Most commonly, $Y(t) = 1$ until right-censoring occurs. Mathematical conditions involving $d\,N(t)$ and $Y(t)$ can be written out to define independence of the censoring process. In practice, however, judgements are made in context on whether withdrawal of an unfailed unit from observation had anything to do with its state.

10.5 Other Fields of Application

10.5.1 Repair and Warranty Data

The unit in this context is an appliance or machine of some sort (refrigerator, photocopier, automobile, pogo stick). Such purchases occasionally go wrong and need to be repaired. If you are lucky this can be done at the seller's expense: apparently, so I am told, when you buy a new car you get a warranty lasting something like three years or 30,000 miles for certain types of repairs. Electrical appliances, likewise, usually come with a guarantee. Data on recurring warranty claims and the like is of concern to manufacturers: they wish to know if production quality has deteriorated over a particular period, for example. Lawless et al. (2009), together with the references therein, give examples of modelling claims processes and more general reliability data.

10.5.2 Sports Data

A wealth of sports data are nowadays available to be downloaded from a variety of Web sites. Many people (mostly men) are welded into the armchair in front of the television for weeks on end: "Is there anything you wish to discuss before the football season starts, dear?" There are even sports statisticians, though mainly concerned with simply quoting precedents for the delight/boredom of radio and television audiences. You know the sort of thing: our cricket statistician tells us that the last time 104 was scored in 15 overs and 153 minutes in any test match was in June 1933, when Smith was out LBW (leg before wicket) . . . Football (soccer), the game that Britain gave to the world and would now rather like back, if that is all right with everybody,

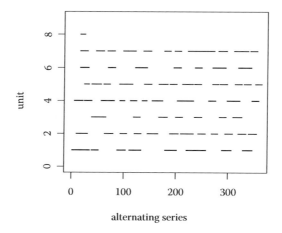

FIGURE 10.4
Alternating processes.

provides a wide range of opportunities for data analysis. In particular, the times to successive goals in a match form a recurrent events process. Dixon and Coles (1997) is one paper among many that apply statistical modelling to such matters of grave concern.

10.5.3 Institution Data

The number of deaths among Prussian cavalry officers from kicks by unruly horses is the classic example for the Poisson distribution. Nowadays, there are certain institutions that have been set up to care for troubled individuals of the human species. Studies with which I have had some involvement concern Young Offender Institutions (disciplinary incidents, Gesch et al., 2002) and psychiatric wards (behavioural incidents, conflict, and self-harm) (Bowers et al., 2009; Stewart et al., 2009). In such places there are seemingly random outbreaks of varying degrees of unrest. Questions to be investigated include whether the frequency of events increases or decreases at particular times or under particular management regimes or under particular treatments, and whether there is serial correlation between outbreaks.

10.5.4 Alternating Periods

Zucchini and MacDonald (2009, Section 16.9) described some data on caterpillar feeding behaviour. Each of a number of caterpillars was monitored over nearly 19 hours, and each minute was classed as *feeding* or *not feeding*. So, the raw data on an individual caterpillar is a long binary chain. Consecutive 1s and 0s can be merged to form sequences of times over which *feeding* and *fasting* are going on. Figure 10.4 shows these for eight caterpillars over the first 360 minutes; caterpillar 8 seems to be on a diet.

The data form a set of processes alternating between two states. Recurrent-events models, such as alternating renewal processes and Markov processes, would be natural contenders here. The approach in Zucchini and MacDonald is via a hidden Markov process (Section 11.1) that determines the developmental stage of the caterpillar. More general situations arise in which there are more than two states and various covariates are recorded.

10.6 Event Counts

It may be that the data record only the number of events over the observation period, not their actual times of occurrence. In that case, given a model for the recurrence times, we have to sum or integrate over these times to obtain the probability distribution of the event counts.

We will assume that a single unit is monitored over the time interval $(0, t)$ and that the occurrence of the non-terminal event does not change the subsequent hazards. This would be the case when the ability of the unit to produce an event depends on its age alone, not on the timing or number of previous events.

10.6.1 Continuous Time: Poisson Process

Cook and Lawless (2007, Section 7.2) give the following analysis: Suppose that events occur in a non-homogeneous Poisson process of rate function $\rho(t)$, and let $\mu(s, t) = \int_s^t \rho(u)du$. The probability of $n = 2$ occurrences, at times t_1 and t_2, over a period $(0, t)$, is then

P{no event in $(0, t_1)$, event at t_1, no event in (t_1, t_2), event at t_2, no event in (t_2, t)}

$$= e^{-\mu(0,t_1)}\rho(t_1)e^{-\mu(t_1,t_2)}\rho(t_2)e^{-\mu(t_2,t)} = e^{-\mu(0,t)}\prod_{j=1}^{n}\rho(t_j).$$

By obvious extension, this last expression applies for n occurrences in general, not just two. To obtain the (marginal) distribution of $N(t)$, the number of events over $(0, t]$, we have to integrate over the t_j:

$$P\{N(t) = n\} = \int_S e^{-\mu(0,t)}\prod_{j=1}^{n}\{\rho(t_j)dt_j\},$$

where the set $S = \{0 < t_1 < \ldots < t_n < t\}$. But $\prod_{j=1}^{n}\{\rho(t_j)dt_j\}$ is symmetric in the t_j so the integral has the same value for all $n!$ orderings of the t_j. Hence,

$$P\{N(t) = n\} = e^{-\mu(0,t)}\int\prod_{j=1}^{n}\{\rho(t_j)dt_j\}/n!,$$

where the integral is now over the whole space $\prod_j \{0 < t_j < t\}$. And so, since $\int_0^t \rho(t_j)\,dt_j = \mu(0, t)$, we obtain

$$P\{N(t) = n\} = e^{-\mu(0,t)}\mu(0, t)^n/n!,$$

showing that $N(t)$ has a Poisson distribution of mean $\mu(0, t)$.

10.6.2 Discrete Time

Suppose that the set of discrete times is $\{0 = \tau_0 < \tau_1 < \tau_2 < \ldots\}$ and that the unit is observed over time period $[\tau_0, \tau_m]$. We will assume that only one event can occur at any one time. Let

$$h_l = P(\text{event at time } \tau_l \mid \text{event history to time } \tau_{l-1})$$

be the discrete hazard, and let $g_l = h_l/(1 - h_l)$ be the corresponding odds ratio. Then, by analogy with the continuous-time case, the probability of $n = 2$ events occurring, at times τ_{j_1} and τ_{j_2}, is

$$\prod_{l=0}^{j_1-1}(1 - h_l) \times h_{j_1} \times \prod_{l=j_1+1}^{j_2-1}(1 - h_l) \times h_{j_2} \times \prod_{l=j_2+1}^{m}(1 - h_l) = \prod_{l=0}^{m}(1 - h_l) \times \prod_{k=1}^{n} g_{j_k}.$$

This last expression holds in general for n taking any value from 0 to m, interpreting $\prod_{k=1}^{0} g_{j_k}$ as 1. To obtain the probability of n occurrences during $[0, \tau_m]$, at unstated times, this expression has to be summed over all $(j_1, \ldots, j_n)$ such that $0 \le j_1 < j_2 < \ldots < j_n \le m$. This is an awkward summation, particularly for larger values of m, but it is essentially the same as the one that occurs when there are tied failure times in partial likelihood (Section 4.3).

Suppose that a random sample of n independently functioning units is observed over time period $[\tau_0, \tau_m]$. Then the mean number of events at time τ_l is nh_l. This may be seen formally as follows: For $0 \le j \le m$ let $e_j = I(\text{event at time } \tau_j)$. Then, the mean number is $nE(e_l)$ and

$$E(e_l) = P(e_l = 1)$$
$$= \sum \{h_0^{e_0}(1 - h_0)^{1-e_0} \ldots h_{l-1}^{e_{l-1}}(1 - h_{l-1})^{1-e_{l-1}}\}$$
$$\times h_l \{h_{l+1}^{e_{l+1}}(1 - h_{l+1})^{1-e_{l+1}} \ldots h_m^{e_m}(1 - h_m)^{1-e_m}\},$$

where the summation is over all binary vectors $(e_0, \ldots, e_{l-1}, e_{l+1}, \ldots, e_m)$. But the right-hand side is equal to

$$\{h_0 + (1 - h_0)\} \times \cdots \times \{h_{l-1} + (1 - h_{l-1})\} \times h_l$$
$$\times \{h_{l+1} + (1 - h_{l+1})\} \times \cdots \times \{h_m + (1 - h_m)\} = h_l.$$

10.6.3 Multivariate Negative Binomial

Many parents will have noticed hyperactivity in children following the intake of fizzy drinks, and so forth. It is no great stretch of the imagination to entertain the idea that what is ingested might affect brain function, though this idea has met with resistance in certain quarters. Gesch et al. (2002) described a study conducted in Aylesbury Young Offenders' Institution. About 200 of the inmates consented to participate in a trial concerning the effect of diet on behaviour. There was a run-in phase, during which the number of incidents of bad behaviour, both minor and major, of each participant was recorded. There followed a trial period during which a randomly selected group took diet supplement pills and the rest took matching placebos. The pills contained nutrients, such as omega-3 fatty acids, that were designed to make up for poor choices from the available menu. The design was double-blind: neither the participants nor those dishing out the pills knew which were which. In short, the statistical analysis showed moderate significance (p=0.027) in support of the hypothesis that some bad behaviour among juveniles can be ascribed to poor diet.

Suppose that a number of independent Poisson processes are associated with a single unit. For example, they could be sequences of disciplinary incidents during consecutive phases by a young offender, or health events experienced by different members of a family during a single phase. Let the Poisson rates be $Z\rho_j$ $(j = 1, \ldots, m)$, where Z is a random effect specific to the unit and $\rho_j = \rho(\mathbf{x}_j)$ is expressed in terms of a set of covariates $\mathbf{x}_j$ attached to the jth component of the unit. For instance, in the prison study described above, $\rho_j = \rho_0 e^{\gamma x_j}$ for phases $j = 1$ and 2, where ρ_0 is the untreated event rate, $x_j = I(j = 2$, active pill) is a binary treatment indicator, and e^{γ} represents a treatment factor. The probability of a set $(r_1, \ldots, r_m)$ of event counts for a single unit is then

$$p(r_1, \ldots, r_m; \mathbf{x}_1, \ldots, \mathbf{x}_m) = \int \prod_{j=1}^{m} \{e^{-z\rho_j} (z\rho_j)^{r_j} / r_j!\} k(z) dz,$$

in which $k(z)$ is the density function of Z, describing its distribution over units. The integral can be explicitly evaluated if a gamma distribution is assumed for Z, and let's not be hoity-toity about such a convenience. So, taking Z to have density $\Gamma(\zeta)^{-1} z^{\zeta-1} e^{-z}$, with shape parameter ζ and scale parameter 1 (the scaling being absorbed into the ρ_j),

$$p(r_1, \ldots, r_m; \mathbf{x}_1, \ldots, \mathbf{x}_m) = \prod_{j=1}^{m} (\rho_j^{r_j} / r_j!) (1 + \rho_+)^{-(\zeta+r_+)} \Gamma(\zeta)^{-1} \Gamma(\zeta + r_+).$$

This is a *multivariate negative binomial distribution* for which further details are given in the Exercises (Section 10.8). Such a model was used in Gesch et al. (2002, though the paper, written in non-mathematical style, does not make this clear), and appears in Cowling et al. (2006) and Cook and Lawless (2007, Section 2.2.3).

10.7 Quasi-Life Tables

A quasi-life table was identified by Baxter (1994) as an array of event counts arising from a number of independent discrete-time renewal processes running concurrently. The counts record only the total numbers of renewals at each time point, not the numbers for the individual processes. For context consider an engineering unit containing a replaceable component. Upon failure the component is replaced and the new one begins operating at the following time point. The renewal processes then result from several units operating independently in tandem, and the event counts are the total numbers of component failures at each time aggregated over the units.

Suppose that a manufacturer or retailer sells a number of units each month to various customers. Say that b_r units are supplied, and begin operating, at time (month) r ($r = 0, 1, 2, \ldots$); b_r is the number of *births* at time r. For an individual unit the component failure record can be represented as a binary sequence, with 1 for a component failure and 0 otherwise. We can then form a binary matrix with one such row per unit. The record for the whole cohort of b_r units up to time t is thus summarised as a $b_r \times (t - r)$ matrix of 0s and 1s. With full information, that is, the whole binary matrix, statistical analyses can be performed without too much trouble. But the situation considered here is when only the column totals of the matrix are available. Let d_{rs} be the sth column total of the matrix, that is, the total number of failures (*deaths*) among the b_r units at time s. For the special case of Poisson processes of renewals statistical analyses were presented by Suzuki et al. (2000) and Karim et al. (2001).

In practice, the unit processes might not be concurrent. Often, a manufacturer will simply record numbers of replacement components supplied each month without identifying the units under repair; even if more detailed records are kept they will often be confidential and not made fully available to the statistician. Baxter (1994) and Tortorella (1996) considered the case where only the totals, $d_{+s} = \sum_{r=0}^{s-1} d_{rs}$, are available for analysis, not the individual d_{rs}. So, not only do we not know which units gave rise to the failures, but we do not know from which cohort of units the failures arose. The likelihood function here is somewhat intractable, and moment-based methods were used in the two cited papers.

10.7.1 Estimation

Suppose that the component lifetime until failure has discrete distribution specified by P(*lifetime* $= j$) $= q_j$ ($j = 1, 2, \ldots$). Denote by a_u the probability of a component failure, not necessarily the first, in a given unit of age u ($u = 1, 2, \ldots$). Homogeneity across cohorts in the a_u is assumed. By the discrete-time renewal equation (Feller, 1962, Section XIII.4):

$$a_u = q_u + \sum_{j=0}^{u-1} q_j a_{u-j} \quad (u = 1, 2, \ldots),$$

in which $q_0 = 0$. The a_u can be estimated from the sample ratios $d_{r,r+u}/b_r$, pooling them over the different cohorts. The preceding equation can then be inverted to yield estimates of the q_u. Baxter (1994) fitted a parametric continuous-time survivor distribution, with survivor function $\bar{F}(t;\theta)$: he matched the q-estimates to the $\bar{F}$-increments by weighted least squares. A more formal analysis of such moment-based estimators was made by Crowder and Stephens (2003), including a derivation of their asymptotic distribution. Also, maximum quasi-likelihood estimators were proposed, and some data sets from Suzuki et al. (2000), Karim et al. (2001), and Tortorella (1996) were re-analysed. A brief summary is now given—for details see the 2003 paper.

The renewal equations can be written succinctly in matrix form as follows. Let $a^{(m)} = (a_1, \ldots, a_m)^{\mathrm{T}}$, $q^{(m)} = (q_1, \ldots, q_m)^{\mathrm{T}}$, and define the $m \times m$ matrices

$$
A_m = \begin{pmatrix}
1 & 0 & 0 & \ldots & 0 & 0 \\
a_1 & 1 & 0 & \ldots & 0 & 0 \\
a_2 & a_1 & 1 & \ldots & 0 & 0 \\
. & . & . & . & . & . \\
a_{m-1} & a_{m-2} & a_{m-3} & \ldots & a_1 & 1
\end{pmatrix}, \quad
Q_m = \begin{pmatrix}
1 & 0 & 0 & \ldots & 0 & 0 \\
-q_1 & 1 & 0 & \ldots & 0 & 0 \\
-q_2 & -q_1 & 1 & \ldots & 0 & 0 \\
. & . & . & . & . & . \\
-q_{m-1} & -q_{m-2} & -q_{m-3} & \ldots & -q_1 & 1
\end{pmatrix}.
$$

Then the equations are, equivalently, $a^{(m)} = A_m q^{(m)}$ or $q^{(m)} = Q_m a^{(m)}$. The matrices A_m and Q_m each have determinant 1, so they are non-singular; moreover, $A_m Q_m = I$.

Consider first the record up to time t from the r cohort alone. Among the b_r units installed at time r there are d_{rs} failures at time s ($s = r, r+1, \ldots, t$). So, based on the binomial distribution of d_{rs}, the estimator $\tilde{a}_m = d_{r,r+m}/b_r$ has mean a_m and variance $a_m(1 - a_m)/b_r$ for $m = 1, \ldots, t - r$. The corresponding estimator for $a^{(m)}$ is $\tilde{a}^{(m)} = (\tilde{a}_1, \ldots, \tilde{a}_m)$: $\tilde{a}^{(m)}$ has mean $a^{(m)}$ and covariance matrix $b_r^{-1} C_m$, where the symmetric $m \times m$ matrix C_m has elements $C_{jj} = a_j(1 - a_j)$ and $C_{jk} = a_j(a_{k-j} - a_k)$ for $j < k$. The estimator for $q^{(m)}$ is $\tilde{q}^{(m)} = \tilde{A}_m^{-1} \tilde{a}^{(m)}$.

Now suppose that we have the records from cohorts 1 to t. Cohorts 1 to $t - m$ each give an estimate for a_m as described above, and we can form a weighted average of these: the best linear unbiased estimate has weights proportional to the b_r. As for the single-cohort case, we can calculate the mean and covariance matrix of the resulting estimator of $a^{(t)}$ and derive the corresponding estimator of $q^{(t)}$.

Suppose now that only the totals $d_{+s} = \sum_{r=0}^{s-1} d_{rs}$ are available, the case considered by Baxter (1994). We have

$$
\mathrm{E}(d_{+s}) = \sum_{r=0}^{s-1} b_r a_{s-r} \quad \text{and} \quad \mathrm{var}(d_{+s}) = \sum_{r=0}^{s-1} b_r a_{s-r}(1 - a_{s-r}),
$$

and for $s < w$, one can derive

$$
\mathrm{cov}(d_{+s}, d_{+w}) = \sum_{r=0}^{s-1} \mathrm{cov}(d_{rs}, d_{rw}) = \sum_{r=0}^{s-1} b_r a_{s-r}(a_{w-s} - a_{w-r}) = \sum_{r=0}^{s-1} b_r C_{s-r,w-r}.
$$

In much the same way as described above, simple moment estimators for the a_j may be written down and their properties investigated. Then, similarly, the corresponding estimator for $q^{(m)}$, together with its properties, follows.

Finally (the word to put a smile on even the most conscientious seminar audient), we consider the case where the distribution q is specified by some parametric family, for example, as a geometric waiting-time distribution with $q_l = (1 - \theta)^{l-1}\theta$ in terms of parameter $\theta \in (0, 1)$. Such an approach can be useful when the b_r are not large, in which case non-parametric estimators and their nominal asymptotic properties might not be very reliable. Moreover, it is straightforward to include covariates in parametric models, for example, to accommodate non-homogeneity across cohorts. Provided that the means and covariances of the d-counts are readily available, one can apply *maximum quasi-likelihood estimation*. Among estimating equations linear in the observations, this is asymptotically optimal (for example, Crowder, 1987).

In a follow-up paper Stephens and Crowder (2004) took the Bayesian approach based on data comprising the counts d_{rs} or just the marginal totals d_{+s}. (The first author came up with some rather nifty McMC techniques for the analyses.) A major issue here was prediction, where, for example, a manufacturer wishes to assess his current exposure to future warranty claims.

10.8 Exercises

1. For a Poisson process of rate ρ let $p_r(t) = P\{N(0, t) = r\}$. Derive

$$p_0(t + dt) = p_0(t)(1 - \rho dt) + o(dt)$$

   and

$$p_r(t + dt) = p_r(t)(1 - \rho dt) + p_{r-1}(t)\rho dt + o(dt) \text{ for } r \geq 1.$$

   Verify that, in the limit $dt \to 0$, these equations yield

$$p_0'(t) = -\rho p_0(t) \text{ and } p_n'(t) = -\rho p_n(t) + \rho p_{n-1}(t).$$

   Last, verify that $p_r(t) = e^{-\rho t}(\rho t)^r / r!$ $(r = 0, 1, \ldots)$ solves these differential-difference equations.

2. Show that, for a Poisson process of rate ρ the inter-event waiting-time distribution is exponential with mean $1/\rho$.

3. Convince yourself that, in the non-homogeneous case, the successive inter-event waiting times are dependent.

4. Consider a Poisson process with rate function $\rho(t)$ and let $\mu(t) = \int_0^t \rho(s)ds$. Show that the time-transformed process $N^*(t) = N\{\mu^{-1}(t)\}$, where μ^{-1} is the inverse function, is homogeneous with rate 1.

5. Work through the details of the preceding problem for cases (a) $\rho(t) = \beta t$, and (b) $\rho(t) = \exp(\alpha + \beta t)$.

6. *Multivariate negative binomial distribution.* Confirm that the r_j have (univariate) negative binomial distributions. Derive $E(r_j) = \zeta\rho_j$, $var(r_j) = \zeta\rho_j(1 + \rho_j)$ and $cov(r_j, r_k) = \zeta\rho_j\rho_k$.

10.9 Hints and Solutions

2. $P(W > w) = P\{N(0, w) = 0\} = p_0(w) = e^{-\rho w}$.

3. $P(W_j > w \mid t_{j-1}) = \exp\{-\int_{t_{j-1}}^{t_{j-1}+w} \rho(s)ds\}$, which depends on t_{j-1}.

4. Let $t^* = \mu^{-1}(t)$. Then, $P\{N^*(t) = r\} = P\{N(t^*) = r\} = e^{-\mu(t^*)}\{\mu(t^*)\}^r / r! = e^{-t}t^r/r!$.

6. *Multivariate negative binomial distribution.* The marginal density of r_j can be calculated as

$$p(r_j) = \int \{e^{-z\rho_j}(z\rho_j)^{r_j}/r_j!\}k(z)dz,$$

which yields the said result. And, noting that $E(Z) = var(Z) = \zeta$,

$$E(r_j) = E\{E(r_j \mid Z)\} = E(Z\rho_j) = \zeta\rho_j,$$
$$var(r_j) = E\{var(r_j \mid Z)\} + var\{E(r_j \mid Z)\} = E(Z\rho_j) + var(Z\rho_j)$$
$$= \zeta\rho_j + \zeta\rho_j^2,$$
$$cov(r_j, r_k) = E\{cov(r_j, r_k \mid Z)\} + cov\{E(r_j \mid Z), E(r_k \mid Z)\}$$
$$= E(0) + cov(Z\rho_j, Z\rho_k) = \zeta(1 + \zeta)\rho_j\rho_k.$$

11

Multi-State Processes

We consider the situation where an individual unit can assume different states at different times. The times at which various states are reached is often the focus, which gives rise to data of the *milestones* variety (Section 19.5). Progressive, or monotone, movement through a sequence of states is the case with most degradation models—see below. In other cases units may move to and fro between states: an example is where a patient's health status varies over time, hopefully improving long-term but suffering periodic setbacks. Such times are often interval censored, for example, when state changes occur between visits to the dentist and a selfless reluctance to trouble her is paramount.

For reference there is a surfeit of books on stochastic processes. Some standard tomes are Cox and Miller (1965), Karlin and Taylor (1975), and Ross (1996). Hougaard (1999) gives a nice review of multi-state models.

11.1 Markov Chain Models

Let S_t denote the state of the unit at time t and assume a finite state-space, with states labelled as $s_0, s_1, \ldots, s_m$. State s_0 is an initial state and state s_m is a terminal, absorbing state. So, assuming that the unit is monitored from *birth*, it starts in state s_0 and then, after a journey visiting none, some or all of the intermediate states, comes to its final resting place in state s_m. In symbols, $S_0 = s_0$ and $S_t = s_m$ for $t \geq T$, T being the time to first reach s_m, that is, the lifetime. In many cases, however, monitoring of a unit begins some way along the line, so that the initial state as recorded in the data is one other than s_0. The situation where there is more than one absorbing state, representing different types of termination, is one of competing risks—see Part III.

11.1.1 Discrete Time

Suppose that S_t can only move at a sequence of specified times, which are conveniently labelled as 1, 2, 3, ... These times might be generated by duty cycles, regular inspections, or scheduled visits to the clinic, and so forth, but are not associated with the progress of S_t. The process is governed by transition probability matrices, $P(t)$, for moving between states at discrete time t:

$$\{P(t)\}_{jk} = P(S_{t+1} = s_k \mid S_t = s_j).$$

In the case of monotone, progressive movement through the states $P(t)$ has non-zero elements only on the diagonal and super-diagonal.

In the time-homogeneous case $P(t) = P$, independent of t. The matrix $P = \{p_{jk}\}$ can be partitioned as

$$P = \begin{pmatrix} R & (I_m - R)1_m \\ 0_m^T & 1 \end{pmatrix},$$

where 1_m is an $m \times 1$ vector of ones, 0_m is an $m \times 1$ vector of zeros, I_m is the $m \times m$ unit matrix, and R is the $m \times m$ matrix of probabilities of transitions among the set of transient states, $\mathcal{R} = (s_0, \ldots, s_{m-1})$; the last row of P corresponds to the absorbing state s_m. The distribution of T_j, the time to absorption from a start in state s_j, can be derived as

$$P(T_j > t) = \sum_{k \in \mathcal{R}} \{R^t\}_{jk} = \{R^t 1_m\}_j, \quad P(T_j = t) = \{(R^{t-1} - R^t)1_m\}_j.$$

Consequent formulae for the mean and variance of T_j are derived in the (Section 11.6, Exercises). The time spent in current state s_j (the *sojourn time*) before moving on has a geometric distribution:

$$P(sojourn\ time > t) = (1 - p_{jj})^t \text{ for } t = 1, 2, \ldots.$$

So, one can predict how long the current state will last, though not with any great precision in view of the relatively large variability of the geometric distribution.

In the non-homogeneous case the survivor function for T_j is $P(T_j > t) = \{\prod_{s=1}^t R(s)1_m\}_j$, obtained by replacing R^t by $\prod_{s=1}^t R(s)$. For the sojourn time in current state j, at current time t_0, replace $(1 - p_{jj})^t$ with $\prod_{u=t_0}^{t_0+t-1}\{1 - p_{jj}(u)\}$.

11.1.2 Continuous Time

We consider the time-homogeneous case here; the inhomogeneous case is rather more tricky. Let $P_{jk}(t) = P(S_{u+t} = s_k \mid S_u = s_j)$ be independent of u (time homogeneity), and make the standard assumptions

$$P_{jk}(dt) = q_{jk}dt + o(dt) \text{ for } j \neq k, \quad \text{and} \quad P_{jj}(dt) = 1 - q_{jj}dt + o(dt).$$

The process is thus governed by *transition intensities* q_{jk} for movement from state s_j to s_k. Starting from state s_j, the process must be in one of the states at a time dt later, so $(1 - q_{jj}dt) + \sum_{k \neq j} q_{jk}dt = 1$, which entails

$$q_{jj} = \sum_{k \neq j} q_{jk};$$

also, $q_{mk} = 0$ for $k \neq m$, since s_m is an absorbing state. Let $p_k(t) = P(state\ s_k\ at\ time\ t)$ with $p_0(0) = 1$ and $\sum_{k=0}^{m} p_k(t) = 1$. Then,

$$p_k(t + dt) = p_k(t)(1 - q_{kk}dt) + \sum_{j \neq k} p_j(t)q_{jk}dt + o(dt),$$

leading to

$$p_k'(t) = -p_k(t)q_{kk} + \sum_{j \neq k} p_j(t)q_{jk}.$$

Let Q be the $m \times m$ matrix with diagonal elements $-q_{jj}$ and off-diagonal elements q_{jk}; Q has row-sums zero and all zeros along its mth row. Then the differential equations can be written in matrix terms as

$$p'(t) = p(t)Q,$$

where $p(t)$ is $1 \times m$ with kth component $p_k(t)$. In similar fashion, conditioning on state s_0 at time 0, we can derive

$$P'(t) = P(t)Q \text{ with } P(0) = I,$$

where I denotes the unit matrix. The solutions can be written symbolically as $p(t) = p(0)e^{tQ}$ and $P(t) = e^{tQ}$, where $e^{tQ} = \sum_{n=0}^{\infty} t^n Q^n / n!$ is the *matrix exponential* function; direct computation of this is not advised (Moler and Van Loan, 2003). The sojourn time in state s_j has an exponential distribution of rate $1 - q_{jj}$, the latter acting as a constant hazard function for *failure*. Aalen (1995) focussed upon the so-called *phase type* distribution of the time to absorption in state s_m.

11.1.3 Hidden Markov Chains

Suppose that $(h_1, h_2, \ldots)$ is a hidden, unobserved sequence of states developing as a time-homogeneous Markov chain with transition probability matrix $P = \{p_{jk}\}$. Suppose also that $(g_1, g_2, \ldots)$ is a sequence of observations whose joint probability distribution is governed by the h_t according to

$$p(g_t \mid g_{t-1}, g_{t-2}, \ldots, h_t, h_{t-1}, h_{t-2}, \ldots) = p(g_t \mid h_t).$$

The g_t may be multivariate of dimension varying with t or even with different probability structures over time, though that would be unusual. The model decrees that, conditionally on the current, latent h_t, the observed outcome g_t ignores all other, historical evidence (just like politicians). The diagram below illustrates the dependence structure for this model, which is an example of a *state-space model*.

$$g_{t-1} \qquad g_t \qquad g_{t+1}$$
$$\uparrow \qquad \uparrow \qquad \uparrow$$
$$\rightarrow h_{t-1} \rightarrow h_t \rightarrow h_{t+1} \rightarrow$$

Diagram for hidden Markov chain process.

The h_t represent the latent process that drives the g_t marker process. The idea is that the observed g_t are sufficiently informative about the unobserved h_t to make decisions about the underlying state of the unit. In the medical context, h_t is the underlying *healthiness* of the patient and g_t would include the reponses to the doctor's questioning. Book-length treatments are given by MacDonald and Zucchini (1997), Zucchini and MacDonald (2009), and Cappe et al. (2005).

An application to defaults within a bond portfolio was made in Giampieri et al. (2005). A hidden Markov chain was used in which the latent variable is the risk state common to all units, such as the current state of the economy, and the number of defaults among the surviving units at any time is binomially distributed. The proposal was tested by simulation and then applied to some real data abstracted from a large data base supplied by Standard & Poor's credit rating agency.

11.1.4 Estimation

There is a wide variety of observational settings. Three main types of multivariate data can be identified among a host of lesser variations. For the first, the system is monitored at given times $t_1, t_2, \ldots$, yielding observations $(S_1, S_2, \ldots)$. In health applications, for example, the observations give feedback on the effects of a treatment. In reliability the reason for collecting the data is typically so that the process can be curtailed before failure occurs, in which case the failure time T is right-censored. The number of observations made will often be random, governed by the approach of S_t to an identified failure level s^*. In other cases, a maximum number of inspections is fixed in advance before withdrawal of the unit. Notwithstanding, failure might occur earlier than expected and then one gets a nice surprise, actually observing T.

The second main type of data is of the *milestones* variety. That is, deterioration levels $(s_1, s_2, \ldots)$ are prespecified and the times $(T_1, T_2, \ldots)$ for the unit to reach these are observed. If the unit is not continuously monitored, some s_j will be missed and then the corresponding T_j will be interval censored.

The third main type of data is when there isn't any data or, to be more precise, when there are no direct observations on S_t. It is quite commonly the case that the actual S_t process is latent, bubbling away underneath but giving rise to so-called *marker processes* that can be observed; hidden Markov processes are like this. A case in point is the healthiness of a patient, which cannot be observed directly but only inferred from a range of diagnostic tests.

Consider first the homogeneous Markov chain in discrete time, and let us just take the simple case in which the times of transitions, and the states occupied, are recorded up to absorption. Suppose that an individual unit starts in state s_0 at time 0, ends in state s_m at time t_l, and that the intermediate transition times and states occupied are $(t_1, \ldots, t_{l-1})$ and $(a_1, \ldots, a_{l-1})$. Then the likelihood function can be derived as

$$p_{s_0 a_1}^{(t_1)} \, p_{a_1 a_2}^{(t_2 - t_1)} \ldots p_{a_{l-1} 1 s_m}^{(t_l - t_{l-1})},$$

where

$$p_{jk}^{(t)} = \mathrm{P}(state\ s_k\ at\ time\ t \mid state\ s_j\ at\ time\ 0) = \{P^t\}_{jk}.$$

Given a sufficient number of transitions among a random sample of units, the elements of P can be estimated from the transition counts; these are in fact, the non-parametric *mle*. If P is parametrised in accordance with some model parametric *mle* can be performed.

The analysis for the homogeneous Markov chain in continuous time, with the same observation regime, is similar: the likelihood is now

$$P_{s_0 a_1}(t_1) \, P_{a_1 a_2}(t_2 - t_1) \ldots P_{a_{l-1} s_m}(t_l - t_{l-1}).$$

More detailed analysis is given in Kalbfleisch and Prentice (2002, Section 8.3), Lawless (2003, Section 11.4), and Cook and Lawless (2007, Section 15.3).

11.2 The Wiener Process

Diffusion processes have continuous sample paths, with continuous state space and continuous time. There are many standard texts on stochastic processes that deal with the mathematics, which is more involved than most of what appears in this book. We will just describe what is probably the most commonly applied of such models, the *Wiener process*, also known as *Brownian motion*, which has the great advantage of tractability. It can take negative values but, if necessary, that can be addressed by working in terms of log S_t rather than S_t itself. Further, its sample paths are not monotone, which can be a bit of a drawback for many applications. However, some real processes can reverse the trend now and again: this is often the case with health and illness.

A standard Wiener process W_t can be defined by the following two properties:

1. it has independent increments, that is, whenever $t_1 < t_2 < t_3 < t_4$, $W_{t_4 - t_3}$ is independent of $W_{t_2 - t_1}$;
2. $W_t - W_s$ has distribution $\mathrm{N}(0, t - s)$ for each $0 \le s < t$.

The definition implies that, as well as being independent, the increments are stationary, the distribution of $W_t - W_s$ depending only on $t - s$.

The distribution of W_t at any given time t is normal; further, the joint distribution of successive observations on the process is multivariate normal. Let T_s denote the time at which the process first reaches level s. Then the distribution of T_s is easily derived as *inverse Gaussian*, which has a simple, explicit expression for its survivor function in terms of Φ, the standard normal distribution function. Hence, likelihood functions can be derived for single observations on either S_t or T_s or a mixture of both. These and other properties of the Wiener process are given below as Exercises (Section 11.6).

A more useful version for many situations is the Wiener process with drift. This is defined as $S_t = \mu_t + \sigma_t W_t$, where μ_t and σ_t are parametrised functions governing the mean and variance of S_t; a particular choice is $\sigma_t = \sigma t^{-1/2}$, which makes $\text{var}(S_t) = \sigma^2$ constant, independent of t.

The process $S_t = \mu_t + \sigma W_t$, with σ independent of time, has independent increments. So, if inspections are scheduled at times $t_1 < t_2 < \ldots$, the increases $\nabla S_j = S_{t_j} - S_{t_{j-1}}$ are independently normally distributed with means $\mu_{t_j} - \mu_{t_{j-1}}$ and variances $\sigma^2(t_j - t_{j-1})$. Hence, the joint distribution for multivariate data, in the form of the set of ∇S_j, is readily accessible. For the case where the variance parameter depends on time, see the Exercises.

For data of the milestones type let $s_1 < s_2 < \ldots$ be the specified levels and let T_j be the first time at which S_t hits level s_j. In general, the intervals $\nabla T_j = T_j - T_{j-1}$ are not independent but, in the time-homogeneous case, $S_t = \mu + \sigma W_t$, the ∇T_j are independent with inverse Gaussian distribution.

Whitmore (1995) considered estimation in the presence of measurement error.

11.3 Wear and Tear and Lack of Care

It has been said that most failures can be traced to an underlying degradation process. I ponder this gloomily as I sit in the dentist's chair. The serious point is that if one can monitor degradation in a system, it might be possible to predict the onset of failure and so take appropriate action; for example, Crowder and Lawless (2007) examined models for predictive maintenance aimed at cost reduction. By *wear* is meant long-term, legitimate usage that gradually makes the unit deteriorate; by *tear* is meant untoward damage; by *lack of care* is meant degradation that develops without any physically applied wear and tear, for example, the accumulation of rust on the neglected paintwork of the car (*mea culpa*).

Cox (1999) has some remarks on the general area; Singpurwalla (1995) set out a framework for a wide class of models; Park and Padgett (2005) proposed models based on geometric Brownian motion and a gamma process for accelerated degradation.

Let S_t be the degradation measure of the system (engineering unit, human subject, and so forth) at time t. For convenience, S_t is often scored so that, as a stochastic process, it has positive drift, often monotone, and with $S_0=0$. When S_t first reaches or exceeds a specified critical level, $s^* > 0$, failure occurs, and this at time T, say. Failure can be hard, when the engine blows up, for example, or soft, when it starts grumbling and a funereal-faced mechanic sadly shakes his head as he starts to write out the bill. There are variations on this particular definition of failure time, and some will be seen in later sections.

In continuous time, the unit might be subject to regular scheduled inspections, or irregular inspections at times not associated with the progress of $\{S_t\}$. In reliability applications a machine might only be monitored when maintenance is performed, or at random times when you remember to check the oil level in the car; in health, the patient's status is only observed at visits to the clinic. In such situations, the recorded waiting times between states are usually interval censored.

Many models have been proposed for degradation processes. Often they are used only to obtain the distribution of the time to ultimate failure of the system. The study of such *first-passage times* or *hitting times*, of a stochastic process reaching a given boundary, forms a sub-topic in univariate rather than multivariate survival analysis. Even for fairly simple processes the extraction of the first-passage distribution can be difficult: see, for example, Durbin (1985) and Cuzick (1981). The statistical literature is dominated by a few tractable cases, some of which will be discussed here. Lee and Whitmore (2006) discussed some models and their statistical application; Balka et al. (2009) gave a nice review.

Example

Consider the wear of a car tire: let S_t be the tread depth at time t. By law, a tire must be replaced when the tread depth is less than some prescribed level. *Time* here would usually be taken to be mileage (continuous). On the other hand, *time* could be taken as the number of revolutions of the wheel (discrete) on the grounds that each point of the tire only experiences wear when it is on the grounds (discounting the odd kick to test the inflation pressure). The loss of tread depth S_t will normally be monitored on a number of occasions, including the once-a-year MOT (Ministry of Transport) inspection and random police checks on the hapless motorist.

11.4 Cumulative Models

The general form is $S_t = S_0 + (d_1 + \cdots + d_{N_t})$, where S_0 is the initial value, d_j is the jth increment, and N_t is the number of increments over the time interval $(0, t]$. Without loss of generality (as they say when losing generality) we will take $S_0 = 0$: in effect, we are assuming that S_0 is known and subtracted from S_t. In any case, even if S_0 is unknown, what follows applies to $S_t - S_0$ rather than to S_t itself. Cumulative damage models have been applied widely in Reliability, but they are clearly relevant to health applications too. The lifetime T is defined as the first passage time of S_t to critical level s^*.

The usual assumption is that the d_j are *iid*, often taking only positive values but not necessarily so, and a Poisson process for N_t is a basic choice. The model can be extended to the case of correlated increments d_j, though the analysis is then more difficult: some progress in obtaining moments is possible in some cases. Cumulative models are relatively tractable but limited in their applicability to processes S_t that have discontinuous sample paths (jumps up and down), though often providing an adequate approximation to the real situation.

If the *mgf* (moment generating function) of d_j is $M_j(z)$, and if the d_j and N_t are all independent, the *mgf* of S_t is given by

$$E(e^{-zS_t}) = E\left\{\exp\left(-z\sum_{j=1}^{N_t} d_j\right)\right\} = E\left[E\left\{\exp\left(-z\sum_{j=1}^{N_t} d_j\right) \mid N_t\right\}\right]$$

$$= E\left\{\prod_{j=1}^{N_t} M_j(z)\right\}.$$

If the d_j are *iid*, $M_j(z) = M(z)$ for all j and, denoting the *pgf* (probability generating function) of N_t by $P_t(z)$, S_t has *mgf*

$$K_t(z) = \sum_{n=0}^{\infty} M(z)^n P(N_t = n) = P_t\{M(z)\}.$$

From this the mean and variance of S_t are easily derived: such moments can be used to construct estimating functions when likelihoods are not tractable.

11.4.1 Compound Poisson Process

Suppose that N_t is Poisson distributed with mean μ_t, an increasing function of time t: $P(N_t = n) = e^{-\mu_t}\mu_t^n/n!$ for $n = 0, 1, 2, \ldots$. This defines a non-homogeneous Poisson process of increments; the special case $\mu_t = \rho t$ corresponds to a homogeneous process of rate ρ. The *pgf* of N_t is $P_t(z) = \exp(-\mu_t + \mu_t z)$ and then the *mgf* of S_t is

$$K_t(z) = \exp\{-\mu_t + \mu_t M(z)\},$$

where $M(z)$ is the *mgf* of the *iid* d_j.

Suppose now that the d_j are independent gamma variates with common density

$$g(x; \xi, \nu) = \Gamma(\nu)^{-1}\xi^{-\nu}x^{\nu-1}e^{-x/\xi};$$

then $E(d_j) = \nu\xi$, $\text{var}(d_j) = \nu\xi^2$. The *mgf* of the d_j is $M(z) = (1 + \xi z)^{-\nu}$, and then the *mgf* of S_t is given as

$$K_t(z) = P_t\{M(z)\} = \exp\{-\mu_t + \mu_t(1 + \xi z)^{-\nu}\}.$$

The corresponding density of S_t can be found using an inversion formula, involving numerical integration in this case. Alternatively, arguing more directly, the distribution of S_t is mixed, having probability mass $e^{-\mu_t} = P(N_t = 0)$ at $S_t = 0$ together with density

$$\sum_{n=1}^{\infty} g(s; \xi, n\nu)(e^{-\mu_t} \mu_t^n/n!)$$

on $0 < s < \infty$; $g(s; \xi, n\nu)$ here is the gamma$(\xi, n\nu)$ density of $\sum_{j=1}^{n} d_j$.

This is all rather unwieldy but the special case of exponentially distributed increments, for which $\nu = 1$ gives more amenable expressions. In the particular context of crack growth, where the d_i are successive increases in the crack length, this assumption has been awarded some experimental justification by Sobczyk and Spencer (1992, Section 23.1.1). In that case S_t has density

$$k_t(s) = \sum_{n=1}^{\infty}\{\Gamma(n)^{-1}\xi^{-n}s^{n-1}e^{-s/\xi}\}(e^{-\mu_t}\mu_t^n/n!)$$

$$= e^{-\mu_t - s/\xi}\,(\xi s/\mu_t)^{-1/2}\,I_1(2\sqrt{\mu_t s/\xi}),$$

where $I_k(.)$ denotes the modified Bessel function of order k (Olver et al., 2010, Section 10.25.2). The survivor function of T follows as

$$P(T > t) = P(S_t < s^*) = \int_0^{s^*} k_t(s)ds,$$

and the density is obtained by differentiating with respect to t.

In the case of a homogeneous process of hits, when $\mu_t = \rho t$, a more tractable expression can be obtained as follows. We have

$$P\left(\sum_{j=1}^{n} d_j < s^*\right) = \int_0^{s^*} g(s; \xi, n\nu)ds = \Gamma(n)^{-1} \int_0^{s^*/\xi} u^{n-1}e^{-u}du$$

$$= 1 - e^{-s^*/\xi}e_{n-1}(s^*/\xi),$$

where $e_n(x) = \sum_{k=0}^{n} x^k/k!$. Hence,

$$P(T > t) = e^{-\rho t} + \sum_{n=1}^{\infty}\{1 - e^{-s^*/\xi}e_{n-1}(s^*/\xi)\}\,e^{-\rho t}(\rho t)^n/n!$$

and the density of T follows by differentiation as $\rho e^{-\rho t - s^*/\xi}\,I_0(2\sqrt{\rho t s^*/\xi})$ (see the Exercises, Section 11.6).

11.4.2 Compound Birth Process

Sobczyk and Spencer (1992, Section 5.6) considered a model for crack growth in which there is an accelerating frequency of increments as the crack grows. Specifically, the waiting times between increments are independent, the nth being distributed as exponential with mean $(n\rho)^{-1}$; denoting the increment occurrence times by $T_1, T_2, \ldots$, the nth waiting time is $T_n - T_{n-1}$. This is a linear birth process, also known as the *Yule-Furry process*, for which

$$P(N_t = n) = e^{-\rho t}(1 - e^{-\rho t})^n;$$

thus, N_t has a geometric distribution with mean $e^{-\rho t} - 1$ (see the Exercises in Section 11.6).

Suppose that the d_j have *mgf* $M(z)$. Then S_t has *mgf*

$$K_t(z) = \sum_{n=0}^{\infty} M(z)^n e^{-\rho t}(1 - e^{-\rho t})^n = e^{-\rho t}\{1 - (1 - e^{-\rho t})M(z)\}^{-1}.$$

For a given form of $M(z)$ this expression can be inverted, most usually numerically, to yield the distribution function of S_t.

Consider the special case where the d_j are exponentially distributed with mean ξ. Then, $M(z) = (1 + \xi z)^{-1}$ and

$$K_t(z) = e^{-\rho t}\{1 - (1 - e^{-\rho t})/(1 + \xi z)\}^{-1} = e^{-\rho t} + (1 - e^{-\rho t})(1 + \xi e^{\rho t}z)^{-1}.$$

Thus, S_t has a mixed distribution with probability mass $e^{-\rho t}$ at $S_t = 0$ and density

$$k_t(s) = (1 - e^{-\rho t})(\xi^{-1}e^{-\rho t})\exp(-\xi^{-1}e^{-\rho t}s)$$

on $0 < s < \infty$ (see the Exercises in Section 11.6). The distribution function of T follows as

$$P(T \le t) = P(S_t \ge s^*) = (1 - e^{-\rho t})\exp(-\xi^{-1}e^{-\rho t}s^*);$$

this is a *Gumbel-type distribution*.

11.4.3 Gamma Process

Suppose that N_t follows a homogeneous Poisson process of rate ρ and that the d_j are *iid* with gamma(ξ, ν) distribution. Then S_t has *mgf*

$$K_t(z) = \exp[-\rho t\{1 - (1 + \xi z)^{-\nu}\}].$$

Consider now the limiting case in which there is an increasing frequency of smaller steps. Let $\rho \to \infty$ and $\nu \to 0$ with $\rho\nu \to \tau$. Thus, the step-size coefficient of variation, $(\nu\xi^2)^{1/2}/(\nu\xi)$, tends to ∞; this is a significant determinant of the nature of the process—see below. The *mgf* of S_t becomes

$$K_t(z) = \exp\{-\rho t(1 - e^{-\nu \log(1+\xi z)})\} \sim \exp[-\rho t\{\nu \log(1 + \xi z)\}]$$
$$\rightarrow \exp\{-\tau t \log(1 + \xi z)\} = (1 + \xi z)^{-\tau t},$$

and so S_t has distribution gamma(ξ, τt); the limit operation has produced a *gamma process* for S_t.

The survivor function of T can be evaluated, since S_t is monotone increasing in t, as

$$P(T > t) = P(S_t < s^*) = \int_0^{s^*} k_t(s)ds,$$

where $k_t(s) = \Gamma(\tau t)^{-1}\xi^{-\tau t}s^{\tau t-1}e^{-s/\xi}$ is the density of S_t. In terms of standard functions,

$$P(T > t) = \Gamma(\tau t)^{-1}\gamma(\tau t, s^*/\xi),$$

where $\gamma(a, z) = \int_0^z u^{a-1}e^{-u}du$ is the *incomplete gamma function* (Olver et al., 2010, Section 8.2).

Since the gamma process has independent increments, statistical analyses for multivariate data are tractable. With scheduled inspections at times $t_1 < t_2 < \ldots$, the random increases $\nabla S_j = S_{t_j} - S_{t_{j-1}}$ over time intervals $\nabla t_j = t_j - t_{j-1}$ are independent with distribution functions $P(\nabla S_j \leq a) = \int_0^a k_{\nabla t_j}(s)ds$. Likewise, for data of the milestones variety, let $s_1 < s_2 < \ldots$ be the specified levels and T_j the times at which they are reached. Then, the random intervals $\nabla T_j = T_j - T_{j-1}$ are independent with survivor functions $P(\nabla T_j > t) = \int_0^{\nabla s_j} k_t(s)ds$, where $\nabla s_j = s_j - s_{j-1}$.

The gamma process is a *jump process*. To see what this entails, go back to the pre-limiting form $S_t = \sum_{j=1}^{N_t} d_j$. For $a > 0$,

$$P(d_j > a) = \int_a^\infty \Gamma(\nu)^{-1}\xi^{-\nu}x^{\nu-1}e^{-x/\xi}dx = q_a,$$

say. For given n,

$$P\{\max(d_1, \ldots, d_n) \leq a\} = (1 - q_a)^n$$

and then

$$P\{\max(d_1, \ldots, d_{N_t}) \leq a\} = \sum_{n=0}^\infty (1 - q_a)^n\, e^{-\rho t}(\rho t)^n/n! = e^{-\rho t q_a}.$$

Now take the limiting case: $\rho \rightarrow \infty$ and $\nu \rightarrow 0$ with $\rho\nu \rightarrow \tau$. As $\nu \rightarrow 0$, $\Gamma(\nu) = \nu^{-1}\Gamma(\nu + 1) \sim \nu^{-1}$, so $q_a \sim \nu \int_a^\infty x^{-1}e^{-x/\xi}dx =: \nu r_{a\xi}$, say. Then,

$$e^{-\rho t q_a} \sim \exp(-\rho t\nu r_{a\xi}) \sim \exp\{-\tau t r_{a\xi}\}.$$

This probability tends monotonically to 0 as $t \rightarrow \infty$ and tends monotonically to 1 as $a \rightarrow \infty$. This means that the probability that none of the d_j exceeds a

tends to zero as $t \to \infty$, so arbitrarily large jumps will eventually occur. As $\rho \to \infty$, $N_t \to_p \infty$ and $\max(d_1, \ldots, d_{N_t})$ becomes the maximum of an infinite number of small-in-probability jumps.

11.4.4 Customer Lifetime Value

A commercial application of cumulative processes is where a company accumulates profit from an ongoing customer. (I dare not think about my own patronage of brewers over the years.) Suppose that a company acquires a customer at time 0 and thereby accumulates income $V_1(t)$ over the period $(0, t)$; the associated cost incurred by the company is $C_1(t)$, say. For simplicity, we assume that $V_1(0) = 0$, $C_1(0) = 0$ and that V_1 and C_1 are increasing processes. Both the customer and the company are free to terminate the association at any time, and suppose that the time to customer-instigated termination is T_1.

In the commercial framework the company will have a policy for dumping poor customers. A customer account will be reviewed, say at time r_1, assuming that $r_1 < T_1$. If the accumulated profit at that time exceeds some given threshold, the customer will be offered an ongoing deal, perhaps on more favourable terms. Specifically, if $V_1(r_1) > v_1$ the new deal has income and cost processes V_2 and C_2, and the second-phase lifetime is T_2, say. We can take $T_2 = 0$ as a possibility, meaning that the new offer has not been taken up. This two-phase framework, comprising (V_1, C_1, T_1) and (V_2, C_2, T_2), can be set out as follows: If $T_1 \leq r_1$ the *CLV* (customer lifetime value) to the company is just $V_1(T_1) - C_1(T_1)$; if $T_1 > r_1$ and $V_1(r_1) \leq v_1$, the *CLV* is $V_1(r_1) - C_1(r_1)$; if $T_1 > r_1$ and $V_1(r_1) > v_1$, the *CLV* is $V_1(r_1) - C_1(r_1) + V_2(T_2) - C_2(T_2)$. In terms of indicator functions,

$$CLV = I(T_1 \leq r_1)\{V_1(T_1) - C_1(T_1)\} + I(T_1 > r_1)[\{V_1(r_1) - C_1(r_1)\}$$
$$+ I(V_1(r_1) > v_1)\{V_2(T_2) - C_2(T_2)\}].$$

Crowder et al. (2007) considered this framework and made certain assumptions about the random variables T_1 and T_2 and the random processes V_1, V_2, C_1, and C_2. Based on these, expressions for the expected value, E(*CLV*), were obtained and computed. The object was to determine values for r_1 and v_1 to optimise E(*CLV*). In addition, a real application was presented.

11.5 Some Other Models and Applications

11.5.1 Empirical Equation Models

A variety of empirical equations has been proposed in various contexts. The physical foundation for such equations ranges from zero (where, for instance, a curve of observations has been described by a power law just for simplicity) to firm-ish (where, for instance, a differential equation based on physical–chemical properties has been derived). The Paris–Erdogan law for crack growth is one such empirical equation: it is defined by $dS_t/dt = kS_t^\gamma$, where S_t is the crack length at time t and k and γ are constants specific to the material. See Kozin and Bogdanov (1981), Sobczyk (1987), and Sobczyk and Spencer (1992) for details. Kozin and Bogdanoff (1983) related the two models discussed in their 1981 paper via a contrived "discretization" of the Paris–Erdogan law to match it to a discrete-time Markov chain model.

To introduce stochastic variation, as observed in experimental results, the constants appearing in such equations can be replaced by random variables. Though such replacement often reproduces the observed variation well, as intended, it does not generally yield additional insight into the physical mechanisms involved. Kozin and Bogdanoff (1981) discussed randomization of the Paris–Erdogan law, in which the parameters and initial conditions are represented as random effects: some data were fitted and goodness of fit discussed, after a fashion. In biostatistics growth curves have a long history: these are based on simple curves, such as polynomials, with coefficients modelled as random effects.

Nelson (1981) presented a case study in which the degradation variable S_t is the breakdown voltage of an insulator, and the explanatory variables are x, the test temperature (absolute), and t, the ageing time: the data form a $4 \times 4 \times 8$ factorial layout (replicates $\times$ temperatures $\times$ ageing times), with a single measurement per test unit, and the model fitted takes $\log S_t$ as $N(m, s^2)$ with $m = a - bte^{-g/x}$ and s^2 constant.

Tomsky (1982) showed how to perform linear regression analysis of repeated-measures data on damage, assuming the damage increments to be independent: the paper is mainly concerned with spelling out the formulae for confidence limits and so forth for engineers.

Lu and Meeker (1993) considered repeated-measures data on crack growth in which S_{ij} is the crack length in the ith experimental unit at time t_{ij} ($i = 1, \ldots, n; j = 1, \ldots, p$). The model adopted was $S_{ij} = \eta(t_{ij}; \beta, b_i) + e_{ij}$, where the e_{ij} are independently distributed as $N(0, \sigma^2)$ and η is a given, possibly nonlinear, function of t_{ij} and parameters β and b_i; b_i is a random coefficient associated with the ith unit. The authors suggested fitting the model to the data by an ad hoc two-stage procedure in which individual-unit curves are first fitted to yield estimates of the b_i, and then the set of estimated b_i are fitted by a multivariate normal distribution. The main aspect of interest is the distribution of T, the time to failure as induced by the S-model. This was calculated by simulation from the S-model with the estimated parameters plugged in. Robinson and Crowder (2000) gave a full likelihood and a Bayesian treatment of the framework.

Tseng et al. (1994) and Chiao and Hamada (1995) tackled data on the loss of luminosity of a fluorescent lamp over time. The model assumed was that S_t (minus log of luminous flux) increases linearly with t, the slope being normally or log-normally distributed over the population of lamps. Repeated measures were made to predict when the luminosity will fall below a specified level. The focus was on conducting factorial experiments to discover the most influential factors in the manufacturing process.

An extension is to introduce a whole stochastic process, rather than just a few random variables, into an empirical equation. In some cases, experimentally observed sample paths tend to cross over each other quite frequently, suggesting that a random process is involved. However, it is often difficult to make analytic progress with models of this type. Tang and Spencer (1989) proposed such a model for fatigue crack length and fitted it to the data of Virkler et al. (1979). Spencer and Tang (1988) adopted a similar model.

Subordinator processes are monotone increasing processes with independent increments and, as such, they provide a wide class of degradation models. Further elaboration can be achieved by introducing random effects. Lawless and Crowder (2004) fitted some repeated-measures, crack-growth data previously subjected to two-stage nonlinear regression by Lu and Meeker (1993). The model assumed was a gamma process extended to accommodate random effects that allow for the random development of the sample paths of different units.

11.5.2 Models for the Stress Process

Let the loading or stress process be U_t. If the damage process S_t has a known relation to the U_t process, and if the latter can be informatively modelled, then the structure of the S_t process can be derived, in principle. Sobczyk (1986) adopted this approach for fatigue crack growth, beginning with a general deterministic crack growth equation and then introducing a random process U_t to represent the loading pattern. Various particular cases were investigated. Suppose, for example, that damage occurs only when U_t exceeds some critical level u^*. Two typical situations are as follows:

1. failure occurs as soon as $U_t > u^*$ for the first time, or for the mth time, that is, on the mth upcrossing;

2. failure occurs as soon as the total time spent above u^* reaches t^*, say, so the failure time T is the smallest value satisfying $\int_0^T I(U_t > u^*)dt = t^*$; a variation on this is when the proportion of time, rather than the total time, is the operative factor—then T is the smallest solution of $T^{-1} \int_0^T I(U_t > u^*)dt = p^*$, p^* being the critical proportion.

Sobczyk and Spencer (1992, Chapter 3) gave some analytic results for the case where U_t is a stationary Gaussian process: they considered numbers of local maxima of U_t, numbers of peaks above u^*, their heights, times between peaks, $u_{1/3}$ (significant wave height, above which 1/3 of peaks rise), and envelope functions. They also considered stationary non-Gaussian loads, for example, generated as nonlinear transforms of Gaussian ones, and nonstationary loading.

Singpurwalla and Youngren (1993) postulated an evolutionary stochastic model in which the hazard function $h_j(t)$ of the jth component in a system has form $h_{0j}(t)U_t$, $h_{0j}(t)$ being a baseline hazard and U_t a process reflecting the environmental stresses. Two particular cases were considered, a gamma process and a shot noise process, and the associated multivariate survivor functions for the failure-time vector were calculated and discussed. One example resulting from the gamma process is the Marshall–Olkin bivariate exponential distribution.

Doksum and Hoyland (1992) considered Wiener processes with transformed time scale $\tau(t)$ to accommodate varying stress via an accelerated-life model; only time to failure was modelled and various forms for $\tau(t)$ were discussed as examples, such as stress increasing stepwise, linearly, and so forth.

Singpurwalla (1995) gave a wider review of hazard processes, concentrating on the associated failure-time distributions.

11.5.3 Stress-Strength Models

Crowder (1991) proposed a model for concrete carbonation. This is a progressive corrosion process in which carbon dioxide from the atmosphere diffuses into the concrete cladding of steel reinforcement, producing acid that eventually causes the latter to rust, expand, and crack the concrete. The model assumed held that Y, the thickness of the concrete cladding at a given point, is log-normally distributed and the CO_2 penetration depth, X_t at time t, has a log-normal distribution scaled by factor $t^{1/2}$; in addition, X_t and Y are independent. *Failure* occurs when X_t (*stress*) reaches Y (*strength*). A more elaborate approach is to have both stress and strength as stochastic processes: failure will occur as soon as the former exceeds the latter.

The problem of reinforcement corrosion was also addressed by Karimi et al. (2005). The spatial variation of the process was tackled using two analytical techniques, one based on random fields and the other on extreme values. Estimates of the probabilities of corrosion onset for various exposure times were calculated and compared.

11.5.4 Other Models and Applications

Bhattacharyya and Fries (1982) considered a certain failure-time model proposed by Birnbaum and Saunders (1969), which is based on a discrete sequence of cumulative damage contributions subsequently approximated by a continuous-time model and first passage to a critical level of damage. They showed that this model is, in fact, an approximation to the standard inverse Gaussian based on the first passage time of a Wiener process.

Some stochastic processes were considered by Desmond (1985) as models for degradation leading to failure in engineering systems: in essence, these processes all have distribution $N(a_t, b_t)$ for S_t, exactly or approximately, and then the resulting failure time distribution is of the Birnbaum-Saunders type. Data analysis was not addressed.

Various elaborations of Markov models were introduced in the book by Bogdanoff and Kozin (1985), such as probabilities of detection of damage, and withdrawal of units in various damage states before failure; they also mentioned data of repeated-measures type, on fatigue crack growth, but did not fit any by the models described earlier; only qualitative descriptions were given and ad hoc methods used for estimation. A more elaborate Markov chain framework in continuous time was considered by Luvalle et al. (1988).

Carey and Koenig (1991) presented a case study, fitting a model to some data on the effect of ageing (at high temperature) on the response characteristics (propagation delay) of an element of a logic circuit in a submarine cable. Their model has two stages: in the first, $S_t - S_0 = q(1 - e^{-\lambda t}) + e_1$, where S_t is the propagation delay for ageing time t, S_0 is the value measured before ageing, and e_1 is $N(0, \sigma_1^2)$; in the second stage, $\log q = a - b/x + e_2$, where x is the ageing temperature and e_2 is $N(0, \sigma_2^2)$. Least squares estimates of q, λ, and σ_1 were obtained by fitting the S-curves in stage 1, and then least squares was applied at stage 2 to estimate a, b, and σ_2, using the stage-1 q estimates as data.

Young (1994) considered a renewal process with exponential lifetimes: the state E_j represents j component-renewals and cases considered, including those of known and unknown numbers of renewals, and censored experiments where j might or might not be observed for unfailed units.

In the production of aluminium an electrical process is used in which the container, a steel box (cell), undergoes serious thermal and mechanical stress. At some point the resulting deterioration becomes too advanced for continued use of the process, at which time the system fails. The data listed in Whitmore et al. (1998) contain failure ages (days) for 17 aluminium reduction cells together with three covariates measured at the time of failure: iron contamination (percent), distortion (inches) and cathode drop (inches). The covariates were regarded as observations on marker processes correlated with an underlying, latent degradation process. Failure occurs when the latter reaches some critical value. Specifically, a four-variate Wiener-process model was applied. Observing marker processes, when the degradation process itself is unobservable, can be useful to predict the time of failure before it happens so that action can be taken to avoid the consequences of actual failure.

11.6 Exercises

11.6.1 Markov Chains

1. *Chapman–Komogorov equations.* Let $p_{jk}^{(r)}$ be the probability, conditional on starting in state j, of being in state k after r steps, so $p_{jk}^{(1)} = \{P\}_{jk}$, the (j, k)th element of P, the transition probability matrix. Show that $p_{jk}^{(r)} = \{P^r\}_{jk}$.

2. Let $\mathcal{R}$ be a particular subset of states (e.g., the transient ones), and let R be the corresponding submatrix of P. Verify that the probability of

a transition from state j to state k in r steps, all within $\mathcal{R}$, is $\{R^r\}_{jk}$. Show that this leads to the survivor function of T_j (time to leave $\mathcal{R}$ after starting in state j) as $P(T_j > t) = \sum_{k \in \mathcal{R}} \{R^t\}_{jk} = R^t 1_m$.

3. Suppose that T takes values $0, 1, \ldots$ with probabilities $p_0, p_1, \ldots$, and let $q_j = P(T > j)$. Recall that $E(T) = \sum_{j=1}^{\infty} q_j$ (Section 2.7). Use this to verify that $E(T_j) = \{(I_m - R)^{-1} 1_m\}_j$.

The variance formula, $\operatorname{var}(T_j) = \{2R(I_m - R)^{-2} 1_m\}_j$ takes a bit more work. Start in state j and consider the first step: you either jump straight to state m or to state k in $\mathcal{R}$, replacing T_j by $T_k + 1$ in the latter case:

$$E(T_j^2) = p_{jm} + \sum_{k \in \mathcal{R}} E\{(T_k + 1)^2\} p_{jk} = \sum_{k \in \mathcal{R}} E(T_k^2) p_{jk} + 2 \sum_{k \in \mathcal{R}} E(T_k) p_{jk} + 1.$$

Denote the vector with jth component $E(T_j)$ by m_1 and that with jth element $E(T_j^2)$ by m_2. Then we have

$$(I_m - R) m_2 = 2R m_1 + 1_m,$$

from which follows

$$m_2 = (I_m - R)^{-1}(2R m_1 + 1_m) = 2R(I_m - R)^{-2} 1_m + m_1,$$

and then $\operatorname{var}(T_j)$ as $(m_2)_j - (m_1)_j^2$.

4. Consider a Markov chain in continuous time. Let $p_k(t)$ and q_{jk} be as defined in Section 11.1, and suppose that only elements q_{jj} and $q_{j,j+1}$ are non-zero for $j = 0, 1, 2, \ldots$ (progressive movement through successive states). Derive

$$\frac{d}{dt}\{e^{q_k t} p_k(t)\} = e^{q_k t} q_{k-1} p_{k-1}(t).$$

Now, $p_0(0) = 1$, and so, taking $q_{-1} = 0$, $p_0(t) = e^{-q_0 t}$. For $k = 1$ we obtain

$$p_1(t) = q_0(q_1 - q_0)^{-1}(e^{-q_0 t} - e^{-q_1 t}),$$

using $p_1(0) = 0$ to obtain the constant of integration. Likewise, the $p_k(t)$ for $k > 1$ can be obtained by recursion.

11.6.2 Wiener Process

1. Verify the following properties.
   a. For $0 \le s < t$, and taking $W_0 = 0$,

   $$\operatorname{var}(W_s) = \operatorname{var}(W_s - W_0) = s,$$
   $$\operatorname{cov}(W_s, W_t) = \operatorname{cov}\{(W_s - W_0), (W_s - W_0) + (W_t - W_s)\} = s + 0.$$

b. The finite-dimensional distributions are multivariate normal, thus making W_t a *Gaussian process*.

c. W_t has the Markov property. For $0 \le s < t < u$ let $\mathcal{H}_{s-} = \{W_r : 0 < r < s\}$ be the history up to time $s-$. Then,

$$
\begin{aligned}
P(W_t \le b \mid W_s = a, \mathcal{H}_{s-}) &= P(W_t - W_s \le b - a \mid W_s = a, \mathcal{H}_{s-}) \\
&= P(W_t - W_s \le b - a) \\
&= P(W_t - W_s \le b - a \mid W_s = a) \\
&= P(W_t \le b \mid W_s = a).
\end{aligned}
$$

d. For independent increments, with $0 < s < t < u$, $\mathrm{var}(W_u - W_s) = \mathrm{var}(W_u - W_t) + \mathrm{var}(W_t - W_s)$. This is satisfied by $\mathrm{var}(W_b - W_a) = b - a$, but not, for example, by $\mathrm{var}(W_b - W_a) = (b - a)^2$.

e. Take $a > 0$ and $W_0 = 0$: the process $\{W_{at}\}$ has the same distribution as $\{a^{1/2} W_t\}$. This is because its finite-dimensional distributions are multivariate normal with the same mean (zero) and covariance structure: for $0 \le s < t$, $\mathrm{cov}(W_{as}, W_{at}) = as = \mathrm{cov}(a^{1/2} W_s, a^{1/2} W_t)$.

2. The sample paths of W_t are continuous but not differentiable: for each t, $\mid W_{t+h} - W_t \mid = O_p(\sqrt{h})$ as $h \to 0$.

   Proof. Let $p_h := P(\mid W_{t+h} - W_t \mid \le a) = 2 \int_0^a (2\pi h)^{-1/2} e^{-s^2/(2h)} ds \sim (\pi h/2)^{-1/2} a$ as $a \to 0$.

   a. Take $a = \sqrt{\pi h/2}$: $p_h \sim 1$, so $\mid W_{t+h} - W_t \mid \to 0$ (in probability) as $h \to 0$, that is W_t is continuous.

   b. Take $a = bh$: $p_h \to 0$, so $h^{-1} \mid W_{t+h} - W_t \mid \to \infty$ (in probability) as $h \to 0$, that is W_t is not differentiable.

   The result can be strengthened, using a refined analysis, to saying that the sample paths are everywhere (at all t-values simultaneously) continuous but not differentiable.

3. The Wiener process is the only process with stationary, independent increments that has continuous sample paths. To see this, let $t_i = i\tau/n$ $(i = 0, \dots, n)$, so that the t_i are equispaced on the time interval $[0, \tau]$. As $n \to \infty$, $\nabla t_i := (t_{i+1} - t_i) \to 0$, so $\nabla W_i := W(t_{i+1}) - W(t_i) \to 0$; further, $\max_i \mid \nabla W_i \mid \to 0$, since W_t is uniformly continuous on the compact set $[0, \tau]$. Hence, $W_\tau = \sum_{i=0}^{n-1} \nabla W_i$ is the sum of n iid random variables of which, as $n \to \infty$, the maximum tends to zero. By the CLT it follows that W_τ is normally distributed.

4. *First passage time.* Let W_t be a standard Wiener process with $W_0 = 0$ and let T_a be the time at which W_t first reaches level $a > 0$. Then, T has inverse Gaussian distribution with distribution function

$$
P(T_a \le t) = 2\,P(W_t \ge a) = 2\{1 - \Phi(a/t^{1/2})\}.
$$

11.6.3 Cumulative Damage Models

1. Take $S_t = \sum_{j=1}^{N_t} d_j$ and derive its mean and variance as $E(S_t) = \mu_t \mu_d$ and $\text{var}(S_t) = \sigma_t^2 \mu_d^2 + \mu_t^2 \sigma_d^2$, where μ_t and σ_t^2 are the mean and variance of N_t and μ_d and σ_d^2 are those of the d_j.

11.6.4 Compound Poisson Process

1. Let $J_n = \int_0^a u^{n-1} e^{-u} du$, where n is a positive integer. Integrate by parts to derive

$$J_n = -a^{n-1} e^{-a} + (n-1) J_{n-1},$$

and then repeatedly to obtain

$$J_n = -e^{-a}(n-1)! \sum_{k=1}^{n-1} a^k/k! + (n-1)!(1 - e^{-a}) = \Gamma(n)\{1 - e^{-a} e_{n-1}(a)\}.$$

2. Obtain the derivative of $P(T > t)$ as

$$\rho e^{-\rho t} + \sum_{n=1}^{\infty} \{1 - e^{-s^*/\xi} e_{n-1}(s^*/\xi)\} \, e^{-\rho t} \{(\rho^{n+1} t^n/n!) - (\rho^n n t^{n-1}/n!)\}.$$

Now use the identity

$$\sum_{n=1}^{\infty} e_{n-1}(a)\{\rho^{n+1} t^n/n! - \rho^n t^{n-1}/(n-1)!\}$$

$$= \sum_{n=1}^{\infty} e_{n-1}(a)\rho^{n+1} t^n/n! - \sum_{n=0}^{\infty} e_n(a)\rho^{n+1} t^n/n!$$

$$= \sum_{n=1}^{\infty} \{e_{n-1}(a) - e_n(a)\}\rho^{n+1} t^n/n! - e_0(a)\rho$$

to verify the formula given for the density of T.

3. Suppose that the d_j are *iid* binary, with $P(d_j = 1) = \pi$, and that N_t is a Poisson process with mean μ_t. Show that S_t has a Poisson distribution with mean $\pi \mu_t$. The original Poisson process has been *thinned* by only registering a proportion π of the damage hits.

11.6.5 Compound Birth Process

1. Let $p_n(t) = P(N_t = n)$ and derive $p_0(t) = e^{-\rho t}$ and

$$p_n'(t) = n\rho p_{n-1}(t) - (n+1)\rho p_n(t) \quad \text{for } n \geq 1.$$

Verify that $p_n(t) = e^{-\rho t}(1 - e^{-\rho t})^n$ solves the differential-difference equations.

2. Verify the form given for $k_t(z)$ by evaluating $\int_0^{\infty} e^{-zs} k_t(s) ds$.

11.6.6 Gamma Process

1. Consider the alternative limiting form in which $\lambda \to \infty$ and $\xi \to 0$ such that $\lambda\xi \to \tau$. (Thus, the step-size coefficient of variation remains constant at $\nu^{-1/2}$.) The mgf of S_t becomes

$$\phi(z) = \exp\{-\lambda t(1 - e^{-\nu \log(1+z\xi)})\} \sim \exp\{-\lambda t(1 - e^{-\nu z\xi})\}$$
$$\sim \exp\{-\lambda t(\nu z\xi)\} \to e^{-\nu\tau tz},$$

which is the mgf of a constant, $\nu\tau t$. In this case $S_t = \nu\tau t$ rises linearly and deterministically with t.

11.7 Hints and Solutions

11.7.1 Markov Chains

1. To get from state j to state k in two steps you have to go to some state in between, say l. So,

$$P(\text{state } k \text{ at time } t+2 \mid \text{state } j \text{ at time } t) = \sum_l p_{jl}\, p_{lk} = \{P^2\}_{jk}.$$

3. $(I_m - R)^{-1} = \sum_{j=0}^{\infty} R^j$ is the *fundamental matrix* (Kemeny and Snell, 1960, Section 3.2).

11.7.2 Wiener Process

1. a. Take $0 < s < t < u$ and $W_0 = 0$: then $W_s - W_0$, $W_t - W_s$ and $W_u - W_t$ are independent normal; hence, via a linear transformation, (W_s, W_t, W_u) is trivariate normal.

4. *Proof.*

$$P(W_t \geq a) = P(W_t \geq a, T_a \leq t) + P(W_t \geq a, T_a > t)$$
$$= P(W_t \geq a \mid T_a \leq t)\, P(T_a \leq t) + 0.$$

But, $P(W_t \geq a \mid T_a \leq t) = \frac{1}{2}$; this is because W_t has zero drift and is therefore equally likely to be above or below level a after being there at time $T_a \leq t$. Hence, remembering that W_t has distribution $N(0, t)$,

$$P(T_a \leq t) = 2\, P(W_t \geq a) = 2\{1 - \Phi(a/t^{1/2})\}.$$

By symmetry, the formula, with a replaced by $\mid a \mid$, also applies for $a < 0$. It also follows, by allowing $t \to \infty$, that $P(T_a < \infty) = 1$, which means that the process will eventually visit each level a. However,

$$E(T_a) = \int_0^{\infty} t\, dP(T_a \leq t) = \cdots = \infty.$$

The distribution of the maximum, $M_t = \max_{0 \le s \le t} W_t$, is also obtained as

$$P(M_t \ge a) = P(T_a \le t) = 2\{1 - \Phi(a/t^{1/2})\}.$$

Cox and Miller (1965, Section 5.7) give the result that for a Wiener process with drift parameter μ and variance parameter σ^2, the first-passage time from 0 to $a > 0$ has survivor function

$$P(T_a > t) = \Phi\left(\frac{a - \mu t}{\sigma t^{1/2}}\right) - \exp\left(\frac{2\mu a}{\sigma^2}\right) \Phi\left(\frac{-a - \mu t}{\sigma t^{1/2}}\right).$$

Thus, as $t \to \infty$, $P(T_a > t)$ tends to 0 when $\mu \ge 0$, and to $1 - \exp(\frac{2\mu a}{\sigma^2})$ when $\mu < 0$.

11.7.3 Cumulative Damage Models

1. $E(S_t) = E\{E(\sum_{j=1}^{N_t} d_j \mid N_t)\} = E(N_t \mu_d).$

$$E(S_t^2) = E\left\{E\left(\sum_{i,j=1}^{N_t} d_i d_j \mid N_t\right)\right\} = E\{N_t(\sigma_d^2 + \mu_d^2) + N_t(N_t - 1)\mu_d^2\}$$
$$= \mu_t(\sigma_d^2 + \mu_d^2) + (\mu_t^2 + \sigma_t^2 - \mu_t)\mu_d^2.$$

11.7.4 Compound Poisson Process

3. For integer r, $P(S_t = r) = \sum_{n=r}^{\infty}\{\binom{n}{r}\pi^r(1 - \pi)^{n-r}\}e^{-\mu_t}\mu_t^n/n! = \{e^{-\mu_t}\mu_t^r\pi^r/r!\}\sum_{n=0}^{\infty}\mu_t^n(1 - \pi)^n/n!.$

11.7.5 Compound Birth Process

1. Start with $p_n(t + dt) = p_{n-1}(t)n\rho dt + p_n(t)\{1 - (n + 1)\rho dt\} + o(dt)$. Then,

$$p_n'(t) = -\rho p_n(t) + e^{-\rho t}n(1 - e^{-\rho t})^{n-1}\rho e^{-\rho t}$$
$$= -\rho p_n(t) + n\rho e^{-\rho t}(1 - e^{-\rho t})^{n-1}\{1 - (1 - e^{-\rho t})\} = \cdots.$$

Part III

Competing Risks

In Part I of this book the single response variable, *time*, was the focus of attention, and in Part II more than one time at a time was tackled. The scenario considered there concerns a single type of event, called *failure* among other things. But in many situations the outcome event can be one of several alternative types, each with its own cause. Now the time has come to take up the cause, as it were, and focus on the type of event that has occurred as well as when it occurred.

In classical competing risks we still only observe a single terminal event and note what type of event. The hair-shirt approach is to say that there is no more to it: we just observe a time and a cause. The more whimsical imagine a whole set of risk processes developing with one of them beating the others to the finish line, and you do not get to see the others come in. (History tends to ignore the runners up.) Both approaches will be considered here.

12

Continuous Failure Times and Their Causes

The bare bones of the competing risks setup are laid out in this chapter. All that is considered is a time and a cause, without the additional, traditional structure of latent failure times; the latter will be addressed subsequently.

12.1 Some Small Data Sets

Many examples of competing risks analyses can be found in the mainstream statistical literature. In this section and in later ones, a few such applications will be described in just enough detail to convey the essentials, that is, the type of data and the methodology employed. More detail can be found in the references cited.

12.1.1 Gubbins

We start with some artificial data, generated here to be in the format required for illustration. Table 12.1 gives the breaking strengths of 55 gubbins. If you're not familiar with the word *gubbins*, it means a small device or gadget, an object of little or no value (now don't get personal). I used to say *widget* until some students asked me what that meant, demonstrating the depressingly wide generation gap; nowadays, it is the thing in the beer can that produces the froth. It is amazing what you can learn from a statistics book! Anyway, I digress. There are two causes of failure here, corresponding to two possible structural weaknesses. The response variable is breaking strength, or loading, rather than failure time, but as far as the statistical setup is concerned, this is merely a change of name. The values are right-censored at 3.0, which explains the first row of the table, labelled *Cause* 0.

12.1.2 Catheter Infection

The data here are taken from Crowder (1997). Like those in Section 1.2, these concern catheter infections. However, here the function of the catheter is to deliver drugs into the patient. The data arise from a study in which the growth of organisms was monitored at two separate sites in the equipment. Day 1 indicates the time at which the catheter was inserted, and the study ran for

TABLE 12.1

Breaking Strengths of Gubbins

Cause 0:	3.0	3.0	3.0	3.0	3.0	3.0	3.0	3.0							
Cause 1:	2.7	2.4	2.2	2.1	2.8	2.1	2.8	2.3	2.1	2.6	2.4	2.7	2.6	2.5	2.3
	2.8	2.1	2.5	2.6	2.4	2.2									
Cause 2:	2.4	1.9	1.8	2.3	2.0	2.4	2.3	1.9	1.8	1.9	2.2	1.9	1.9	2.3	2.4
	1.9	1.8	2.4	1.8	2.3	2.4	1.9	1.8	2.0	2.4	2.3				

one week. For each patient, two samples (one from each site) for analysis were collected on days 2 to 7, yielding failure times 1 to 6. There were 334 patients, of whom 250 had their catheters removed prematurely for reasons unconnected with infection.

The figures in Table 12.2 are head counts: for instance, 5 patients had an infection at site 2 on day 3, 11 patients had infections at both sites on day 1, and 9 patients lasted till day 5 before having their catheters removed for other reasons. The data provide an example in which the failure time is a discrete variable.

12.1.3 Superalloy Testing

The data in Table 12.3 have been taken from Problem 5.5 of Nelson (1982). They arise from a "low-cycle fatigue test of 384 specimens of a superalloy at a particular strain range and high temperature." The numbers of cycles to failure are grouped on a log_{10}-scale into intervals of width 0.05, and the lower endpoint of each interval is shown in the table. Failure is defined by the occurrence of a surface defect (type 1) or an interior defect (type 2).

The data provide an example of *interval censoring*; this will be tackled in Section 16.5.

12.2 Basic Probability Functions: Continuous Time

In classical competing risks the observed outcome comprises T, the time to failure, and C, the *cause*, *type*, or *mode* of failure. The failure time T is taken to be a continuous variate for the present, and the cause C can take one of a

TABLE 12.2

Catheter Infection at Two Sites

Time (days)	1	2	3	4	5	6
Site 1	2	2	1	0	0	0
Site 2	7	7	5	3	2	2
Both sites	11	20	15	3	3	1
Removed	35	104	55	19	9	28

Source: Crowder, M.J., 1997, A Test for Independence of Competing Risks with Discrete Failure Times, *Lifetime Data Analysis*, 3, 215–223. Reproduced by kind permission from Springer Science+Business Media, B.V.

TABLE 12.3

Log Cycles to Failure: Numbers of Specimens with Defects of Types 1 and 2

Lower End	No. Defects Type 1	No. Defects Type 2	Lower End	No. Defects Type 1	No. Defects Type 2	Lower End	No. Defects Type 1	No. Defects Type 2
3.55	0	3	4.00	0	16	4.45	7	2
3.60	0	6	4.05	3	12	4.50	4	3
3.65	0	18	4.10	3	4	4.55	2	1
3.70	1	23	4.15	2	3	4.60	1	0
3.75	3	47	4.20	5	10	4.65	1	0
3.80	1	44	4.25	2	3	4.70	0	0
3.85	0	57	4.30	3	4	4.75	1	0
3.90	1	37	4.35	7	7	4.80	0	0
3.95	0	20	4.40	12	4	4.85	1	0

Source: Nelson, W.B., 1982, *Applied Life Data Analysis*, Wiley, New York, Problem 5.5. Reproduced by permission of John Wiley & Sons, Inc.

fixed (small) number of values labelled $1, \ldots, p$. The basic probability framework is thus a bivariate distribution in which one component, C, is discrete and the other, T, is continuous. It will be assumed for now that to every failure can be assigned one and only one cause from the given set of p causes. To pick a nit, they are called *risks* before failure and *causes* afterward: the risks compete to be the cause.

For example, in Medicine C might be the cause of death and T the age at death. In Reliability C might identify the faulty component in a system and T the running time from start-up to breakdown. In Engineering our subject is more likely to be referred to as the reliability of non-repairable series systems. In Manufacturing T might be the usage (e.g., mileage) and C the cause of breakdown of a machine (e.g., vehicle). In Economics T might be the time spent on producing the forecast and C the reason for its turning out to be wrong.

The identifiable probabilistic aspect of the model is the joint distribution of C and T. This can be specified in terms of the so-called *sub-distribution functions* $F(j, t) = P(C = j, T \leq t)$, or equivalently by the *sub-survivor functions* $\bar{F}(j, t) = P(C = j, T > t)$. These functions are related by $F(j, t) + \bar{F}(j, t) = p_j$, where

$$p_j = P(C = j) = F(j, \infty) = \bar{F}(j, 0)$$

gives the marginal distribution of C. Thus, $F(j, t)$ is not a proper distribution function because it reaches only the value p_j instead of 1 at $t = \infty$. We assume implicitly that $p_j > 0$ and $\sum_{j=1}^{p} p_j = 1$. Note that $\bar{F}(j, t)$ is not, in general, equal to the probability that $T > t$ for failures of type j; that probability is a conditional one, $P(T > t \mid C = j)$, expressible as $\bar{F}(j, t)/p_j$. The *sub-density function* $f(j, t)$ for continuous T is $-d\bar{F}(j, t)/dt$.

The marginal survivor function and marginal density of T can be calculated from

$$\bar{F}(t) = \sum_{j=1}^{p} \bar{F}(j,t) \text{ and } f(t) = -d\bar{F}(t)/dt = \sum_{j=1}^{p} f(j,t).$$

The related conditional probabilities are P(*time t* | *cause j*) $= f(j,t)/p_j$ and P(*cause j* | *time t*) $= f(j,t)/f(t)$. These all have their uses in different contexts. For instance, $f(j,t)/p_j$ gives the distribution of age at death from cause j, which is of vital interest in medical applications (and of growing interest to me as I get older). Likewise, for a system of age t the probability of ultimate failure from cause j is

$$P(C = j \mid T > t) = \bar{F}(j,t)/\bar{F}(t).$$

12.2.1 Exponential Mixture

Consider the simple form $\pi_j e^{-t/\xi_j}$ for $\bar{F}(j,t)$ ($j = 1,\ldots,p$), where $\xi_j > 0$, $\pi_j > 0$ and $\pi_1 + \cdots + \pi_p = 1$. The marginal survivor function of T is $\bar{F}(t) = \sum_{j=1}^{p} \pi_j e^{-t/\xi_j}$ and the marginal probabilities p_j for C are the π_j. The conditional distributions are defined by

$$P(T > t \mid C = j) = e^{-t/\xi_j} \text{ and } P(C = j \mid T = t) = \pi_j \xi_j^{-1} e^{-t/\xi_j} \bigg/ \sum_{k=1}^{p} \pi_k \xi_k^{-1} e^{-t/\xi_k}.$$

The model can thus be interpreted as a mixture of exponentials: the mixture proportions are the π_j, and, for a unit in the jth group, the failure time distribution is exponential with mean ξ_j. Note that C and T are not independent unless the ξ_j are all equal. By differentiation, the sub-densities are $f(j,t) = \pi_j \xi_j^{-1} e^{-t/\xi_j}$.

To think about how such a model might be applied in practice, consider the case $p = 2$. Suppose that $n_1 = 20$ failures of type 1 and $n_2 = 10$ of type 2 were recorded, and $n_0 = 15$ failure times were *right-censored* at fixed time t_0; that is, all that is known about these 15 times is that they exceeded t_0. On this information alone we might be tempted to estimate π_1 as 20/30 and π_2 as 10/30. Later, after watching *Match of the Day*, it might occur to us that one team seemed to take much longer to get going than the other. So the 15 censored times might all have turned out to be of type 2 had the game gone to extra time. In this case we would estimate π_1 as 20/45 and π_2 as 25/45, quite different from the previous estimates. The fact that there are twice as many failures of type 1 as of type 2 within time t_0 has as much to do with the relative sizes of ξ_1 and ξ_2 as it does with those of π_1 and π_2. To make any sense out of the data, we should really try to account for both these effects. This is just what the likelihood function does, anticipating Section 13.1. The relevant probabilities are

$$P(T > t_0) = \bar{F}(t_0) = \sum_{j=1}^{2} \pi_j e^{-t_0/\xi_j}$$

for the n_0 right-censored observations, and

$$P(C = j \mid T \le t_0) = F(j, t_0)/F(t_0) = \pi_j(1 - e^{-t_0/\xi_j})/\left\{1 - \sum_{k=1}^{2} \pi_k e^{-t_0/\xi_k}\right\}$$

for the n_j observed failures of type j. Estimates based on the likelihood function

$$\{\bar{F}(t_0)\}^{n_0} \{F(1, t_0)/F(t_0)\}^{n_1} \{F(2, t_0)/F(t_0)\}^{n_2}$$

will make provision for all three independent parameters (π_1, ξ_1, ξ_2). In practice, we would not use this likelihood function based only on (n_0, n_1, n_2): we would use a fuller one based on the observed times also, but the point about disentangling the parameters holds just the same.

12.3 Hazard Functions

Various hazard functions, which describe probabilities of imminent disaster, are associated with the competing risks setup. The hazard function for failure from cause j, in the presence of all risks 1 to p, is defined in similar fashion to the univariate version in Section 2.1:

$$h(j, t) = \lim_{\delta \downarrow 0} P(C = j, T \le t + \delta \mid T > t) = \lim_{\delta \downarrow 0} \delta^{-1}\{\bar{F}(j, t) - \bar{F}(j, t + \delta)\}/\bar{F}(t)$$

$$= f(j, t)/\bar{F}(t).$$

The overal hazard function is $h(t) = \sum_{j=1}^{p} h(j, t)$. Some authors call the $h(j, t)$ *cause-specific* hazard functions, while others have used this term for the marginal hazards in the latent failure times setup (Section 14.1). The $F(j, t)$ are called *sub-distribution functions*, the $\bar{F}(j, t)$ are called *sub-survivor functions*, and the $f(j, t)$ are called *sub-density functions*. But what then should we call the $h(j, t)$? No contest—they are the *sub-hazard functions*.

12.3.1 Exponential Mixture

The sub-hazards for this mixture of exponentials (Section 12.2) are

$$h(j, t) = \pi_j \xi_j^{-1} e^{-t/\xi_j} \left/ \sum_{k=1}^{p} \pi_k e^{-t/\xi_k} \right.$$

and the overall hazard is

$$h(t) = \sum_{k=1}^{p} \pi_k \xi_k^{-1} e^{-t/\xi_k} \Bigg/ \sum_{k=1}^{p} \pi_k e^{-t/\xi_k}.$$

Unless the ξ_j are all equal these functions are not constant over time, this in spite of the fact that the underlying distributions are exponential.

12.3.2 Lemma (Kimball, 1969)

In applications it is often of interest to assess the consequences of changes in certain risks. In engineering the load profile might be altered so that some components of the system are placed under increased stress. In Medicine a treatment might improve the chances of surviving a particular disease. In our family cruelly inflicting home math practice on the children is designed to decrease the hazard of failure in that particular subject.

Suppose that the sub-hazard function $h(j, t)$ is increased from its previous level over the time interval $I = (a, b)$ and that the other sub-hazards are unchanged. It is reasonable to conjecture that the overall probability of failure in I is increased and that the relative probability of failure from a cause other than j in I is decreased. This is proved to be true in the following lemma that serves as an example of this type of calculation. The opposite conclusions would hold if $h(j, t)$ were decreased over I, of course.

Suppose that $h(c, t)$ is increased on $t \in I = (a, b)$ for $c = j$ only. Then (1) P_I, the probability of failure in I, conditional on survival to enter I, is increased, and (2) P_{Ik}, the probability of failure in I from cause k, conditional on entry to I, is decreased for $k \neq j$.

Proof   We have

$$P_I = \mathrm{P}(T \leq b \mid T > a) = \{\bar{F}(a) - \bar{F}(b)\}/\bar{F}(a) = 1 - \exp\left\{-\int_a^b h(t)dt\right\}$$

Since $h(t) = \sum_j h(j, t)$, the specified increase results in a decrease in the exponential term, and so (1) is verified. For (2) we have

$$P_{Ik} = \mathrm{P}(C = k, T \leq b \mid T > a) = \{\bar{F}(k, a) - \bar{F}(k, b)\}/\bar{F}(a)$$

$$= \int_a^b f(k, t)dt/\bar{F}(a) = \int_a^b h(k, t)\{\bar{F}(t)/\bar{F}(a)\}dt.$$

Now, $h(k, t)$ is unchanged from its previous level, whereas $\bar{F}(t)/\bar{F}(a) = \exp\{-\int_a^t h(s)ds\}$ is decreased, by the same argument as used for (1). Hence the result.

12.3.3 Weibull Sub-Hazards

Stochastic models for competing risks can be specified directly in terms of the sub-hazards, rather than the sub-survivor functions or sub-densities, as in the following example.

Suppose that the form $\phi_j \xi_j^{-\phi_j} t^{\phi_j - 1}$ is adopted for $h(j, t)$. (This form is just used here as an illustration. A discussion of ways of choosing suitable hazard forms will be given later in Section 13.5.) Then the overall hazard function is $h(t) = \sum_{j=1}^{p} \phi_j \xi_j^{-\phi_j} t^{\phi_j - 1}$. The marginal survivor function of T follows as

$$\bar{F}(t) = \exp\left\{ -\int_0^t h(s) ds \right\} = \exp\left\{ -\sum_{j=1}^{p} (t/\xi_j)^{\phi_j} \right\},$$

which shows that T has the distribution of $\min(T_1, \ldots, T_p)$, the T_j being independent $Weibull(\xi_j, \phi_j)$ variates. The sub-densities $f(j, t)$ can be calculated as $h(j, t)\bar{F}(t)$, but integration of this to find $\bar{F}(j, t)$ is generally intractable.

Note that this model is not the same as the one obtained by extending the exponential mixture of Section 12.2 to its Weibull counterpart, which has sub-survivor functions of form $\pi_j \exp\{-(t/\xi_j)^{\phi_j}\}$ for $j = 1, \ldots, p$.

12.4 Proportional Hazards

The relative risk of failure from cause j at time t is $h(j, t)/h(t)$. If this is independent of t for each j then *proportional hazards* are said to obtain. The phrase is used here with a different meaning from that of the PH model (Section 3.3). In the present case, as time marches on the relative risks of the various causes of failure stay in step, no one increasing its share of the overall risk, though the latter might rise or fall. This is beginning to sound a bit like independence of cause and time of failure, and it is proved to be so in the theorem below.

In many areas of application proportional hazards would probably be the exception rather than the rule. For instance, in the health sciences the relative risk of cot death and senile dementia might be expected to vary with age. The assumption, at least in piecewise fashion along the time scale, has been attributed to Chiang (1961) by David (1970). Seal (1977, Section 3) indicated that it goes back much further.

12.4.1 Weibull Sub-Hazards

Proportional hazards obtains only if the ϕ_j are all equal, say to ϕ. In this case,

$$\bar{F}(t) = \exp(-\xi^+ t^\phi), \quad f(j, t) = h(j, t)\bar{F}(t) = \phi \xi_j^{-\phi} t^{\phi-1} \exp(-\xi^+ t^\phi),$$
$$\bar{F}(j, t) = \pi_j \exp(-\xi^+ t^\phi),$$

where $\xi^+ = \sum_{j=1}^{p} \xi_j^{-\phi}$ and $\pi_j = \xi_j^{-\phi}/\xi^+$. This is a mixture model in which $p_j = P(C = j) = \pi_j$ and

$$P(T > t \mid C = j) = P(T > t) = \exp(\xi^+ t^\phi).$$

12.4.2 Theorem (Elandt–Johnson, 1976; David and Moeschberger, 1978; Kochar and Proschan, 1991)

The following conditions are equivalent:

1. proportional hazards obtains;
2. the time and cause of failure are independent;
3. $h(j, t)/h(k, t)$ is independent of t for all j and k.

In this case $h(j, t) = p_j h(t)$ or, equivalently, $f(j, t) = p_j f(t)$ or $F(j, t) = p_j F(t)$.

Proof   For (1) and (2) we have

$$P(cause\ j \mid time\ t) = f(j, t)/f(t) = h(j, t)/h(t);$$

independence of t must obtain for both sides of this equation or for neither. The last statement of the theorem follows since the left-hand side equals p_j in this case. For (1) and (3) note that

$$h(j, t)/h(t) = h(j, t) \bigg/ \sum_{k=1}^{p} h(k, t)$$

is independent of t for all j if and only if (3) holds.

Part (2) of the theorem says, for instance, that failure during some particular period does not make it any more or less likely to be from cause j than failure in some other period.

Another hazard function that could be defined is

$$f(j, t)/\bar{F}(j, t) = -d\log \bar{F}(j, t)/dt.$$

This looks like the univariate form (Section 2.1); it is the hazard for failures from cause j only. However, as will be seen, it is the $h(j, t)$ that arise naturally in competing risks. The conditioning event $\{T > t\}$ used in $P(C = j, T \leq t+\delta \mid T > t)$ to define $h(j, t)$ is more relevant in practice than the conditioning event $\{C = j, T > t\}$ used when the denominator is $\bar{F}(j, t)$ instead of $\bar{F}(t)$: at time t you know that failure has not yet occurred, but you probably do not know what the eventual cause of failure will turn out to be.

12.5 Regression Models

As in Section 3.3 the basic probability functions are now developed to cope with covariates. Here we see how those general classes of regression models might be adapted for competing risks. For generally dependent risks, it is the sub-distributions that determine the structure, so it is in terms of these that the models are framed. In what follows, ψ_{xj} is a positive function of the vector $\mathbf{x}$ of explanatory variables, for example, $\psi_{xj} = \exp(\mathbf{x}^{\mathsf{T}} \beta_j)$ or $\psi_{xj} = \exp(\mathbf{x}_j^{\mathsf{T}} \beta)$.

12.5.1 Proportional Hazards (PH)

The univariate PH model has $h(t; \mathbf{x}) = \psi_x h_0(t)$, where h_0 is some baseline hazard function. A natural adaptation for competing risks can be made in terms of the sub-hazard functions as

$$h(j, t; \mathbf{x}) = \psi_{xj} h_0(j, t).$$

A second stage can be imposed, taking proportional hazards now in the competing risks sense, to replace $h_0(j, t)$ by $p_j h_0(t)$:

$$h(j, t; \mathbf{x}) = \psi_{xj} p_j h_0(t).$$

In a statistical analysis, with parametrically specified h-functions, one can test for these successive restrictions on the sub-hazard models in the usual parametric way.

12.5.2 Accelerated Life (AL)

This model specifies the form $\bar{F}(t; \mathbf{x}) = \bar{F}_0(\psi_x t)$ for a univariate survivor function, with $\bar{F}_0$ some baseline survivor function. A natural analogue for competing risks is

$$\bar{F}(j, t; \mathbf{x}) = \bar{F}_0(j, \psi_{xj} t)$$

in terms of baseline sub-survivor functions $\bar{F}_0(j, t)$. A second, proportional hazards, stage can be incorporated as above to yield

$$\bar{F}(j, t; \mathbf{x}) = \bar{F}_{00}(\psi_{xj} t)^{p_j}$$

for some $\bar{F}_{00}$.

12.5.3 Proportional Odds (PO)

In the univariate version of this model, the odds on the event $\{T \leq t\}$ are expressed as

$$\{1 - \bar{F}(t; \mathbf{x})\}/\bar{F}(t; \mathbf{x}) = \psi_{xj}\{1 - \bar{F}_0(t)\}/\bar{F}_0(t).$$

A natural competing risks version of this is

$$\{1 - \bar{F}(j, t; \mathbf{x})\} / \bar{F}(j, t; \mathbf{x}) = \psi_{xj}\{1 - \bar{F}_0(j, t)\} / \bar{F}_0(j, t).$$

A second, proportional hazards, stage can be incorporated as above, replacing $\bar{F}_0(j, t)$ here with $\bar{F}_{00}(t)^{p_j}$.

12.5.4 Mean Residual Life (MRL)

The univariate version has $m(t) = \mathrm{E}(T - t \mid T > t)$ modelled as $m(t; \mathbf{x}) = \psi_x m_0(t)$. To extend this to competing risks we can define *sub-mrl* functions as

$$m(j, t) = \int_t^\infty (y - t) f(j, y \mid y > t) dy = \bar{F}(t)^{-1} \int_t^\infty \bar{F}(j, y) dy;$$

and $m(t) = \sum_{j=1}^p m(j, t)$. The inverse relation, expressing the $\bar{F}(j, t)$ in terms of the $m(j, t)$ is

$$\bar{F}(j, t) / \bar{F}(t) = m(t)^{-1}\{m'(t) + 1\} m(j, t) - m'(j, t).$$

The proportional *mrl* model can be extended straightforwardly to competing risks as

$$m(j, t; \mathbf{x}) = \psi_{xj} m_0(j, t).$$

12.6 Examples

12.6.1 Exponential Mixture

A regression model for the exponential mixture can be specified in terms of log-linear forms as (1) $\log \xi_{ij} = \mathbf{x}_i^{\mathrm{T}} \beta_j$, and (2) $\log \pi_{ij} = \kappa_i + \mathbf{x}_i^{\mathrm{T}} \gamma_j$. In (2) κ_i is determined by the requirement that $\pi_{i+} = \pi_{i1} + \cdots + \pi_{ip} = 1$ as

$$\kappa_i = -\log \sum_{j=1}^p \exp\left(\mathbf{x}_i^{\mathrm{T}} \gamma_j\right),$$

so the model for π_{ij} is expressible as

$$\pi_{ij} = \exp\left(\mathbf{x}_i^{\mathrm{T}} \gamma_j\right) \bigg/ \sum_{l=1}^p \exp\left(\mathbf{x}_i^{\mathrm{T}} \gamma_l\right);$$

an identifiability constraint, such as $\gamma_p = 0$, is required.

If $\mathbf{x}_i$ is of dimension q, the full parameter set, comprising vectors $\beta_1, \ldots, \beta_p$ and $\gamma_1, \ldots, \gamma_{p-1}$, will be of dimension $(2p - 1)q$. This will often be too large for the limited set of data at hand. In that case, further constraints will be necessary to reduce the number of parameters to be estimated. A typical reduction, applicable in the usual case where x_i has first component 1, is to take $\beta_j = (\beta_{0j}, \beta)$; here, β_{0j} is a scalar *intercept* parameter, allowed to vary over j, and β is a vector of regression coefficients of length $q - 1$, the same for all j. A similar reduction applied to the γ_j is not sensible: it would effectively remove all the γ-regression coefficients, since they would be divided out in the expression above for π_{ij}, just leaving the intercepts γ_{0j}. An alternative reduction is to set selected components of the β_j and γ_j to zero. This is applicable when the corresponding components of x_i are known to have no effect on the associated λ or π.

The regression version here is an accelerated life model provided that $\pi_{ij} = \pi_j$ for all i; in this case $\bar{F}_0(j, t)$ can be taken as $\pi_j e^{-t}$ with $\psi_{ij} = \exp(x_i^T \beta_j)$.

Example (Larson and Dinse, 1985)

These authors used a model similar to the exponential mixture. They justified it as being appropriate for situations in which the eventual failure cause is determined at time zero by some random choice. An example that they quote is where, following treatment for a terminal disease, a patient might be cured or not and this can only be determined at some future date. Their model, which allows for a vector x of explanatory variables, is as follows:

1. $P(C = j; \mathbf{x}) = p_{xj} = \phi_{xj} / \sum_{l=1}^{p} \phi_{xl}$, where $\phi_{xj} = \exp(\mathbf{x}^T \beta_j)$ and $\beta_p = 0$ for identifiability;

2. $P(T > t \mid C = j; \mathbf{x}) = \exp\{-\psi_{xj} G_j(t)\}$, where $G_j(t) = \int_0^t g_j(s)ds$, $g_j(s)$ is piecewise constant in s and $\psi_{xj} = \exp(\mathbf{x}^T \gamma_j)$.

Thus, log-linear models are used for ϕ_{xj} and ψ_{xj}, with regression coefficients β_j and γ_j, respectively. From these specifications follow

$$\bar{F}(j, t) = p_{xj} \exp\{-\psi_{xj} G_j(t)\}, \quad f(j, t) = p_{xj} \psi_{xj} g_j(t) \exp\{-\psi_{xj} G_j(t)\},$$

and hence $\bar{F}(t)$, $h(j, t)$, and so forth.

Larson and Dinse computed maximum likelihood estimates via an expectation-maximisation (EM) algorithm that replaces right-censored failure times with unknown cause by a set of suitably-weighted hypothetical ones with known causes. They applied their analysis to some heart-transplant data with two causes of death: transplant rejection and other causes. There were $n_1 = 29$ rejection cases, $n_2 = 12$ deaths from other causes, and $n_0 = 24$ patients still alive at the end of the study. The explanatory variables comprised a mismatch score (comparing donor–recipient tissue compatibility), age, and waiting time (from admission to the program to transplantation). They also compared their parametric model-based analysis of these data with a semi-parametric analysis.

12.7 Exercises

1. The exponential mixture model has $\bar{F}(j, t) = \pi_j e^{-t/\xi_j}$. What is the probability that a unit of current age t will eventually succumb to cause j? How does this vary with t? What if the ξ_j are all equal?

2. Extend the analysis of Question 1 to the corresponding Weibull version.

3. Repeat Question 1 for the Weibull sub-hazards model with $\phi_j = \phi$ (PH version).

4. Investigate a Pareto mixture model, like the exponential mixture model.

5. MRL model: Derive the relation shown, expressing $\bar{F}(j, t)$ in terms of the $m(j, t)$.

6. Suppose that the risks are time limited, so that risk j becomes zero at time v_j (where some v_j can be infinite to give the usual case). How might the exponential mixture be modified to reflect such a situation? Investigate the consequences.

12.8 Hints and Solutions

1. $P(C = j \mid T > t) = \bar{F}(j, t)/\bar{F}(t) = \pi_j e^{t/\xi_j} / \sum_{k=1}^{p} \pi_k e^{-t/\xi_k}$. For $t = 0$ this probability is π_j. For large t it is approximately $\pi_j e^{-t/\xi_j} / \pi_m e^{-t/\xi_m}$, where $\xi_m = \max(\xi_j)$, and this $\to 0$ for $\xi_j < \xi_m$. If the ξ_j are equal, the probability tends to 1.

5. First derive

$$m'(j, t)\bar{F}(t) + m(j, t)\bar{F}'(t) = -\bar{F}(j, t),$$

by differentiation of the integral definition of $m(j, t)$. Then, using the relation

$$\bar{F}'(t) = -\bar{F}(t)\{m'(t) + 1\}/m(t)$$

from Section 5.3, obtain

$$\bar{F}(j, t)/\bar{F}(t) = m(t)^{-1}\{m'(t) + 1\}m(j, t) - m'(j, t).$$

6. A brute-force modification is to take $\bar{F}(j, t) = \pi_j e^{-t/\xi_j} I(t \leq v_j)$, using the indicator function.

13

Continuous Time: Parametric Inference

13.1 The Likelihood for Competing Risks

13.1.1 Forms of the Likelihood Function

Suppose that data $\{(c_i, t_i) : i = 1, \ldots, n\}$ are available, and that parametric models have been specified for the sub-densities $f(j, t)$. Then the likelihood function is given by

$$L = \prod_{obs} f(c_i, t_i) \times \prod_{cens} \bar{F}(t_i).$$

Here $\prod_{obs}$ denotes the product over cases for which failure has been observed and $\prod_{cens}$ denotes the product over right-censored failure times of cases where failure has not yet been observed. The latter can conveniently be indicated in the data by $c_i = 0$ (in accordance with the *censoring indicator* of Section 3.2), and t_i is then recorded as the time at which observation ceased. The censoring mechanism is assumed here to operate independently of the failure mechanisms.

The likelihood can be expressed in terms of the sub-hazard functions: using $f(j, t) = h(j, t)\bar{F}(t)$, L can be written as

$$L = \prod_{obs} h(c_i, t_i) \times \prod_{i=1}^{n} \bar{F}(t_i);$$

further, in the second term here $\bar{F}(t_i)$ can be written as $\exp\{-\int_0^{t_i} h(s)ds\}$ with $h(s) = \sum_{j=1}^{p} h(j, s)$. Thus, an alternative specification for the model can be made entirely in terms of the $h(j, t)$ rather than the $f(j, t)$.

Another way of writing the likelihood function is

$$L = \prod_{i=1}^{n} \{h(c_i, t_i)^{d_i} \bar{F}(t_i)\},$$

where d_i is the censoring indicator for case i: $d_i = 1$ if death or failure has been observed, $d_i = 0$ if the failure time is right-censored; in short, $d_i = I(c_i \neq 0)$ in terms of the indicator function. Again, this can be re-expressed as $L = L_1 L_2 \ldots L_p$ with

$$L_j = \prod^{j} h(j, t_i) \times \prod_{i=1}^{n} \exp\left\{ -\int_0^{t_i} h(j, s)ds \right\},$$

where $\prod^j$ denotes the product over $\{i : c_i = j\}$. Here, L_j accounts for the contributions involving the sub-hazard $h(j, t)$. If, as is often the case, the model has been parametrized so that the $h(j, t)$ have non-overlapping parameter sets, such a partition of L can be useful. Numerically, it might be much easier to maximize each L_j separately, since there are fewer parameters to handle at any one time. Sampford (1952) made this point in connection with his *multistimulus* model. Theoretically, the parameter sets will be *orthogonal*, which can facilitate inference considerably (Cox and Reid, 1987). Thus, for example, if we are particularly interested in one of the failure causes, say the jth, the parameters governing $h(j, t)$ can be estimated and examined without reference to those of the other failure causes. In fact, one does not even have to specify parametric forms for the other $h(k, t)$ at all. Obviously, the greater the number of observations contributing to L_j, the better the estimates of the parameters of $h(j, t)$. Observed failures due to an identified cause are more useful in this respect than right-censored failure times. From the point of view of Statistics, then, censoring is a fate worse than death.

13.1.2   Uncertainty about C

In some cases it is not altogether clear after a system failure exactly what went wrong. For instance, after a machine breakdown the fault might be narrowed down only to one of several possibilities. This is commonly the case where failure is destructive in some way, such as in a plane crash. Again, relatives might refuse to allow a post-mortem, thus leaving the cause of death uncertain. In such cases, and provided that the observed time and set of possible causes gives no additional information about which individual cause was the culprit, the likelihood contribution for an observed failure is just modified from $f(j, t)$ to $\sum f(j, t)$, with summation over the set of possible indices j. Clearly, it is likely that inference will suffer as the amount of uncertainty in the data increases. In the extreme, where the failure cause is not documented at all, the likelihood involves only the function $\bar{F}(t)$ and its derivative; that is, inference is possible only about the marginal distribution of T. In this case we are back to univariate survival analysis.

13.1.3   Uncertainty about T

The instant of failure can be missed because the night watchman was asleep or the recording equipment was on the blink (malfunctioning). A system might be inspected only periodically, a breakdown in the meantime causing only a minor, unnoticed disaster. All that is known is that it happened between times t_1 and t_2, say. The likelihood contribution for a failure from cause j during the time interval (t_1, t_2) is $\bar{F}(j, t_1) - \bar{F}(j, t_2)$. It can even happen that the failure cause is known ahead of the failure, the time being right-censored (e.g., Dinse, 1982); then t_2 is taken as ∞. (There are always people who knew just what was going to happen.) If there is also uncertainty about the cause of failure, this difference must be summed over the possible indices j.

13.1.4 Maximum Likelihood Estimates

We consider first a very simple, and therefore atypical, example just to illustrate the derivation of maximum likelihood estimators.

13.1.4.1 *Weibull Sub-Hazards*

This model has $h(j, t) = \phi_j \xi_j^{-\phi_j} t^{\phi_j - 1}$ and $\bar{F}(t) = \exp\{-\sum_{j=1}^p (t/\xi_j)^{\phi_j}\}$. The log-likelihood contributions take the form

$$\log L_j = n_j \log \phi_j - n_j \phi_j \log \xi_j + (\phi_j - 1) \sum^j \log t_i - \xi_j^{-\phi_j} s_j,$$

where n_j is the number of observed failures from cause j, $\sum^j$ denotes summation over $\{i : c_i = j\}$ and $s_j = \sum_{i=1}^n t_i^{\phi_j}$. Assuming that the ξ_j and ϕ_j are all distinct,

$$\partial \log L / \partial \xi_j = \partial \log L_j / \partial \xi_j = -n_j \phi_j / \xi_j + \phi_j \xi_j^{-\phi_j - 1} s_j,$$

and so the *mle* $\hat{\xi}_j$ can be written down explicitly in terms of ϕ_j as $(s_j/n_j)^{1/\phi_j}$. That for ϕ_j requires numerical maximization of $\log L_j$. In the particular exponential case, $\phi_j = 1$, $\hat{\xi}_j = t_+/n_j$, where $t_+ = t_1 + \cdots + t_n$ is the *total time on test*, including censored times (Section 3.2).

13.1.4.2 *Exponential Mixture*

Simplifications, such as the ability to solve likelihood equations explicitly for some parameters, are not typically possible for competing risks. This is illustrated by the complicated likelihood function that results from a simple model in the following example.

The parameter set of this regression model given in Section 12.6 comprises $(\beta_1, \ldots, \beta_p)$ and $(\gamma_1, \ldots, \gamma_p)$, and the corresponding log-likelihood function is

$$\log L = \sum_{obs} \log(\pi_{ic_i} \xi_{ic_i}^{-1} e^{-t_i/\xi_{ic_i}}) + \sum_{cens} \log\left(\prod_{k=1}^p \pi_{ik} e^{-t_i/\xi_{ik}}\right)$$

$$= \sum_{obs} \left\{\kappa_i + \mathbf{x}_i^T \gamma_{c_i} - \mathbf{x}_i^T \beta_{c_i} - t_i e^{-\mathbf{x}_i^T \beta_{c_i}}\right\} + \sum_{cens} \log\left\{\prod_{k=1}^p \exp(\kappa_i + \mathbf{x}_i^T \gamma_k + t_i e^{\mathbf{x}_i^T \beta_k})\right\}.$$

The derivatives of $\log L$ with respect to the parameters are rather unwieldy algebraic expressions, and explicit formulae for the *mle* are not obtained.

There is plenty of scope for getting knee deep in algebra when investigating particular models in the likelihood function; see, for example, David and Moeschberger (1978). Without mentioning any names, some of the literature puts one in mind of the sons of toil shifting tons of soil. In this book such algebraic heroics will be avoided. We will take the view that, in computing terms, one just needs some code that will return a value for the likelihood function on call. The rest can be left to the professionals, the numerical analysts and computer programmers who have developed powerful methods for function optimisation over the last quarter of a century. The general problem of computing maximum likelihood estimates numerically is discussed in more detail in Appendix D.

13.2 Model Checking

The methods outlined in Section 3.4 can be extended to cope with competing risks. We will just consider model enhancement and uniform residuals here.

13.2.1 Goodness of Fit

As an example of model enhancement one might fit Weibull sub-distributions, extending the exponential mixture model, with $\bar{F}(j, t) = \pi_j \exp\{-(t/\xi_j)^{\phi_j}\}$, and then test to see whether they can be downgraded to exponentials, with $\phi_j = 1$.

Various comparisons can be made of estimated quantities with their observed counterparts, for example, $p_j = \bar{F}(j, 0)$ (the proportion of failures from cause j), $\bar{F}(j, t)/p_j$ (the survivor function of failures from cause j), and $\bar{F}(t)$ (the survivor function for all failures). In the survivor function plots, right-censored observations might have to be accommodated by appropriate modification of the empirical survivor functions.

13.2.2 Uniform Residuals

Residuals can be computed for competing risks data by an extension of the methods used for univariate survival analysis. Thus, the survivor function for failures of type j is the conditional one, $p_j^{-1}\bar{F}(j, t)$; this is just $\bar{F}(t)$ in the case of proportional hazards. Let the observed failure times of type j be t_{ij} ($i = 1, \ldots, n_j$) and define $e_{ij} = p_{ij}^{-1}\bar{F}_i(j, t_{ij})$; this is just a *probability integral transform* of t_{ij}, using the survivor function instead of the distribution function. The subscript i on $\bar{F}_i$ and p_{ij} indicates that these may be functions of explanatory variables $\mathbf{x}_i$, so that the ordering among the e_{ij} might not be the same as that originally among the t_{ij}. With estimates inserted for the parameters, the $\hat{e}_{ij}$ should resemble a random sample of size n_j from $U(0, 1)$, the uniform distribution on (0,1). However, censoring, as usual, muddies the water. Of the n_0 censored failure times t_{i0}, some would eventually have turned out to be failures from cause j. If we knew which ones, we could transform them to $e_{i0j} = p_{ij}^{-1}\bar{F}_i(j, t_{i0})$ and put them in with the uncensored e_{ij}, after making the adjustment of replacing e_{i0j} by $(e_{i0j} + 1)/2$; this is obtained by adding the mean residual lifetime (Section 12.5), $(1 - e_{i0j})/2$, to e_{i0j}. Since we do not know which ones, we could arbitrarily select the most likely ones, that is, those for which q_{ij} is largest among the q_{ik} $(k = 1, \ldots, p)$, where

$$q_{ij} = P_i(C = j \mid T > t_{i0}) = \bar{F}_i(j, t_{i0})/\bar{F}_i(t_{i0}).$$

In this way we would obtain an approximation to the full set of residuals from failures of type j. More favourably, this arbitrary selection from the censored values can be avoided if the e_{i0j} $(i = 1, \ldots, n_0)$ are all larger than the e_{ij}. This would certainly obtain, for instance, if there were no explanatory variables and all failure times were censored at the same value. Then the e_{ij} are the first n_j order statistics from a uniform sample of size $m_j = n_j + q_j n_0$, where q_j is the proportion of censored values that would eventually be type j failures. If we knew q_j, we could then plot residuals as the points $\{i/(m_j + 1), \hat{e}_{ij}\}$, $i/(m_j + 1)$ being the expected value of e_{ij}. What we can do is to insert an estimate of q_j, an obvious one being $q_{j1} + \cdots + q_{jn_0}$. If there is only a small amount of overlap between the e_{i0j} and the e_{ij}, we could reduce the cut-off point to the lowest e_{i0j} value and treat the e_{ij}s beyond as censored. Evidently, the situation is not ideal for generating residuals, and in practice, some ad hoc measures will often be needed. Though such ad-hockery is sometimes derided it is often just a form of common sense, that quality of judgement equally derided by the snooty.

13.3 Inference

The discussion of Section 3.1 applies equally here to competing risks. Regarding the Bayesian approach, the literature does not seem to be overburdened, though there would not appear to be any special technical problems with the setup. However, just as the likelihood equations are usually intractable, necessitating numerical solutions, so are the Bayesian manipulations analytically impossible for all but the very simplest models. And here follows an atypically tractable example.

Example (Bancroft and Dunsmore, 1976)

Consider the model with sub-survivor functions $\bar{F}(j, t) = p_j e^{-\lambda_+ t}$, where $p_j = \lambda_j/\lambda_+$ and $\lambda_+ = \lambda_1 + \cdots + \lambda_p$. This is a special case of the exponential mixture. The likelihood function for data $\{(c_i, t_i) : i = 1, \ldots, n\}$ is

$$L(\lambda) = \prod_{c_i \neq 0} f(c_i, t_i) \times \prod_{c_i = 0} \bar{F}(t_i) = \prod_{c_i \neq 0} \{\lambda_{c_i} e^{-\lambda_+ t_i}\} \times \prod_{c_i = 0} e^{-\lambda_+ t_i} = \left\{\prod_{j=1}^{p} \lambda_j^{n_j}\right\} e^{-\lambda_+ t_+},$$

where $\lambda = (\lambda_1, \ldots, \lambda_p)$, $t_+ = t_1 + \cdots + t_n$ and n_j is the number of failures of type j in the sample. For a Bayesian analysis we need a joint prior for the λ_j. A natural conjugate prior has the form of independent gamma distributions:

$$f(\lambda) \propto \prod_{j=1}^{p} \{\lambda_j^{\gamma_j - 1} e^{-\nu_j \lambda_j}\}.$$

The joint posterior distribution for the λ_j is now given by

$$f(\lambda \mid data) \propto f(\lambda) L(\lambda) \propto \prod_{j=1}^{p} \{\lambda_j^{\gamma_j + n_j - 1} e^{-(\nu_j + t_+)\lambda_j}\},$$

representing independent gamma distributions for the λ_j with modified shape and scale parameters. Bancroft and Dunsmore derived the predictive density for a future observation (C, T):

$$f(c, t \mid data) = \int f(c, t \mid \lambda, data) f(\lambda \mid data) d\lambda$$

$$\propto \int \lambda_c e^{-\lambda_+ t} \prod_{j=1}^{p} \{\lambda_j^{\gamma_j + n_j - 1} e^{-(\nu_j + t_+)\lambda_j}\} d\lambda$$

$$= \prod_{j=1}^{p} \{(t + \nu_j + t_+)^{-a_{cj}} \Gamma(a_{cj})\},$$

where $\lambda_0 = 1$ and $a_{cj} = \delta_{cj} + \gamma_j + n_j$, δ_{cj} being the *Kronecker delta* taking value 1 if $c = j$ and 0 otherwise. If $\nu_j = \nu$ for all j, C and T are conditionally independent, given the data, with

$$P(C = c \mid data) \propto \prod_{j=1}^{p} \Gamma(a_{cj}),$$

which leads to

$$P(C = c \mid data) = (\gamma_c + n_c)/(\gamma_+ + n_+),$$

and T having Pareto density

$$f(t \mid data) = (a_+ - 1)^{-1}(t + \nu + t_+)^{-a_+}$$

on $(0, \infty)$, where $a_+ = 1 + \gamma_+ + n_+$, $\gamma_+ = \gamma_1 + \cdots + \gamma_p$ and $n_+ = n_1 + \cdots + n_p$.

13.4 Some Applications

In this section some numerical examples are presented. Complete, definitive analyses are not catalogued—that would require a whole report for each one. Rather, the aim is to illustrate the various methods described above, showing the sorts of things that can be done and how. In addition, one or two other areas of application are briefly described.

13.4.1 Gubbins

These data (Section 12.1) will be fitted first with the exponential mixture model. The model, with $p = 2$, has three independent parameters: π, with $\pi_1 = \pi$ and $\pi_2 = 1 - \pi$, and ξ_1 and ξ_2. To fit it we use `fitmodel`, which calls `wblmix` to return the negative log-likelihood function. The latter function is set up for the more general Weibull mixture model, the parameters are expressed as $\pi_j = 1/(1 + e^{-\theta_{1j}})$, $\xi_j = e^{\theta_{2j}}$ and $\phi_j = e^{\theta_{3j}}$, and the overall parameter set used by the function is

$$\theta = (\theta_{11}, \ldots, \theta_{1,p-1}, \theta_{21}, \ldots, \theta_{2p}, \theta_{31}, \ldots, \theta_{3p});$$

the missing θ_{1p} is expressed internally as $1 - \sum_{j=1}^{p-1} \theta_{1j}$, and if this is negative a huge penalty value is returned to `optim` (see the CRC Press Web site noted in the Preface). The R-code is

```
#gubbins data (Sec 13.4)
n0=8; n1=21; n2=26; nd=n0+n1+n2; gb0=rep(3.0,n0);
gb1=c(2.7,2.4,2.2,2.1,2.8,2.1,2.8,2.3,2.1,2.6,2.4,2.7,2.6,2.5,2.3,
      2.8,2.1,2.5,2.6,2.4,2.2);
gb2=c(2.4,1.9,1.8,2.3,2.0,2.4,2.3,1.9,1.8,1.9,2.2,1.9,2.3,2.4,
      1.9,1.8,2.4,1.8,2.3,2.4,1.9,1.8,2.0,2.4,2.3);
cb=c(rep(0,n0),rep(1,n1),rep(2,n2)); #censoring indicator
dx1=cbind(c(gb0,gb1,gb2),cb); dx1;
#fit exponential model
mc=2; kt=1; kc=2; mdl=1; opt=0; adt=c(nd,mc,kt,kc,mdl,opt,0);
np=3; par0=c(0,0,0); par1=fitmodel(wblmix,np,par0,dx1,adt);
```

To fit the more general Weibull model we just change to five parameters and `mdl=2`:

```
#fit Weibull model
mdl=2; adt=c(nd,mc,kt,kc,mdl,opt,0);
np=5; par0=c(0,0,0,0,0); par1=fitmodel(wblmix,np,par0,dx1,adt);
```

The maximised log-likelihoods are -126.8653 under the exponential mixture and -59.52443 under the Weibull. The log-likelihood ratio statistic for the enhanced model yields $\chi_2^2 = 2(126.8653 - 59.52443) = 134.68$ and $p = 2.4 \times 10^{-96}$: it's not looking good for the exponential fit.

We proceed with the Weibull mixture model. The estimated parameters and their standard errors can be transformed back using

```
se=c(0.2701,0.0320,0.0208,0.1838,0.1579); wblmixb(mdl,mc,par1,se);
```

in which the printed standard errors have been copied and pasted into the code. This produces $\hat{\pi} = 0.53$ (0.13), $\hat{\xi}_1 = 2.83$ (0.09), $\hat{\xi}_2 = 2.20$ (0.05), $\hat{\phi}_1 = 6.83$ (1.25) and $\hat{\phi}_2 = 9.96$ (1.57), with the standard errors shown in the accompanying parentheses; the latter have been obtained via the *delta method* (Exercises), and we make the usual provisos for standard errors accompanying range-limited parameters. We will not pursue this further: the data set is limited, so more searching tests for model fit are not likely to be productive, and we have bigger fish to fry.

13.4.2 Survival Times of Mice

Hoel's (1972) data, reproduced in Table 13.1, arose from a laboratory experiment in which male mice of strain RFM were given radiation dose 300 rads at five to six weeks old. There are two groups of mice: conventional lab environment (Group 1) and germ-free environment (Group 2). The survival times are measured in days and the causes of death are 1 (thymic lymphoma), 2 (reticulum cell sarcoma), and 3 (other). Hoel developed a model that incorporates individual immunity and fitted it to the data by maximum likelihood.

We will not take Cause 3 as representing independent censoring, as might be the usual option, because it is not clear that this would be valid here: it will just be treated as a third cause. The Weibull mixture model will be applied, allowing for possible differences between groups in the parameters. The full parameter set is $\theta = (\theta_1, \ldots, \theta_{16})$, in which (π_1, π_2, π_3) is determined by (θ_1, θ_2) for Group 1 and by (θ_3, θ_4) for Group 2; (ξ_1, ξ_2, ξ_3) is determined by $(\theta_5, \theta_6, \theta_7)$ for Group 1 and by $(\theta_8, \theta_9, \theta_{10})$ for Group 2; (ϕ_1, ϕ_2, ϕ_3) is determined by $(\theta_{11}, \theta_{12}, \theta_{13})$ for Group 1 and by $(\theta_{14}, \theta_{15}, \theta_{16})$ for Group 2. We apply exponential transformations for ξ and ϕ, as for the Gubbins data in the preceding

TABLE 13.1

Survival Times of Mice

	Group 1														
Cause 1	159	189	191	198	200	207	220	235	245	250	256	261	265	266	280
	343	356	383	403	414	428	432								
Cause 2	317	318	399	495	525	536	549	552	554	557	558	571	586	594	596
	605	612	621	628	631	636	643	647	648	649	661	663	666	670	695
	697	700	705	712	713	738	748	753							
Cause 3	40	42	51	62	163	179	206	222	228	252	259	282	324	333	341
	366	385	407	420	431	441	461	462	482	517	517	524	564	567	586
	619	620	621	622	647	651	686	761	763						
	Group 2														
Cause 1	158	192	193	194	195	202	212	215	229	230	237	240	244	247	259
	300	301	321	337	415	434	444	485	496	529	537	624	707	800	
Cause 2	430	590	606	638	655	679	691	693	696	747	752	760	778	821	986
Cause 3	136	246	255	376	421	565	616	617	652	655	658	660	662	675	681
	734	736	737	757	769	777	800	806	825	855	857	864	868	870	870
	873	882	895	910	934	942	1015	1019							

Source: Hoel, D.G., 1972, A Representation of Mortality Data by Competing Risks, *Biometrics*, 28, 475–488. Reproduced by permission of the International Biometric Society.

application. For the π parameters we will now use a full logit transformation for a change: for each group, $\pi_j = e^{\theta_j}/(1 + \sum_{k=1}^{p-1} e^{\theta_k})$ for $j = 1, \ldots, p-1$ and $\pi_p = 1/(1 + \sum_{k=1}^{p-1} e^{\theta_k})$. The R-code is

```
#mice data (Sec 13.4)
dx1=read.table('hoel.dat',header=T); attach(dx1); #dx1;
dx2=as.numeric(unlist(dx1)); nd=181; dim(dx2)=c(nd,3);
dx2[1:nd,1]=dx2[1:nd,1]/1000; #dx2
#fit model
p=3; ng=2; kt=1; kc=2; kx=3; mdl=1; opt=0; #see hoell
adt=c(nd,p,ng,kt,kc,kx,mdl,opt,0);
np=16; par0=rep(0,np); #llkd=hoell(par0,dx2,adt);
par1=fitmodel(hoell,np,par0,dx2,adt);
pars=hoela(mdl,p,ng,par1); pars; #untransformed parameters
```

The full model gives maximised log-likelihood -94.52742. In a univariate Weibull distribution, $\log \xi$ is a location parameter and ϕ is a scale parameter for $\log T$. With normal linear models, one often entertains homogeneity of the scale parameters, here over the two groups. Under this constraint, that (ϕ_1, ϕ_2, ϕ_3) is the same for the two groups, the maximised log-likelihood is -183.6220. (This can be achieved by temporarily setting these parameters equal in hoell, remembering to also set np=13 in the main code.) It is hardly worth bothering, but the log-likelihood ratio test gives $\chi_3^2 = 178.19$, yielding a p-value that could probably give way to Planck's constant. Let us agree that the data are telling us loudly that separate ϕ-parameters must be retained for the two groups. Similar assessments suggest that the other parameters also differ between groups.

The full (untransformed) *mle* are as follows: $\hat{\pi}_1 = (0.22, 0.38, 0.39)$, $\hat{\pi}_2 = (0.35, 0.18, 0.46)$, $\hat{\xi}_1 = (0.31, 0.65, 0.46)$, $\hat{\xi}_2 = (0.39, 0.75, 0.79)$, $\hat{\phi}_1 = (3.59, 8.05, 2.07)$, and $\hat{\phi}_2 = (2.20, 6.13, 4.39)$. The main differences seem to be a lower probability of death from Cause 2 in Group 2, and longer average survival times. The ϕ parameters determine finer aspects of the shape of the survival distributions; for instance, as ϕ increases beyond 1.0 in an ordinary Weibull distribution, the density becomes more symmetric.

For goodness of fit, index plots of uniform residuals and individual log-likelihood contributions are shown in Figure 13.1 for the two groups. The R-code is

```
#residual plots (with par1 pasted in from mle fit)
mdl=1; opt=1; adt=c(nd,p,ng,kt,kc,kx,mdl,opt,0);
par1=c(-0.5725726,-0.02603007,-0.2703272,-0.9295114,
       -1.164870,-0.4336928,-0.7723763,-0.9396544,-0.2859788,-0.2330752,
       1.278570,2.085393,0.7291779,0.7905226,1.813396,1.479963);
res=hoell(par1,dx2,adt); n1=99; n2=82; #group sizes
ra1=res[1:n1]; ra2=res[(n1+1):nd];
rb1=res[(nd+1):(nd+n1)]; rb2=res[(nd+n1+1):(2*nd)];
par(mfcol=c(2,2));
pll=indxplot(n1,ra1,1,'uniform residuals',''); title(main='Group 1',sub='');
pll=indxplot(n1,rb1,1,'log-likelihoods','');
pll=indxplot(n2,ra2,1,'uniform residuals',''); title(main='Group 2',sub='');
pll=indxplot(n2,rb2,1,'log-likelihoods','');
rtn=moran(ra1,n1,8,1); rtn=moran(ra2,n2,8,1);
```

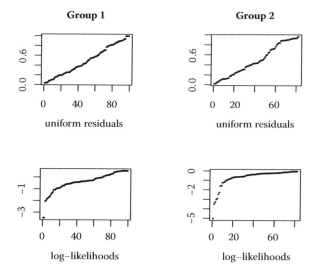

FIGURE 13.1
Index plots of uniform residuals and individual log-likelihood contributions.

There are a couple of lowliers in the log-likelihood plots but the overall picture does not seem too far out of line. However, honesty forces me to reveal that *Moran tests* (Section 3.4) tell a different story: the chi-square values are $\chi^2_{99} = 475.39$ and $\chi^2_{82} = 425.50$ for Groups 1 and 2, giving vanishingly small p-values. So, assuming that the sample sizes here are not of the order that would make any test significant, we must accept that the Weibull mixture is a poor fit as far as the residuals are concerned. Whether the fit is good enough for useful medical conclusions to be drawn, I'm not qualified to judge.

13.4.3 Fracture Toughness

In assessing the safety of nuclear power plants, various tests are routinely carried out. In one such experiment specimens of irradiated weld metal are subjected to an increasing load until the test is terminated due to cracking and/or fracture or for some other unrelated reason. The specimen is subsequently examined and certain measurements are made. Crowder (1995) listed data on 91 such specimens. The recorded variables are specimen width, temperature (at which the experiment on the specimen was conducted), final crack length (which can be zero), and toughness (a measure of the work done in tearing the specimen apart). The competing risks aspect here is the type of termination: cleavage/rupture (with or without cracking), ductile tearing (test terminated after crack appearance but before cleavage), and interrupted (meaning censored).

The situation had been recognised as one of competing risks by the engineers. Thus, a basic sub-survivor function $\bar{F}(c, k) = P(C = c, K > k)$ can be set up for C (termination type) and K (fracture toughness); the covariates identified by the engineers were specimen width, temperature, and crack length Δa. However, the last one is, like fracture toughness, a random outcome rather than a fixed covariate. Moreover, both K and Δa are often censored, and the latter is often zero. For the three types of termination we have the following information:

Cleavage: 49 specimens, 37 with $\Delta a = 0$, K and Δa both observed.

Ductile tearing: 37 specimens, K and $\Delta a > 0$ both right-censored (test terminated prior to cleavage).

Interrupted: 5 specimens, 3 with $\Delta a = 0$, K and Δa both right-censored.

A model with the following components was proposed to accommodate the particular aspects of the data:

Marginal distribution of Δa:
$$P(\Delta a = 0) = p_a \text{ and } P(\Delta a > y_a) = (1 - p_a)\bar{F}_a(y_a) \text{ for } y_a \geq 0.$$
Conditional distribution of K given Δa:
$$P(K > y_K \mid \Delta a = y_a) = \bar{F}_K(y_K \mid y_a) \text{ for } y_a \geq 0.$$

The corresponding densities are $f_K(y_k \mid y_a) = -\partial \bar{F}_K(y_k \mid y_a)/\partial y_k$ and $f_a(y_a) = -\partial \bar{F}_a(y_a)/\partial y_a$, and then likelihood contributions can be assembled:

P(*cleavage at* $K = u$ *with* $\Delta a = 0$) $= p_a f_K(u \mid 0)$.

P(*cleavage at* $K = u$ *with* $\Delta a = v > 0$) $= (1 - p_a)f_K(u \mid v)f_a(v)$.

P(*ductile tearing with* $K = u$ *and* $\Delta a = v > 0$) $=$
$(1 - p_a)\int_v^\infty \bar{F}_K(u \mid y)f_a(y)dy$.

P(*interrupted at* $K = u$ *with* $\Delta a = 0$) $= p_a \bar{F}_K(u \mid 0)$
$+ (1 - p_a)\int_0^\infty \bar{F}_K(u \mid y)f_a(y)dy$.

P(*interrupted at* $K = u$ *with* $\Delta a = v > 0$) $= (1 - p_a)\int_v^\infty \bar{F}_K(u \mid y)f_a(y)dy$.

Parametric forms for the survivor functions $\bar{F}_a$ and $\bar{F}_K$ were adopted, with parameters permitted to depend on the covariates, and the fitted model was used to draw conclusions. No doubt you will have found this rather brief outline sufficiently enthralling to encourage you to go and look up the details in the paper cited.

13.4.4 Length of Hospital Stay

Survival models are often generated from first-passage times of stochastic processes, in particular, in the statistics of degradation as described in Section 6.3. Horrocks and Thompson (2012) gave a nice extension of this kind of construction to competing risks. The framework concerned the length of hospital stay of a patient, with one of two possible outcomes, cure or death. There is an interest in relating the time and outcome to various covariates. Their model is based on a Wiener process with drift, representing the underlying health status of the patient over time. The time is up when the process first reaches an upper barrier (cure) or a lower one (death).

Consider a Wiener process with parameters μ (drift) and σ^2 (volatility), and starting at zero; the barriers are at $u > 0$ and $l < 0$. With the constraint $\sigma = 1, u$ and l are treated as scale parameters to be estimated; all three parameters, μ, u, and l, can be modelled in terms of covariates. Formulae are given in the cited paper for the two sub-density and sub-survivor functions as infinite series; the associated first-passage-two-barrier distribution was dubbed FP2B. The terms in the series rapidly become negligible so that computation can proceed based on truncated versions.

The paper continued with an application to publicly available data on hospital discharges in Utah during 1996. Although the authors took care to point out that their analysis is for illustrative purposes only, they did draw some interesting conclusions. A mixture model was also investigated, in which μ becomes a normally distributed random effect. Finally, the model was extended to accommodate a third outcome, transfer to another institution. For this, an intermediate barrier is inserted between 0 and u, transfers occurring when the patient's health status has improved sufficiently.

13.5 Some Examples of Hazard Modelling

The usual broad classification of hazard functions is into three types: (1) constant over time, as would be appropriate in a constant environment subject only to random shocks; (2) increasing, as in the case of a steadily deteriorating system; and (3) decreasing, as for systems that wear in and become less liable to failure as time goes on. The univariate Weibull distribution is the usual start for discussing different types of behaviour of hazard functions, and we shall honour tradition. Thus, $h(t) = \phi\xi^{-1}(t/\xi)^{\phi-1}$ is the hazard function corresponding to the Weibull survivor function $\bar{F}(t) = \exp\{-(t/\xi)^\phi\}$. It is (1) constant for $\phi = 1$, corresponding to an exponential distribution, (2) increasing for $\phi > 1$, and (3) decreasing for $\phi < 1$.

The construction of customized hazard functions, designed to suit the application at hand, has received some attention. In standard survival analysis we have to consider one hazard function, and this causes enough problems. In the competing risks context we have to consider simultaneously a whole set of them, the $h(j, t)$ for $j = 1, \ldots, p$.

13.5.1 Exponential Mixture

The sub-survivor functions are $\bar{F}(j, t) = \pi_j e^{-t/\xi_j}$ and the sub-hazards are

$$h(j, t) = \pi_j \xi_j^{-1} e^{-t/\xi_j} \bigg/ \sum_{k=1}^{p} \pi_k e^{-t/\xi_k}.$$

Let $\xi_a = \min(\xi_1, \dots, \xi_p)$ and $\xi_b = \max(\xi_1, \dots, \xi_p)$. Then $h(a, t) = O$ $(e^{-(\xi_a^{-1} - \xi_b^{-1})t})$ as $t \to \infty$, so $\int_0^\infty h(a, s)ds < \infty$. This makes the ath component of the independent-risks proxy distribution improper, that is, $\bar{F}_a^*(\infty) > 0$. Now, none of this matters in the least unless the situation calls for an interpretation in terms of latent failure times. Then, if the risks were actually independent, or even if only the partial Makeham condition $h_a(t) = h(a, t)$ held (Section 14.5), we would be forcing $\bar{F}_a(\infty) > 0$, and so unwittingly assuming T_a to be an improper failure time, that is, $P(T_a = \infty) > 0$. In contrast, with independent exponential latent lifetimes, $\bar{F}_j(t) = e^{-t/\xi_j}$ and $h(j, t) = h_j(t) = \xi_j^{-1}$.

That this kind of accident can happen was pointed out by Nadas (1970) in the context of independent risks. We do not have an entirely free hand in constructing models for the sub-distributions. If there is to be a latent lifetime structure in the background then there are dangers in not deriving the sub-distributions from an underlying joint latent failure time distribution.

13.5.2 Gumbel's Bivariate Exponential

The sub-hazard functions for this distribution (Section 6.2) are $h(j, t) = \lambda_j + \nu t$. In terms of the broad classification made above, this represents the sum of a constant hazard, λ_j, and an increasing one, νt, remembering that $\nu > 0$. This form might be appropriate for a system in which the jth component both deteriorates and is subject to random shocks over time.

13.5.3 Bivariate Makeham Distribution

From Section 7.5 the bivariate survivor function has the form

$$\bar{G}(\mathbf{t}) = \exp\{-\lambda_1 t_1 - \lambda_2 t_2 - \lambda_{12} \max(t_1, t_2) + \psi_1 \phi_1^{-1}(1 - e^{\phi_1 t_1}) + \psi_2 \phi_2^{-1}(1 - e^{\phi_2 t_2}).$$

This is an example of customising hazard functions: the sub-hazards,

$$h(j, t) = \lambda_j + \psi_j e^{\phi_j t} \qquad h(\{1, 2\}, t) = \lambda_{12},$$

are constructed to have the intended functional behaviour. Here T can be expressed as $\min(R_1, R_2, R_{12}, W_1, W_2)$, where the Rs are independent, exponentially distributed times to failure of types $\{1\}$, $\{2\}$, and $\{1, 2\}$; and W_1 and W_2 are Gompertz distributed. Arnold and Brockett (1983) showed that the bivariate latent lifetime distribution is parametrically identifiable by the sub-hazard functions, that is, by competing-risks data.

13.5.4 Kimber and Grace: The Dream Team

This data set has been sportingly provided for our entertainment by Alan Kimber, my colleague for many happy years at Surrey University. Alan played cricket at a high level for a good number of years, somewhat higher than my own level in football. His paper (Kimber and Hanson, 1993) applied techniques of survival analysis to batting scores in cricket. The failure time is taken as the number of runs scored until *out*, right-censored if *not out*. The

TABLE 13.2

Alan Kimber's Batting Data for Guildford: $(t, c) = $ (Runs, How Out)

1981	2 b	0 b	22 c	120 b	61 r	5 b	32 n	24 b	27 r	
1982	55 n	84 b	43 b	67 c	6 c	49 s	0 c	13 b	12 c	39 c
	7 c	51 c	1 b	18 l	16 c	10 c	12 c	11 b	2 c	30 n
	42 c	77 c	53 b	17 r	34 n	35 n	6 c	40 n	41 n	
1983	32 n	16 b	28 b	36 b	20 c	1 r	3 n	95 n	3 n	23 c
	10 b	5 s	56 n	0 c	11 b	9 c	22 b	12 c	34 b	40 n
	41 n									
1984	19 c	25 c	1 c	20 l	29 c	41 b	0 n	40 n	0 n	4 n
	1 n	20 n	46 c	11 c	35 c	23 c	39 b	17 c	42 n	
1985	12 s	15 l	0 l	6 c	31 n	4 l	1 n	23 c	0 n	6 n
	11 b	19 l	9 c	41 c	0 r	31 c				
1986	13 l	43 b	11 l	25 c	7 n	43 c	8 n	21 r	4 b	42 c
	6 n	27 n	29 c	1 b						
1987	11 n	28 n	37 n	6 b	51 n	17 l	0 b	3 c	0 c	14 n
	0 b	25 c	10 l	7 c	3 c					
1988	11 c	58 c	2 b	36 c						
1989	42 n	9 c	5 c	2 c	13 c	12 b	52 n	21 b		
1990	25 c	8 b	100 n	6 c	13 c	14 c	48 n	23 c	12 c	

Source: Kimber, A.C. and Hansford, A.R., 1993, A Statistical Analysis of Batting in Cricket, *J. Roy. Statist. Soc.* A 156, 443–455. Reproduced by permission of Alan Kimber.

runs may be regarded as a proxy for the time at the wicket, or as reflecting better than clock time the wear and tear on the batsman. (This might account for Kimber's ever-youthful appearance.) Hazard modelling is natural in this context: each time the batsman faces the bowler, there is a clear risk of mistake. In fact, it is cricketing folklore that the hazard varies critically with runs scored; for example, there might be an increased risk early on, until the batsman gets his "eye in," and then, if he gets that far, he runs into the "nervous nineties," when the anxiety to reach a century dominates. If, beside the number of runs, we take note of the cause of failure, that is, how the batsman was dismissed, the data become of competing risks type. The officially admitted causes of failure here are b = bowled, c = caught, l = lbw (leg before wicket), s = stumped, r = run out, n = not out. Table 13.2 gives Kimber's batting record during his time playing for Guildford, 1981 to 1990. (He maintains that his century in 1990 would have been augmented but for the intervention of tea!)

Models with constant hazard, and with increased hazard when the number of runs is less than 10 or in the 90s, were fitted. Without wishing to be in any way judgemental, it has to be said in a caring way that there is not enough data in the 90s to address that particular anxiety. The appropriate likelihood function for discrete failure times is given in Section 16.4: here τ_l can take values $0, 1, 2, \ldots$, $\bar{F}(\tau_l; \mathbf{x}_i)$ is expressed as $\prod_{s=0}^{l}\{1 - h(\tau_s)\}$ and $f(c_i, \tau_l; \mathbf{x}_i)$ as $h(c_i, \tau_l)\bar{F}(\tau_{l-1}; \mathbf{x}_i)$, there being no explanatory variables. In turn, for the constant-hazard model $h(c, \tau_l)$ is expressed as

$$h_c = \exp(-\theta_c) \Big/ \left\{ 1 + \sum_{j=1}^{5} \exp(-\theta_j) \right\} \quad (c = 1, \ldots, 5),$$

the θ_c being parameters of unrestricted range. The resulting log-likelihood is -593.68607. For the modified hazards model h_c is replaced by $h_c \exp(\theta_6)$ whenever τ_l was below 10 or in the 90s, the exponential factor representing the anxiety effect. The log-likelihood for this model is -593.67800, giving log-likelihood ratio chi-square $\chi_1^2 = 0.016$ ($p \approx 0.90$), the estimated anxiety factor actually being very slightly less than 1.0. Either Kimber has nerves of steel or he does not know what the score is. Just for interest, the sub-hazard estimates come out as 0.009, 0.017, 0.003, 0.001, 0.002, with standard errors suggesting that no two are equal; their sum, the overall hazard, is 0.032. The sub-hazard for Cause 2 is as large as the rest put together, illustrating the batsman's typical generosity in always being ready to give a catch to the fielders.

Another grand old man of English cricket was Dr. W.G. Grace. His scores from matches in the latter half of the 19th century are to be found in Lodge (1990). In Grace's case, possibly because the data are more extensive, the anxiety factor does show moderate significance ($\chi_1^2 = 3.90$, $p < 0.05$), its estimate being 1.35; the sub-hazard estimates show that Grace's preferred mode of departure was to be caught out, being bowled coming a fairly close second.

13.5.5 A Clinical Trial (Lagakos, 1978)

The data arose from a lung-cancer clinical trial: there are 194 cases with 83 deaths from Cause 1 (local spread of disease) and 44 from Cause 2 (metastatic spread), the remaining 67 times being right-censored (Cause 0). Three explanatory variables were also recorded: activity performance, a binary indicator of physical state (x_1), a binary treatment indicator (x_2), and age in years (x_3).

Lagakos (1978) listed the data to which he fitted a model with constant sub-hazards, $h(j, t) = \xi_j$ ($j = 1, 2$). The three explanatory variables were accommodated via a log-linear regression model, $\log \xi_j = x^T \beta_j$, in which $\beta_j = (\beta_{0j}, \beta_{1j}, \beta_{2j}, \beta_{3j})^T$ and x has first component 1 so that β_{0j} is the intercept term. Maximum likelihood estimation was applied, goodness of fit was assessed from residual plots, and various hypothesis tests were performed.

Lagakos's model is a special case of Weibull sub-hazards (Section 12.3), which we will now adopt: the sub-hazards are $h(j, t) = \phi_j \xi_j^{-\phi_j} t^{\phi_j - 1}$. For computation we have used R-function `fitmodel` with `wblshz` to return the negative log-likelihood (listed on the CRC Press Web site noted in the Preface). The R-code for reading the data in and re-formatting it, and then fitting the full model, is

```
#Lagakos data (Sec 13.5)
dx1=read.table('lagakos1.dat',header=T); attach(dx1);
dx2=as.numeric(unlist(dx1)); nd=194; dim(dx2)=c(nd,5);
dx2[1:nd,1]=time/100; dx2[1:nd,5]=age/100; #rescale
#Weibull sub-hazards model
mc=2; kt=1; kc=2; mx=4; kx=c(0,3,4,5);
mdl=0; np=mc*mx+mc; opt=0; iwr=0;
par0=runif(np,0,1); adt=c(nd,mc,kt,kc,mx,kx,mdl,opt,iwr);
par1=fitmodel(wblshz,np,par0,dx2,adt);
```

The constrained model, in which $\phi_j = 1$ ($j = 1, 2$), gives the same results as those quoted by Lagakos, with maximised log-likelihood -45.5702.
For the unconstrained, full-model version we obtain -35.2115. The resulting log-likelihood ratio chi-square value, 20.72 (with 2 degrees of freedom,
$p = 3 \times 10^{-5}$), indicates that the constraint is unreasonable. The ϕ estimates
are both greater than 1, giving sub-densities that are zero at $t = 0$ and so
are humped rather than being of monotone decreasing, negative exponential
shape. Presumably, observed times are too thin on the ground near $t = 0$ for
an exponential distribution.

The estimated values of ϕ_1 and ϕ_2 in the full model are quite close, which
suggests that the underlying values might be equal. This hypothesis turns out
to be tenable: the maximised log-likelihood under this constraint is -35.2927,
giving chi-square 0.16 (1 degree of freedom, $p = 0.69$) in comparison with
the full model. Thus, we now adopt this model, with ϕ_1 and ϕ_2 both equal
to ϕ, say. For convenience we will refer to this as Model 1, the previous, full,
unconstrained model being Model 0. The estimates and their standard errors,
based on the data in which the times and ages have been scaled by a factor
0.01, are as follows:

$$\hat{\beta}_1 = (-1.28, -0.47, -0.32, 1.26), \quad se(\hat{\beta}_1) = (0.50, 0.17, 0.20, 0.84),$$
$$\hat{\beta}_2 = (0.50, -0.43, -0.32, -0.96), \quad se(\hat{\beta}_2) = (0.75, 0.23, 0.28, 1.18),$$
$$\hat{\phi} = 1.37, \quad se(\hat{\phi}) = 0.09.$$

Model 1 has proportional hazards: in this case $\bar{F}(t) = \exp(-\xi^+ t^\phi)$, where
$\xi^+ = \sum_{j=1}^p \xi_j^{-\phi}$, and

$$\bar{F}(j, t) = \int_t^\infty h(j, s)\bar{F}(s)ds = \int_t^\infty \phi \xi_j^{-\phi} s^{\phi-1} e^{-\xi^+ s^\phi} ds = \pi_j \bar{F}(t),$$

where $\pi_j = \xi_j^{-\phi}/\xi^+$.

Further questions of interest, centred on the values of the regression coefficients β, were considered by Lagakos. Judging by the standard errors given
above, one or more of the βs might be zero, in which case the corresponding explanatory variable has no effect on the ξ_j value. In fact, leaving aside
the intercept terms, β_{01} and β_{02}, with only one exception none of the normal
deviate ratios ($\hat{\beta}$ value/standard error) exceeds the magic threshold, 1.96;
the exception is $\hat{\beta}_{11}$, for which the ratio is $-0.47/0.17$. As a first shot at model
reduction, let us re-fit with $\beta_{21}, \beta_{31}, \beta_{22}$, and β_{32} all set to zero. This yields maximised log-likelihood -38.6441; on comparison with Model 1 (proportional
hazards), $\chi_4^2 = 6.70$ ($p = 0.15$). So, it seems that the evidence for influence
of the second and third x variables is weak. The parameter estimates for this
reduced model are

$$\hat{\beta}_1 = (-0.80, -0.38), \quad se(\hat{\beta}_1) = (0.11, 0.16),$$
$$\hat{\beta}_2 = (-0.33, -0.38), \quad se(\hat{\beta}_2) = (0.15, 0.23),$$
$$\hat{\phi} = 1.35, \quad\quad\quad se(\hat{\phi}) = 0.09.$$

The estimates for these remaining parameters have not undergone drastic revision from the previous values, which is reassuring. Whether we should retain β_{12} is a matter for discussion. Its normal deviate ratio, $0.38/0.23 = 1.65$, is not large. On the other hand, $\hat{\beta}_{11}$ and $\hat{\beta}_{12}$ look pretty similar (they differ in the third decimal place). In the parameter covariance matrix the estimates are 273.474×10^{-4} for $\text{var}(\hat{\beta}_{11})$, 515.368×10^{-4} for $\text{var}(\hat{\beta}_{12})$, and 0.761×10^{-4} for $\text{cov}(\hat{\beta}_{11}, \hat{\beta}_{12})$; thus, the estimated variance of the difference, $\hat{\beta}_{11} - \hat{\beta}_{12}$, is

$$10^{-4}(273.474 - 2 \times 0.761 + 515.368) = 0.079.$$

This yields normal deviate $(0.37906 - 0.37738)/\sqrt{0.079} = 0.006$. I would not wish to appear for the case against equality (or for diversity) here.

So, the effect of x_1 (patient's ability to walk) appears to be real, and the same for the two causes of death, but any effects that x_2 (treatment) and x_3 (age) might have do not show up strongly in the data. These conclusions are in line with those of Lagakos.

13.6 Masked Systems

In many engineering systems the components are grouped into modules. It is common practice nowadays for repairs to be performed simply by replacing a failed module without bothering to identify the failed component within the module. This is said to be more economical. (But, for whom, pray? I cannot be alone in the experience of trying to replace a very minor component on my car and being told that I have to buy the whole sealed unit at vast expense.) In such cases, the failed component will only be identified as one of several. In addition to the work reviewed in more detail below, relevant papers include those by Gross (1970), Miyakawa (1984), Usher and Hodgson (1988), Usher and Guess (1989), Doganaksoy (1991), Guess et al. (1991), and Lin and Guess (1994).

In a parametric analysis involving masked systems, the likelihood can be simply modified as described in Section 13.1. However, it is clear that if two components, say j_1 and j_2, always appear together in a module, their separate contributions will not be estimable. This is because in the likelihood function the sub-density function $f(c_i, t_i)$ will be replaced with $\sum_{j \in c_i} f(j, t_i)$, with summation over the components j in module c_i, and $\bar{F}(t_i)$ will be expressible as $\sum_{j=1}^{p} \bar{F}(j, t_i)$, with

$$\bar{F}(j, t_i) = \int_{t_i}^{\infty} \sum_{j \in c_i} f(j, s) ds;$$

thus, $f(j_1, t)$ and $f(j_2, t)$ only appear together in combination as $f(j_1, t) + f(j_2, t)$ and are therefore separately unidentifiable. If the separate hazards of the different components are to be estimated, then we need either sufficient distribution of them around different modules, so that the combinations can be disentangled, or some assumptions about the sub-densities.

Albert and Baxter (1995) showed how to apply the expectation-maximisation (EM) algorithm to maximize the likelihood for such situations, taking the identifiers of the failing components to be missing data. They mentioned the identifiability problem when two components always appear together in the same module, but said that this was "unlikely in practice."

Reiser et al. (1995) treated a situation in which a subset of components can be identified post-mortem that contains the guilty component, implicitly assuming that disentanglement is possible in the data. For illustration, they gave some three-component data in which all six non-empty subsets of $\{1, 2, 3\}$ appeared in the list of identified failure causes. Such data arise when the breakdown can be diagnosed better in some failed units than in others. The authors were careful to make explicit the assumption that the observed time T and set S of possible causes gives no additional information about which individual cause C in S was the source of the trouble: in probability terms, their assumption is that, for some fixed $j' \in s$,

A1: $P(S = s \mid C = j, T = t) = P(S = s \mid C = j', T = t)$ for all $j \in s$.

This would not be true, for instance, in a two-component system where complete destruction of the system and, with it, any causal evidence are more likely to be due to failure of the second component than of the first.

The contribution of a failure time t with diagnosed cause-subset s to the likelihood function is, under assumption A1,

$$P(S = s, T = t) = \sum_{j \in s} P(C = j, T = t) \, P(S = s \mid C = j, T = t)$$

$$= P(S = s \mid C = j', T = t) \sum_{j \in s} f(j, t),$$

writing $f(j, t)$ for $P(C = j, T = t)$; assumption A1 removes the terms $P(S = s \mid C = j, T = t)$ and replaces them with the single factor $P(S = s \mid C = j', T = t)$ outside the summation. Hence, the likelihood function becomes

$$L = \prod_{obs} \left\{ P(S = s_i \mid C = j', T = t_i) \sum_{j \in s_i} f(j, t_i) \right\} \times \prod_c \bar{F}(t_i).$$

Reiser et al. (1995) pointed out that, provided that the factor $\prod_{obs} P(S = s_i \mid C = j', T = t_i)$ does not involve the parameters of interest, it may be treated effectively as a constant and omitted from the calculations. They went on to present a Bayesian analysis in which, for simplicity, the component lifetimes are independent exponential variates, and using the usual invariant improper prior for the exponential rate parameters.

The approach outlined in the previous paragraph was extended by Guttman et al. (1995) to the case of *dependent masking* in which assumption A1 is replaced by

A2: $P(S = s \mid C = j, T = t) = a_j P(S = s \mid C = j', T = t)$ for all $j \in s$,

where $j' \in s$ is fixed and a_j does not depend on t. Assumption A2 is that the relative likelihoods of masking, given different component failures, do not change over time. It reduces to A1 when $a_j = 1$ for all j; if $a_j = 0$, failure of component j cannot yield the observation $S = s$. Under A2 the likelihood contributions have the form

$$P(S = s, T = t) = \sum_{j \in s} P(C = j, T = t) \, P(S = s \mid C = j, T = t)$$
$$= P(S = s \mid C = j', T = t) \sum_{j \in s} a_j f(j, t).$$

Guttman et al. considered a two-component system, so there is only one quantity a_j to deal with. They applied a Bayesian analysis with component lifetimes and prior as in Reiser et al. (1995). Further work is presented by Craiu and Reiser (2006).

13.7 Exercises

13.7.1 Applications

1. Use the delta method to obtain the standard errors of the transformed parameters in the Nelson data. These are $\pi = 1/(1 + e^{-\theta})$, $\xi = e^{\theta}$ and $\xi = e^{\theta}$.

2. Develop a Pareto mixture model along the lines of the Weibull one. The Pareto distribution has longer tails than those of the exponential and Weibull, and so might serve as a useful alternative. The span of failure times in the Nelson data looks huge. So, how about writing some code (like that used for the Weibull-mixture fit) and trying it out?

3. Use the delta method to obtain the standard errors of the transformed π-parameters in the Hoel data.

4. In view of the lack of fit of the Weibull mixture model to the Hoel data, try out the Pareto mixture (assuming that you took up the challenge in Question 2).

14

Latent Lifetimes

The specification of competing risks via a set of latent lifetimes is covered in this chapter. This is the traditional route, often with the additional assumption of independence between these concealed lifetimes ("born to blush unseen, and waste their sweetness on the desert air"). It should be mentioned at the outset that this approach carries a health warning: there is a certain problem of model identifiability that will be examined in detail in Chapter 17.

14.1 Basic Probability Functions

In the traditional approach to competing risks it is assumed that there is a potential failure time associated with each of the p risks to which the system is exposed. Thus, T_j represents the time to system failure from cause j ($j = 1, \ldots, p$): the smallest T_j determines the time T to overall system failure, and its index C is the cause of failure, that is, $T = \min(T_1, \ldots, T_p) = T_C$. Once the system has failed, the remaining lifetimes are lost to observation, in a sense they cease to exist, if they ever did. For example, in an electric circuit with p components in series, the circuit will be broken as soon as any one component fails: T is then equal to the failure time of the component that fails first, and C identifies the failing component.

The vector $\mathbf{T} = (T_1, \ldots, T_p)$ will have a joint survivor function $\bar{G}(\mathbf{t}) = P(\mathbf{T} > \mathbf{t})$ and marginal survivor functions $\bar{G}_j(t) = P(T_j > t)$. In the older, actuarial and demographic, terminology the sub-survivor functions $\bar{F}(j, t)$ were called the *crude survival functions* and the marginal $\bar{G}_j(t)$ the *net survival functions*. It will be assumed for the present that the T_j are continuous and that ties cannot occur, that is, $P(T_j = T_k) = 0$ for $j \neq k$, otherwise C is not so simply defined (Section 17.2). If the joint survivor function $\bar{G}(\mathbf{t})$ of $\mathbf{T}$ has known form, then $\bar{F}(t)$ can be evaluated as $\bar{G}(t\mathbf{1}_p)$, where $\mathbf{1}_p = (1, \ldots, 1)$ is of length p.

14.1.1 Tsiatis's Lemma (Tsiatis, 1975)

The sub-densities can be calculated directly from the joint survivor function of the latent failure times as

$$f(j, t) = \left[-\partial \bar{G}(\mathbf{t}) / \partial t_j \right]_{t\mathbf{1}_p},$$

the notation $[\ldots]_{t1_p}$ indicating that the enclosed function is to be evaluated at $\mathbf{t} = t\mathbf{1}_p = (t, \ldots, t)$.

Proof. By definition

$$f(j, t) = \lim_{\delta \downarrow 0} \delta^{-1} \{ \bar{F}(j, t) - \bar{F}(j, t + \delta) \}$$

$$= \lim_{\delta \downarrow 0} \delta^{-1} P\{ t < T_j \le t + \delta, \cap_{k \neq j} (T_k > T_j) \}.$$

But,

$$P\{ t < T_j \le t + \delta, \cap_{k \neq j} (T_k > t + \delta) \} \le P\{ t < T_j \le t + \delta, \cap_{k \neq j} (T_k > T_j) \}$$
$$\le P\{ t < T_j \le t + \delta, \cap_{k \neq j} (T_k > t) \}.$$

Divide by δ and then let $\delta \downarrow 0$ to obtain the result.

It follows immediately from the lemma that the sub-hazards can also be calculated directly from $\bar{G}(\mathbf{t})$ as

$$h(j, t) = f(j, t) / \bar{F}(t) = [-\partial \log \bar{G}(\mathbf{t}) / \partial t_j]_{t1_p}.$$

14.1.2 Gumbel's Bivariate Exponential

Gumbel's distribution was introduced in Section 6.2: it has joint survivor function

$$G(\mathbf{t}) = \exp(-\lambda_1 t_1 - \lambda_2 t_2 - \nu t_1 t_2).$$

The marginal survivor function of T is $\bar{F}(t) = \exp(-\lambda_+ t - \nu t^2)$, where $\lambda_+ = \lambda_1 + \lambda_2$. For $j = 1, 2$, we have the sub-densities

$$f(j, t) = \left[(\lambda_j + \nu t_{3-j}) \bar{G}(\mathbf{t}) \right]_{(t,t)} = (\lambda_j + \nu t) \bar{F}(t)$$

and the sub-survivor functions

$$\bar{F}(j, t) = \int_t^\infty (\lambda_j + \nu s) \exp(-\lambda_+ s - \nu s^2) \, ds$$

$$= \frac{1}{2} [-\exp(-\lambda_+ s - \nu s^2)]_t^\infty + \frac{1}{2} (\lambda_j - \lambda_{3-j}) \int_t^\infty \exp(-\lambda_+ s - \nu s^2) \, ds$$

$$= \frac{1}{2} \exp(-\lambda_+ t - \nu t^2) + \frac{1}{2} (\lambda_j - \lambda_{3-j}) (\pi/\nu)^{1/2}$$
$$\times \exp\left(\lambda_+^2 / 4\nu \right) \Phi\{ -(2\nu)^{1/2} (t + \lambda_+ / 2\nu) \},$$

where Φ is the standard normal distribution function. We seem to have started with an exponential distribution and ended up with the normal—competing risks is full of surprises! If $\lambda_1 = \lambda_2$ we have the simplified form

$$\bar{F}(j, t) = \frac{1}{2} \exp(-\lambda_+ t - \nu t^2).$$

The sub-hazard functions are $h(j, t) = \lambda_j + vt$ and the overall system hazard function is $h(t) = \lambda_+ + 2vt$. If $v = 0$ the latent failure times are independent and we have a constant hazard rate $h(t) = \lambda_+$, as might be expected from exponential distributions. The effect of dependence is to make the overall hazard increase linearly in time, as might not be expected. Proportional hazards obtains if and only if $(\lambda_1 + vt)/(\lambda_2 + vt)$ is independent of t, that is, if either $\lambda_1 = \lambda_2$ or $v = 0$.

As a complement to Tsiatis's lemma, the sub-densities can be calculated as follows from the joint density $g(t)$ of T, which is useful for some purposes. Substituting the expression

$$\bar{G}(\mathbf{t}) = \int_{t_1}^{\infty} ds_1 \dots \int_{t_p}^{\infty} ds_p \{g(\mathbf{s})\}$$

into the lemma, we obtain

$$f(j, t) = \int_t^{\infty} ds_1 \dots \int_t^{\infty} ds_{j-1} \int_t^{\infty} ds_{j+1} \dots \int_t^{\infty} ds_p \{g(\mathbf{s})\};$$

thus, $g(\mathbf{t})$ is integrated over (t, ∞) with respect to each coordinate except the jth. For $p = 2$ it yields $f(j, t) = \int_t^{\infty} g(\mathbf{r}_j) ds$, where $\mathbf{r}_1 = (t, s)$ and $\mathbf{r}_2 = (s, t)$.

Figure 14.1 shows which probabilities are determined by the sub-survivor functions and marginal survivor functions in the bivariate case. Thus, $\bar{F}(2, t)$ and $\bar{G}_2(t)$ determine probabilities in horizontal strips either side of the line $t_1 = t_2$; $\bar{G}_2(t)$ here denotes the marginal survivor function of T_2, by definition $P(T_2 > t)$. Similarly, $\bar{F}(1, t)$ and $\bar{G}_1(t)$ determine the content of vertical strips. The integral formula for $f(j, t)$ given above for the bivariate case is illustrated by the probability mass b in Figure 14.1 when $s_2 \to s_1$: $f(2, t) = \int_t^{\infty} g(\mathbf{r}) ds$ with $\mathbf{r} = (s, t)$.

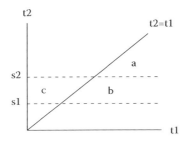

FIGURE 14.1
Probability content of various regions for bivariate case $a = \bar{F}(2, s_2)$; $b = \bar{F}(2, s_1) - \bar{F}(2, s_2)$; $c = \bar{G}_2(s_1) - \bar{G}_2(s_2) - b$.

14.2 Some Examples

One or two well-known multivariate failure time distributions will be examined from the point of view of competing risks.

14.2.1 Freund's Bivariate Exponential

This model was described in Section 7.3. Applying Tsiatis's lemma, the sub-densities, survivor, and hazard functions are derived as

$$f(j, t) = \lambda_j e^{-\lambda_+ t}, \quad \bar{F}(j, t) = (\lambda_j/\lambda_+) e^{-\lambda_+ t}, \quad \bar{F}(t) = e^{-\lambda_+ t}, \quad h(j, t) = \lambda_j.$$

Notice that these functions involve (λ_1, λ_2) but not (μ_1, μ_2); this is because they describe only the time to first failure. In consequence, the likelihood function of Section 13.1 does not involve (μ_1, μ_2), and so no information is forthcoming on these parameters from standard competing risks data. However, full information can be obtained if the system can be observed in operation after a first failure, which is just the situation for which the model is constructed.

The fact that $h(j, t) = \lambda_j$, independent of t, for this distribution should come as no surprise because of the way in which the distribution is set up. During the time before the first failure, which is what the $h(j, t)$ describe, the component failure times are independent exponential variates. During this phase the sub-hazard functions are equal to the marginal ones, and these are constant.

A special case of this model was proposed by Gross et al. (1971). They considered the symmetric case, $\lambda_1 = \lambda_2$ and $\mu_1 = \mu_2$, as appropriate when an individual can survive the loss of one of a pair of organs such as the lungs or kidneys.

14.2.2 Frailty Models

Following on from Section 8.1,

$$\bar{G}(\mathbf{t}) = P(\mathbf{T} > \mathbf{t}) = \int_0^\infty e^{-zs} dK(z) = L_K(s),$$

where $s = \sum_{j=1}^p H_j(t)$. In general terms we can derive $\bar{F}(t)$ from $\bar{G}(t\mathbf{1}_p)$ as $L_K(s_t)$, where $s_t = H_1(t) + \cdots + H_p(t)$, and

$$f(j, t) = [-\partial L_K(s)/\partial t_j]_{t\mathbf{1}_p} = -h_j(t) L'_K(s_t),$$

with $L'_K(s) = dL_K(s)/ds$. The sub-hazard functions are then given by

$$h(j, t) = -h_j(t) L'_K(s_t)/L_K(s_t) = -h_j(t) \, d \log L_K(s_t)/ds_t.$$

We will revisit the MB and MW (multivariate Burr and multivariate Weibull) distributions from Section 8.2. For these the integrated hazard $H_j(t)$ takes

Weibull form $(t/\xi_j)^{\phi_j}$, and we define $s = \sum_{j=1}^{p}(t_j/\xi_j)^{\phi_j}$ and $s_t = \sum_{j=1}^{p}(t/\xi_j)^{\phi_j}$ (spot the difference).

14.2.3 Multivariate Burr (MB)

This distribution has multivariate survivor function $\bar{G}(\mathbf{t}) = (1+s)^{-\nu}$. The sub-densities are then

$$f(j,t) = \nu\phi_j\xi_j^{-\phi_j}t^{\phi_j-1}(1+s_t)^{-\nu-1},$$

and $\bar{F}(t) = (1+s_t)^{-\nu}$. Calculation of $\bar{F}(j,t)$, as the integral of $f(j,t)$, is tractable only if $\phi_j = \phi$ for all j; in this case $s_t = \xi^+t^\phi$ and $\bar{F}(j,t) = (1/\xi^+)(1+s_t)^{-\nu}$, where $\xi^+ = \sum_{j=1}^{p}\xi_j^{-\phi}$. The sub-hazard functions are given by

$$h(j,t) = \nu\phi_j\xi_j^{-\phi_j}t^{\phi_j-1}(1+s_t)^{-1};$$

they are proportional if $\phi_j = \phi$ for all j. The sub-hazards here have a richer structure than in the previous examples. For $\phi_j < 1$, $h(j,t)$ is infinite at $t = 0$ and decreasing thereafter. For $\phi_j > 1$ its behaviour is less transparent: $h(j,t) \sim \nu\phi_j\xi_j^{-\phi_j}t^{\phi_j-1}$ for small t, and

$$h(j,t) \sim \left(\nu\phi_j\xi_j^{-\phi_j}/\xi_m^{-\phi_m}\right)t^{\phi_j-\phi_m-1}$$

for large t, where m is the index of the largest ϕ_k; thus, $h(j,t)$ is initially increasing from zero at $t = 0$, and eventually behaves as $t^{\phi_j-\phi_m-1}$ with $\phi_j - \phi_m - 1 \leq -1$.

Figure 14.2a shows some plots. In particular, the sub-hazard functions rise at first but eventually fall back to earth. This rather limits the application of this model: in most situations hazard functions tend to increase with age.

14.2.4 Multivariate Weibull (MW)

The multivariate survivor function for this distribution is $\bar{G}(\mathbf{t}) = \exp(-s^\nu)$. The sub-densities are then

$$f(j,t) = \nu\phi_j\xi_j^{-\phi_j}t^{\phi_j-1}s_t^{\nu-1}\exp\left(-s_t^\nu\right),$$

and $\bar{F}(t) = \exp(-s_t^\nu)$. Calculation of $\bar{F}(j,t)$, as the integral of $f(j,t)$, is tractable only if $\phi_j = \phi$ for all j; in this case $s_t = \xi^+t^\phi$ and $\bar{F}(j,t) = (1/\xi^+)\exp(-s_t^\nu)$, where $\xi^+ = \sum_{j=1}^{p}\xi_j^{-\phi}$. The sub-hazard functions are given by

$$h(j,t) = \nu\phi_j\xi_j^{-\phi_j}t^{\phi_j-1}s_t^{\nu-1};$$

they are proportional if $\phi_j = \phi$ for all j. For $\phi_j < 1$, $h(j,t)$ is infinite at $t = 0$ and decreasing thereafter, remembering that $0 < \nu < 1$. For $\phi_j > 1$,

$$h(j,t) = \left(\nu\phi_j\xi_j^{-\phi_j}/\xi_l^{(1-\nu)\phi_l}\right)t^{\phi_j-1-(1-\nu)\phi_l}$$

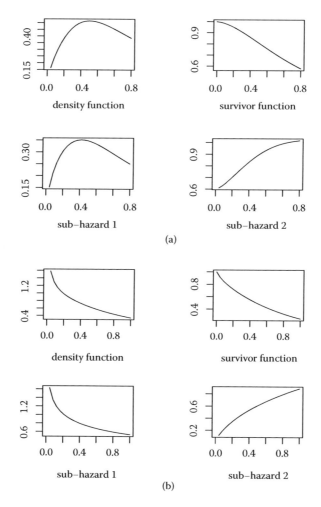

FIGURE 14.2
(a) MB distribution: overall density and survivor functions, and sub-hazard functions; (b) MW distribution: overall density and survivor functions, and sub-hazard functions.

for small t, where ϕ_l is the least among the ϕ_k, and

$$h(j,t) \sim \left(v\phi_j \xi_j^{-\phi_j}/\xi_m^{(1-v)\phi_m}\right)t^{\phi_j-1-(1-v)\phi_m}$$

for large t, where ϕ_m is the maximum of the ϕ_j. Thus, for $\phi_j > 1$, $h(j,t)$ behaves like $t^{\phi_j-1-(1-v)\phi_l}$ near $t = 0$ and like $t^{\phi_j-1-(1-v)\phi_m}$ as $t \to \infty$, where both $\phi_j - 1 - (1-v)\phi_l$ and $\phi_j - 1 - (1-v)\phi_m$ could be positive or negative. Thus, for example, the fabled *bathtub shape* (down-along-up) is not attainable because it would require $\phi_j - 1 - (1-v)\phi_l < 0$ and $\phi_j - 1 - (1-v)\phi_m > 0$, which would entail $\phi_l > \phi_m$.

   Figure 14.2b shows some plots. The sub-hazards here can rise or fall with age, so the model is more flexible in this respect than the multivariate Burr.

14.2.5 A Stochastic Process Model

Yashin et al. (1986) proposed the following setup in connection with human mortality. There is an unobserved physiological ageing process, $\{z(t) : t \geq 0\}$, determined by a stochastic differential equation:

$$dz(t) = \{a_0(t) + a_1(t)z(t)\}dt + b(t)W(t),$$

where $W(t)$ is a standard *Wiener process* or *Brownian motion* of specified dimension. In the terminology, $z(t)$ is an *Ito process* with *drift coefficient* $a_0(t) + a_1(t)z(t)$ and *diffusion coefficient* $b(t)$, and Yashin et al. gave arguments in support of these model specifications for the context. The essence of the competing risks model here is that the latent failure times are conditionally independent given the current history, $\mathcal{H}_t = \{z(s) : 0 \leq s \leq t\}$, of the process: specifically,

$$P(\mathbf{T} > \mathbf{t} \mid \mathcal{H}_t) = \prod_{j=1}^{p} P(T_j > t_j \mid \mathcal{H}_t).$$

Further, it is assumed that $P(T_j > t_j \mid \mathcal{H}_t) = h_j(t_j \mid z_t)$, a marginal hazard depending on $\mathcal{H}_t$ only through the current value z_t. To obtain the marginal hazard functions, $h_j(t_j) = d \log G_j(t_j)/dt_j$, z_t must be integrated out. The various other functions now follow, and Yashin et al. gave some examples, further development, and an application to some mortality data.

14.2.6 Other Applications

Moeschberger and David (1971) wrote down the forms of the likelihood functions for generally dependent risks for observed failure times and for grouped and censored failure times. They went on to consider in detail independent Weibull and exponential latent failure times, giving formulae for the likelihood function and its derivatives. Explanatory variables were not considered. They applied the Weibull model to Boag's (1949) data.

Moeschberger (1974) derived the forms of the likelihood functions for bivariate normal and Weibull distributions. The former does not yield explicit expressions for the sub-densities, they involve the normal distribution function integral; in any case, the normal distribution does not provide a wholly convincing model since lifetimes cannot be negative and their distributions are usually positively skewed. The latter was taken to be of the Marshall–Olkin type (Section 7.4). Moeschberger discussed various properties of the Weibull model and fitted it, by maximum likelihood, to some of Nelson's (1970) data.

Farragi and Korn (1996) considered a particular family of frailty models previously suggested by Oakes (1989). For competing risks in the clinical context they proposed a version in which the medical treatment affects only one of the latent failure times.

14.3 Further Aspects

14.3.1 Latent Failure Times versus Hazard Functions

As indicated above, the traditional approach to modelling competing risks is via a multivariate latent failure time distribution $\bar{G}(\mathbf{t})$, moreover, one in which the T_j are independent. The modern approach is more hazard conscious: the framework for modelling and inference is based on the sub-hazard functions.

Prentice et al. (1978), reiterated in Kalbfleish and Prentice (1980, Section 7.1.2), argued strongly against the traditional approach. Their attack was based partly on the non-identifiability of $\bar{G}(\mathbf{t})$ (Section 17.1). They also pointed out that the traditional aim of isolated inference about a subset of components of **T** makes unjustifiable assumptions about the effect of removing the other risks, even if this were possible. Finally, they cast doubts on the physical existence of hypothetical latent failure times in many contexts. Thus, one should set up models only for observable phenomena, a kind of "What you see is what you set" doctrine.

The important arguments of Prentice et al. (1978) deserve examination. The lack of identifiability is softened to some extent in the regression case (Section 17.5). That removal of some risks might change the circumstances applies equally to hazard models, in fact to any circumstance in statistics where one is trying to extrapolate beyond what has been observed. Finally, the one centering upon the lack of physical meaning of latent failure times seems to be arguable in some cases. A case for the defence, of using a multivariate failure time distribution, has been put by Crowder (1994). Other support for such heresy is given by Aalen (1995), maybe the models should go further than the data, and that the amount of speculation this entails may be fruitful.

In chewing over the arguments, what is perhaps the main recommendation for the hazard-based approach has not yet been mentioned. This is that models for processes evolving over time can be developed much more naturally in terms of hazards than multivariate survivor functions. Thus, one can deal with quite complex situations that would be difficult, even intractable, from the traditional point of view. A vast amount of modern methodology and interpretation has now been developed and implemented around hazard functions.

14.3.2 Marginals versus Sub-Distributions

Historically, the main aim of competing risks analysis was seen as the estimation of the marginal distributions, that is, taking data in which the risks act together and trying to infer how some of them would act in isolation. In the older terminology (Section 14.1), this is an attempt to make inferences about the *net risks* from observations on the *crude risks*. In more modern terminology it is like investigative reporting, uncovering private lives or, rather, lifetimes. In algebraic terms it amounts to deriving the $\bar{G}_j(t)$ from the $\bar{F}(j, t)$. It will be seen in Chapter 17 that this cannot be done unless $\bar{G}(\mathbf{t})$ is restricted in some way, perhaps by specifying a fully parametric model for it. Without making such assumptions the best that can be done is to derive some bounds for $\bar{G}_j(t)$ in terms of the $\bar{F}(j, t)$. The question of whether it is sensible to want to estimate the $\bar{G}_j(t)$ in the first place is another matter. It is only natural to want to focus on the cause of failure of major interest, and make inferences and predictions free of the nuisance aspects. For example, the doctor would probably want to give advice about the illness in question, without having to bring in the possibility of being run over by a bus. However, the counter-argument is that the $\bar{G}_j(t)$ do not describe events that physically occur; they only describe failures from isolated causes in situations in which all the other risks have been removed somehow. It is the $\bar{F}(j, t)$, not the $\bar{G}_j(t)$, that are relevant to the real situation, that is, to failures from cause j that can actually occur. Prentice et al. (1978) emphasized these points in their vigorous criticism of the traditional approach to competing risks.

In certain special circumstances it might be possible to observe the effect of failure cause j alone, that is, with other causes absent. In that case the *marginal hazard rate* $h_j(t) = -d \log \bar{G}_j(t)/dt$ is relevant; $h_j(t)$ is also known as the *marginal intensity function*. However, it might happen that the removal of the other causes of failure materially changes the circumstances. In that case it is not valid to assume that when T_j is observed in isolation its distribution is the same as the former marginal T_j-distribution derived from the joint distribution. This has long been recognized (for example, Makeham, 1874; Cornfield, 1957).

Elandt-Johnson (1976) made a thoughtful distinction in this connection. She differentiated between (1) simply ignoring some causes of failure, in which case one sets the corresponding t_j to zero in $\bar{G}(\mathbf{t})$ and works with the marginal survivor function of the rest, and (2) eliminating or postponing some causes of failure by improved care or maintenance, in which case one should work with the conditional survivor function of the rest given that the corresponding T_j are effectively infinite. Clearly, (1) and (2) will in general give different answers, and she gave examples to illustrate this.

An obvious inequality connecting the observable and unobservable functions is $\bar{F}(j, t) \leq \bar{G}_j(t)$, that is, $P(C = j, T_j > t) \leq P(T_j > t)$. The following theorem gives more refined ones.

14.3.3 Peterson's Bounds (Peterson, 1976)

1. Let $v = \max\{t_1, \ldots, t_p\}$. Then

$$\sum_{j=1}^{p} \bar{F}(j, v) \leq \bar{G}(\mathbf{t}) \leq \sum_{j=1}^{p} \bar{F}(j, t_j).$$

2. For $t > 0$ and each k,

$$\sum_{j=1}^{p} \bar{F}(j, t) \leq \bar{G}_k(t) \leq \bar{F}(k, t) + (1 - p_k).$$

Proof

1. Note that $\bar{G}(\mathbf{t}) = \sum_{j=1}^{p} P(C = j, \mathbf{T} > \mathbf{t})$, and that the right hand side is bounded by:

$$rhs \geq \sum_{j} P(C = j, \mathbf{T} > v\mathbf{1}_p) = \sum_{j} P(C = j, T > v) = \sum \bar{F}(j, v);$$

$$rhs \leq \sum_{j} P(C = j, T_j > t_j) = \sum_{j} \bar{F}(j, t_j).$$

2. Set $t_k = t$ and all other $t_j = 0$ in (i).

Peterson (1976) remarked that, although the bounds cannot be improved, because they are attainable by particular distributions, they are generally rather wide, not restricting $\bar{G}(\mathbf{t})$ and $\bar{G}_j(t_j)$ with any great precision. Thus, it is possible to have a pretty good picture of the $\bar{F}(j, t)$ without this picture's telling you much about $\bar{G}(\mathbf{t})$ or the $G_j(t_j)$. We shall return to this identifiability aspect in Chapter 17. Klein and Moeschberger (1988) showed how the bounds can be made tighter if certain additional information is available. More recently, Bedford and Meilijson (1998) have given more refined bounds for the marginals, and other developments have been made by Nair (1993) and Karia and Deshpande (1997).

Rachev and Yakovlev (1985) studied the inverse problem to Peterson's, deriving bounds for the sub-survivor functions $\bar{F}(j, t)$ in terms of the marginal $\bar{G}_j(t)$. One can envisage a situation where the components can be tested individually to predict their behaviour when incorporated into a system. From the marginals they derived bounds for the probabilities $p_j = \bar{F}(j, 0)$ in a two-component system, and for the covariance $\text{cov}(T_1, T_2)$. They applied their bounds in a certain model for the effects and interaction of two injuries, and derived confidence intervals for the p_j. They assumed for this that empirical estimates of the functions $\bar{G}_1(t)$, $\bar{G}_2(t)$, and $\bar{F}(t)$ are available.

14.4 Independent Risks

The classical assumption is that the risks act independently, that is, that the latent failure times T_j are independent. This would appear to be especially dubious in medical applications, which is, ironically, the context in which competing risks developed. The complex biochemical interactions between different disease processes and physiological conditions would seem to rule out any hope of independence at the outset, making such an assumption not even worthy of barroom debate. On the other hand, in reliability theory a lot of work involving complex systems is based on an assumption of stochastic independence, this being justified by the physically independent functioning of components. Even in such situations there might be effective dependence between risks. This can be due to load sharing between components, or randomly fluctuating loads, or other shared common factors such as working environment; previous exposure to wear and tear; and quality of materials, manufacture, and maintenance. To assume independence one must be sure that a failure of one type has absolutely no bearing at all on the likelihood of failure of any other type, not even through some indirect link.

In spite of the preceding diatribe it is useful to study the special properties of independent-risks systems. This is (a) to understand better the classical approach, (b) to disentangle properties that rely on independence from ones that do not, and (c) to identify the effects of a lack of independence. First, it is clear that when independent risks obtains the set of marginals $\{\bar{G}_j(t)\}$ determines the joint $\bar{G}(\mathbf{t})$ as $\prod_{j=1}^{p} \bar{G}_j(t_j)$. Thereby all other probabilistic aspects of the setup have also been determined. For instance, Tsiatis's lemma yields

$$f(j, t) = g_j(t) \prod_{k \neq j} \bar{G}_k(t) = \{g_j(t)/\bar{G}_j(t)\} \prod_{k=1}^{p} \bar{G}_k(t) = h_j(t)\bar{F}(t)$$

in this case, where $g_j(t) = -d\bar{G}_j(t)/dt$ is the marginal density of T_j.

Example (Lagakos, 1978)

Lagakos' model (Section 13.5) was based on independent exponential latent failure times for the two causes of death: he took $\bar{G}(\mathbf{t}) = \bar{G}_1(t_1)\bar{G}_2(t_2)$, with $\bar{G}_j(t) = e^{-\lambda_j t}$ ($j = 1, 2$). From $f(j, t) = h_j(t)\bar{F}(t)$ then follows

$$f(j, t) = \lambda_j e^{-\lambda_+ t}, \quad \bar{F}(j, t) = (\lambda_j/\lambda_+)e^{-\lambda_+ t},$$

where $\lambda_+ = \lambda_1 + \lambda_2$.

The following theorem gives some other implications of independent risks: in particular, it shows that the unobservable $\bar{G}_j(t)$ can then be derived explicitly from the observable $\bar{F}(j, t)$.

14.4.1 Gail's Theorem (Gail, 1975)

The implications (1) $\Rightarrow$ (2) $\Rightarrow$ (3) $\Rightarrow$ (4) hold for the following propositions:
  1. *independent risks obtains;*
  2. *$h(j, t) = h_j(t)$ for each j and $t > 0$;*
  3. *the set of sub-survivor functions $\bar{F}(j, t)$ determines the set of marginals $\bar{G}_j(t)$, explicitly,*

$$\bar{G}_j(t) = \exp\left\{-\int_0^t h(j, s)ds\right\};$$

  4. *$\bar{G}(t\mathbf{1}_p) = \prod_{j=1}^{p} \bar{G}_j(t)$.*

Proof Under (1), $\log \bar{G}(t) = \sum_{j=1}^{p} \log \bar{G}_j(t_j)$ so, differentiating with respect to t_j and then setting $\mathbf{t} = t\mathbf{1}_p$,

$$[-\partial \log \bar{G}(\mathbf{t})/\partial t_j]_{t\mathbf{1}_p} = -d \log \bar{G}_j(t)/dt$$

for each j. By Tsiatis's lemma (Section 14.1), the left-hand side here is equal to

$$f(j, t)/\bar{G}(t\mathbf{1}_p) = f(j, t)/\bar{F}(t) = h(j, t),$$

and the right-hand side is $h_j(t)$, so (2) is verified. Integration with respect to t now yields

$$\log \bar{G}_j(t) = -\int_0^t h(j, s) ds,$$

in which the $h(j, s)$ are determined by the $\bar{F}(j, s)$ via

$$h(j, s) = -d\bar{F}(j, s)/ds \div \sum_{j=1}^p \bar{F}(j, s).$$

Thus, (3) is verified. Lastly, (4) follows from

$$\prod_{j=1}^p \bar{G}_j(t) = \exp\left\{ -\sum_{j=1}^p \int_0^t h(j, s) ds \right\} = \exp\left\{ -\int_0^t h(s) ds \right\} = \bar{F}(t) = \bar{G}(t\mathbf{1}_p).$$

Part (4) of the theorem looks a bit like independence of the T_j but isn't because that would require equality of $\bar{G}(\mathbf{t})$ and $\prod \bar{G}_j(t_j)$ for all $\mathbf{t}$. What it does say is, in effect, that $\bar{R}(t\mathbf{1}_p) = 1$, where $\bar{R}(\mathbf{t})$ is the dependence measure defined in Section 6.2; this is independence on the diagonal $\mathbf{t} = t\mathbf{1}_p$.

Example (Berkson and Elveback, 1960)

The authors considered two independent risks with exponential failure times. They gave some basic formulae, discussed estimation, including maximum likelihood, and applied the methods to a prospective study relating smoking habits to time and cause of death among 200,000 U.S. males. Their model can be written in terms of the sub-distribution functions as

$$F(j, t) = (\alpha_j/\alpha_+)(1 - e^{-\alpha_+ t}),$$

for non-smokers, and

$$F(j, t) = (\beta_j/\beta_+)(1 - e^{-\beta_+ t}),$$

for smokers; $j = 1$ for death from lung cancer, $j = 2$ for death from other causes, $\alpha_+ = \alpha_1 + \alpha_2$ and $\beta_+ = \beta_1 + \beta_2$. The comparison of particular interest is between α_j and β_j.

Example (Klein and Basu, 1981)

The authors considered Weibull accelerated-life tests. Their model is one of independent risks with marginal hazard functions

$$h_j(t; \mathbf{x}) = \psi_{xj} h_0(t),$$

where $h_0(t) = \phi_j t^{\phi_j - 1}$ has the Weibull hazard form with shape parameter ϕ_j, the scale parameter being absorbed in $\psi_{xj} = \exp(\mathbf{x}^\mathsf{T}\beta_j)$, which combines the vector $\mathbf{x}$ of explanatory variables with the regression coefficients β_j. At first sight this looks more like a univariate proportional hazards specification than an accelerated life one, but the two coincide for the Weibull distribution. The authors computed likelihood functions and derivatives explicitly for Type 1 (fixed time), Type 2 (fixed number of failures), and progressively (hybrid scheme) censored samples.

14.4.2 Other Applications

Ebrahimi (1996) extended parametric modelling with independent latent failure times to the case of uncertainly-observed failure causes.

Gasemyr and Natvig (1994) outlined a Bayesian approach to estimating latent lifetimes based on autopsy data.

Herman and Patel (1971) assumed independent risks, for which they summarized the basic probability functions. They went on to consider, for two risks, parametric maximum likelihood estimation for Type 1 (fixed time) censored samples, and specialized their formulae to exponential and Weibull latent failure times. They re-analyzed Mendenhall and Hader's (1958) data based on the exponential model.

14.5 The Makeham Assumption

Condition (2) of Gail's theorem (Section 14.4) is well known in the competing risks business. Gail (1975) called it the *Makeham assumption*, after the 1874 paper, and we shall follow suit. Cornfield (1957) was also concerned with this assumption, and Elandt-Johnson (1976) referred to it as the identity of forces of mortality. It says that the two hazard functions for risk j, the one in the presence of the other risks and the other in their hypothetical absence, are equal. Even when the risks are not independent it can still hold for some j, as demonstrated explicitly by the example of Williams and Lagakos (1977, Section 3). According to the theorem given below in Section 17.3 this can obtain for all j when the risks are not independent: one can start with an independent-risks model, for which, necessarily, $h(j, t) = h_j(t)$ for all j, and then modify $\bar{G}(\mathbf{t})$ with infinite variety without disturbing any of the $h(j, t)$ or $h_j(t)$.

The Makeham assumption is closely related to the dependence measure $\bar{R}(\mathbf{t})$ via

$$[\partial \log \bar{R}(\mathbf{t}) / \partial t_j]_{t1_p} = h_j(t) - h(j, t).$$

A further implication is that

$$f(j, t) = \{g_j(t) / \bar{G}_j(t)\} \bar{F}(t) = g_j(t) \left\{ \prod_{k \neq j} \bar{G}_k(t) \right\} \bar{R}(t1_p) = g_j(t) \left\{ \prod_{k \neq j} \bar{G}_k(t) \right\}.$$

This says that $f(j, t)$ is deceptively expressible in the same form that it would have in the independent-risks case.

Under the Makeham assumption the likelihood contribution L_j in Section 13.1 involves only the jth marginal survivor function $\bar{G}_j(t)$, remembering that $h_j(t) = -d \log \bar{G}_j(t)/dt$. However, even in this case, it is the $\bar{F}(j, t)$, not the $\bar{G}_j(t)$, that are of real-world relevance, as discussed in Section 14.3. In the special case of independence the function $\bar{F}(j, t)$ is given by

$$\int_t^\infty f(j, s)ds = \int_t^\infty g_j(s) \prod_{k \neq j} \bar{G}_k(s)ds,$$

from Tsiatis's lemma (Section 14.1). Thus, even here, to make inferences about observable occurrences of failures of type j one still needs to use all the $\bar{G}_j(t)$. The point is that observation of a failure of type j necessarily involves the prior non-occurrence of failures of all other types.

Part (3) of Gail's theorem (Section 14.4.1) states that, under independent risks, the set of observable distributions $\bar{F}(j, t)$ and the set of marginal distributions $\bar{G}_j(t)$ each give complete descriptions of the setup. That $(1) \Rightarrow (3)$ has been proved by Berman (1963) and Nadas (1970); these papers also give formula (3) in slightly different forms.

14.5.1 Proportional Hazards

For the case of independent risks, the equivalence of proportional hazards and independence of time and cause of failure (Section 12.4) has been shown in various versions by Allen (1963), Sethuraman (1965), and Nadas (1970a).

The following result has been given for the special case of independent risks by Armitage (1959) and Allen (1963). As now shown, it actually holds under the weaker, Makeham assumption.

Lemma (Crowder, 1994)

Under the Makeham assumption proportional hazards obtains if and only if $\bar{G}_j(t) = \bar{F}(t)^{p_j}$.

Proof Proportional hazards means $h(j, t) = p_j h(t)$. Under the Makeham assumption, that $h(j, t) = h_j(t)$, this is $h_j(t) = p_j h(t)$, i.e.,

$$d \log \bar{G}_j(t)/dt = p_j \, d \log \bar{F}(t)/dt.$$

On integration, this gives the result.

The relation in the lemma, $\bar{G}_j(t) = \bar{F}(t)^{p_j}$, is reminiscent of that in the theorem of Section 12.4, $\bar{F}(j, t) = p_j \bar{F}(t)$. However, that one is rather more fundamental in the sense that it is equivalent to proportional hazards whether or not the Makeham assumption holds, in fact, whether or not there are any latent failure times associated with the setup at all.

A consequence of the lemma is that in the independent-risks case $\bar{G}(\mathbf{t})$ can be expressed as $\prod_{j=1}^{p} \bar{F}(t_j)^{p_j}$, that is, solely in terms of the function $\bar{F}(t)$. For statistical modelling this means that we only have one univariate survivor function $\bar{F}(t)$ to worry about, rather than a whole set of $\bar{F}(j, t)$ or a multivariate $\bar{G}(\mathbf{t})$. Suppose, for example, that each $\bar{F}(j, t)$ requires q parameters when modelled separately. Then the full set would require pq parameters that would be reduced to $q + p - 1$ under proportional hazards, the extra $p - 1$ parameters being the p_js.

David (1970) noted that the relation $\bar{F}(t) = \bar{G}_j(t)^{1/p_j}$ has the same form as one that represents the survivor function of the minimum of an integer number, $1/p_j$, of independent variates each with survivor function $\bar{G}_j(t)$. This led him to suggest the use of extreme value distributions, in particular, Weibull with $\bar{F}(t) = \exp\{-(t/\xi)^\phi\}$. In a later article, David (1974) reviewed some basic theory and considered parametric maximum likelihood estimation, in particular for independent Weibull latent failure times. He also discussed the Marshall–Olkin system (Section 7.4) and gave a generalisation.

14.6 A Risk-Removal Model

Hoel (1972) developed a model that allows individuals in the population to have their own immunity profiles. Each individual has a set of risk indicators $(r_1, \ldots, r_p)$ such that $r_j = 1$ if he is vulnerable to risk j and $r_j = 0$ if not, the proportion vulnerable being q_j in the population. Technically, the latent failure time T_j is represented as S_j/r_j, where the S_j and r_j are all independent, S_j has distribution function $G_{S_j}(t)$, and $P(r_j = 1) = q_j$. As usual $T = \min(T_1, \ldots, T_p) = T_C$. In effect, the T_j are being turned into improper random variables in the sense that $P(T_j > t) \to 1 - q_j$, not 0, as $t \to \infty$. Hoel took one of the q_j to be equal to 1 to make T finite with probability 1—no immortality here, then. The r_j are not directly observed: they are random effects that vary from case to case. If the r_j were known, one could just condition out the T_j for which $r_j = 0$, as suggested by Elandt-Johnson (1976, see Section 14.3 above). The q_j are parameters of the model.

The marginal survivor function of T_j can be obtained as

$$\bar{G}_j(t) = P(T_j > t) = P(S_j > tr_j) = P(S_j > t)q_j + P(S_j > 0)(1 - q_j)$$
$$= 1 - q_j G_{S_j}(t).$$

Hence,

$$\bar{F}(t) = \prod_{j=1}^{p} \bar{G}_j(t) = \prod_{j=1}^{p}\{1 - q_j G_{S_j}(t)\},$$

$$h(j, t) = h_j(t) = g_j(t)/\bar{G}_j(t) = q_j g_{S_j}(t)/\{1 - q_j G_{S_j}(t)\},$$

where $g_{S_j}(t) = dG_{S_j}(t)/dt$ is the density of S_j, and $f(j, t) = h(j, t)\bar{F}(t)$.

Hoel applied his model to two groups of mortality data, each with three risks; we borrowed his data earlier in Section 13.4. For $G_{S_j}(t)$ he used the Makeham–Gompertz form $1 - \exp\{-\phi_j(1 - e^{\lambda_j t})\}$. There were no explanatory variables, other than the grouping, and he employed parametric maximum likelihood estimation and used some goodness-of-fit indicators and plots.

A dependent-risks version of Hoel's model can be derived as follows. We have

$$\bar{G}(\mathbf{t}) = P\{\cap_{j=1}^{p}(S_j > r_j t_j)\} = \sum_r g(\mathbf{r})\bar{G}_S(r_1 t_1, \ldots, r_p t_p),$$

where the summation $\sum_r$ is over all binary vectors $\mathbf{r} = (r_1, \ldots, r_p)$, $\bar{G}_S(\mathbf{s})$ is the joint survivor function of $\mathbf{S} = (S_1, \ldots, S_p)$, and $g(\mathbf{r})$ is the joint probability function of $\mathbf{r}$ given by

$$g(\mathbf{r}) = \prod_{j=1}^{p} \{q_j^{r_j}(1 - q_j)^{1-r_j}\}.$$

For example, for $p = 2$

$$\bar{G}(\mathbf{t}) = \bar{G}_S(t_1, t_2)q_1 q_2 + \bar{G}_S(0, t_2)(1 - q_1)q_2 + \bar{G}_S(t_1, 0)q_1(1 - q_2)$$
$$+ \bar{G}_S(0, 0)(1 - q_1)(1 - q_2),$$

yielding marginals

$$\bar{G}_1(t_1) = q_1\bar{G}_S(t_1, 0) + (1 - q_1) \text{ and } \bar{G}_2(t_2) = q_2\bar{G}_S(0, t_2) + (1 - q_2).$$

For independent S_j, $\bar{G}(\mathbf{t})$ reduces to $\prod_{j=1}^{p}\{q_j G_{S_j}(t_j) + (1 - q_j)\}$, in accordance with $\bar{G}_j(t) = 1 - q_j G_{S_j}(t)$. The associated probability functions are

$$\bar{F}(t) = \bar{G}(t\mathbf{1}_p) = \sum_r g(\mathbf{r})\bar{G}_S(t\mathbf{r})$$

and

$$f(j, t) = [-\partial\bar{G}(\mathbf{t})/\partial t_j]_{t\mathbf{1}_p} = \sum_r g(\mathbf{r})r_j[-\partial\bar{G}_S(\mathbf{s})/\partial s_j]_{t\mathbf{r}}.$$

14.7 A Degradation Process

Consider a system of independently functioning components, the jth being subject to a degradation process $\{Y_{jt} : t \geq 0\}$ with $Y_{j0} = 0$. As soon as Y_{jt} reaches level $y_j^* > 0$ the jth component fails, thereby bringing the whole system crashing down around our ears. Let T_j be the time to failure of component j, $T_j = \inf\{t : Y_{tj} > y_j^*\}$. Then the system failure time is $T = \min(T_1, \ldots, T_p)$

and the system lifetime has survivor function

$$P(T > t) = \prod_{j=1}^{p} P(T_j > t).$$

14.7.1 Wiener Process

The following result can be extracted from that quoted in Section 11.6:

$$P(T_j > a) = \Phi\left(\frac{y_j^* - \mu_j t}{\sigma_j t^{1/2}}\right) - \exp\left(\frac{2\mu_j y_j^*}{\sigma_j^2}\right) \Phi\left(\frac{-y_j^* - \mu_j t}{\sigma_j t^{1/2}}\right);$$

μ_j and σ_j are the drift and variance parameters of $\{Y_{jt}\}$.

14.7.2 Compound Poisson and Compound Birth Processes

The relevant formulae for the component lifetime survivor functions are given in Section 11.4.

14.7.3 Gamma Process

The survivor function for the jth component lifetime can be written as

$$P(T_j > t) = \Gamma(\tau_j t)^{-1} \gamma(\tau_j t, y_j^*/\xi_j),$$

from the formula given in Section 11.4; ξ_j and τ_j are scale and shape parameters for the process $\{Y_{jt}\}$.

For the case where the components do not function independently, an appropriate multivariate joint distribution is needed for the T_j. This is likely to make life much more difficult.

14.8 Exercises

1. Consider Freund's bivariate exponential distribution (Sections 7.3 and 14.2). Under what parametric conditions are the risks independent? How about the Makeham assumption?

2. Repeat Question 1 for the MB and MW distributions (Sections 7.3 and 14.2).

3. An extension of Gumbel's bivariate exponential distribution. Replace t_j with $\phi_j(t)$ for specified functions ϕ_j ($j = 1, 2$). Then the joint survivor function is

$$\bar{G}(t_1, t_2) = \exp\{-\lambda_1 \phi_1(t_1) - \lambda_2 \phi_2(t_2) - \nu \phi_1(t_1)\phi_2(t_2)\}.$$

Derive the joint density function as

$$g(t_1, t_2) = \bar{G}(t_1, t_2)\phi_1'(t_1)\phi_2'(t_2) \left[-v + \{\lambda_1 + v\phi_2(t_2)\}\{\lambda_2 + v\phi_1(t_1)\} \right].$$

What conditions do ϕ_1 and ϕ_2 need to satisfy for this to be a valid bivariate distribution? Write down the survivor function $\bar{F}(t)$ and derive the sub-densities $f(j, t)$ and sub-hazards $h(j, t)$. Show that proportional hazards obtains when $\phi_1(t) = (\lambda_2/v)(e^{vt} - 1)$ and $\phi_2(t) = (\lambda_1/v)(e^{vt} - 1)$. How do you think this result was obtained?

14.9 Hints and Solutions

1. Independence: $\mu_j = \lambda_j$ ($j = 1, 2$). Makeham: $h(2, t) = \lambda_2$,

$$h_2(t) = \{\lambda_1\mu_2 e^{-\mu_2 t} + (\lambda_2 - \mu_2)\lambda_+ e^{-\lambda_+ t}\}/\{\lambda_1 e^{-\mu_2 t} + (\lambda_2 - \mu_2)e^{-\lambda_+ t}\} :$$

equal if $\lambda_2 = \mu_2$.

3. $g(t_1, t_2) = \partial^2 \bar{G}(t_1, t_2)/\partial t_1 \partial t_2 = \cdots$.
Valid if $\bar{G}(0, 0) = 1$ and $g(t_1, t_2) \geq 0$ everywhere, which is so if $\phi_j(0) = 0$, $\phi_j(t) \geq 0$ and $v \leq \lambda_1 \lambda_2$.
Survivor function $\bar{F}(t) = \bar{G}(t, t)$, sub-densities $f(1, t) = \bar{G}(t, t)\{\lambda_1 + v\phi_2(t)\}\phi_1'(t)$, sub-hazards $h(1, t) = f(1, t)/\bar{F}(t)$.
Proportional hazards if $h(1, t)/h(2, t)$ independent of t, which is so under the given ϕ_j. How? Write down the condition and solve for the ϕ_j.

15

Continuous Time: Non- and Semi-Parametric Methods

15.1 The Kaplan–Meier Estimator

Suppose that the observed failure times, from all causes, are $t_1 < t_2 < \cdots < t_m$, and take $t_0 = 0$ and $t_{m+1} = \infty$. Denote by R_{jl} the set of r_{jl} individuals who fail from cause j at time t_l; then r_{jl} is zero and R_{jl} is empty for $l = 0$ and $l > m$. Also, unless there are tied failures from different causes at t_l, only one of $r_{1l}, \ldots, r_{pl}$ will be non-zero. Let $R_l = R_{1l} \cup \ldots \cup R_{pl}$ and denote by S_l the set of s_l individuals whose failure times, t_{ls} ($s = 1, \ldots, s_l$), are right-censored during $[t_l, t_{l+1})$.

We take the form of the log-likelihood function as in Section 13.1:

$$L = \prod_{obs} f(c_i, t_i) \times \prod_{cens} \bar{F}(t_i).$$

For the purpose here, write $f(c, t)dt$ as $\nabla \bar{F}(c, t)$, which denotes $\bar{F}(c, t-) - \bar{F}(c, t)$. Then,

$$L = \prod_{l=1}^{m} \prod_{i \in R_l} \nabla \bar{F}(c_i, t_i) \times \prod_{l=0}^{m} \prod_{i \in S_l} \bar{F}(t_{ls}).$$

The maximum likelihood estimate of $\nabla \bar{F}(c_i, t_l)$ cannot be zero, since this would make L zero, so $\hat{\bar{F}}(c_i, t_l)$ must be discontinuous at t_l whenever $i \in R_l$. Also, for L to be maximised it is necessary to take $\hat{\bar{F}}(t_{ls}) = \hat{\bar{F}}(t_l)$ for each individual $i \in R_l$ (as in Section 4.1). For this, one must take $\hat{\bar{F}}(j, t_{ls}) = \hat{\bar{F}}(j, t_l)$ for each j; this is because $\bar{F}(t) = \sum_j \bar{F}(j, t)$ and each $\bar{F}(j, t)$ is monotone decreasing in t. In particular, $\hat{\bar{F}}(t_{0s}) = \hat{\bar{F}}(t_0) = 1$, so the second product term in L effectively begins at $l = 1$ rather than at $l = 0$. Hence, we have to maximize

$$\prod_{l=1}^{m} \left[\prod_{j=1}^{p} \{\nabla \bar{F}(j, t_l)\}^{r_{jl}} \times \{\bar{F}(t_l)\}^{s_l} \right],$$

where $\bar{F}(j, t)$ has a discontinuity at t_l if $r_{jl} > 0$. As in Section 4.1 the discontinuities in the *mle* of the $\bar{F}(j, t)$ make them discrete subsurvivor functions, and so we re-express the functions in terms of discrete sub-hazards:

$$\nabla \bar{F}(j, t_l) = h(j, t_l)\bar{F}(t_l-), \quad \bar{F}(t_l) = \prod_{s=1}^{l}\{1 - h(t_s)\},$$

where $h(t) = \sum_j h(j, t)$. Now the $h(j, t_l)$ have to be found to maximize

$$\prod_{l=1}^{m}\left\{\prod_{j=1}^{p}\left[h(j, t_l)\prod_{s=1}^{l-1}\{1 - h(t_s)\}\right]^{r_{jl}} \times \left[\prod_{s=1}^{l}\{1 - h(t_s)\}\right]^{s_l}\right\}$$

$$= \prod_{l=1}^{m}\left[\prod_{j=1}^{p}h(j, t_l)^{r_{jl}} \times \{1 - h(t_l)\}^{q_l - r_l}\right],$$

where $r_l = r_{1l} + \cdots + r_{pl}$ and

$$q_l = (r_l + s_l) + \cdots + (r_m + s_m).$$

The *mle* are

$$\hat{h}(j, t_l) = r_{jl}/q_l, \quad \hat{h}(t_l) = r_l/q_l,$$

and those for $\bar{F}(t)$, p_c and $\bar{F}(j, t)$ follow from the formulae

$$\bar{F}(t_l) = \prod_{s=1}^{l}\{1 - h(t_s)\}, \quad f(c, t_l) = h(c, t_l)\bar{F}(t_{l-1}),$$

$$p_c = \sum_{l=1}^{m} f(c, t_l), \quad \bar{F}(c, t_l) = \sum_{s=l+1}^{m} f(c, s).$$

If $s_m > 0$, $\hat{h}(t_m) < 1$ and so $\hat{\bar{F}}(t_m) > 0$. Then $\hat{\bar{F}}(t)$ is undefined for $t > t_m$, which implies the same for the consequent estimates. For example, in this case

$$\hat{p}_+ := \sum_{l=1}^{m} \hat{f}(t_l) = 1 - \hat{\bar{F}}(t_m) < 1 :$$

a pragmatic solution is to divide the $\hat{p}_c$ by $\hat{p}_+$, so forcing the estimates to sum to 1.

Plots of the estimated functions can be made as described previously. Note the distinction between cumulative and integrated hazard functions: in the case of continuous failure times, $-\log \bar{F}(t)$ is the integrated overall hazard function, not the cumulative one; again, $-\log \bar{F}(j, t)$ is neither the integrated nor the cumulative sub-hazard function.

TABLE 15.1

Estimates of Hazard and Survivor Functions

l	t_l	$\hat{h}_{1l}$	$\hat{h}_{2l}$	$\hat{h}_{3l}$	$\hat{h}_l$	$\hat{F}(1, t_l)$	$\hat{F}(2, t_l)$	$\hat{F}(3, t_l)$	$\hat{F}(t_l)$
1	0.040	0.000	0.000	0.010	0.000	0.222	0.384	0.384	0.990
2	0.042	0.000	0.000	0.010	0.010	0.222	0.384	0.374	0.980
3	0.051	0.000	0.000	0.010	0.010	0.222	0.384	0.364	0.970
4	0.062	0.000	0.000	0.010	0.010	0.222	0.384	0.354	0.960
5	0.159	0.011	0.000	0.000	0.011	0.212	0.384	0.354	0.949
...	...	...	...	...	...	...	...	...	...
91	0.738	0.000	0.200	0.000	0.200	0.000	0.020	0.020	0.040
92	0.748	0.000	0.250	0.000	0.250	0.000	0.010	0.020	0.030
93	0.753	0.000	0.333	0.000	0.333	0.000	0.000	0.020	0.020
94	0.761	0.000	0.000	0.500	0.500	0.000	0.000	0.010	0.010
95	0.763	0.000	0.000	1.000	1.000	0.000	0.000	0.000	0.000

15.1.1 Survival Times of Mice

The estimates are derived here for Hoel's (1972) data (Section 13.4). Some results of the computations are given in Table 15.1 for Group 1 of the data; the survival times have been scaled down by a factor of 1000. For more detailed dissection of the computations refer to Table 16.2 in the following chapter. The sample size here is 99 and there are 4 tied times, so the full table has 95 rows of which only the first 5 and the last 5 are shown here.

Plots of the log-survivor and cumulative hazard functions are shown in Figure 15.1 for Groups 1 and 2 separately. The R-code used to produce the

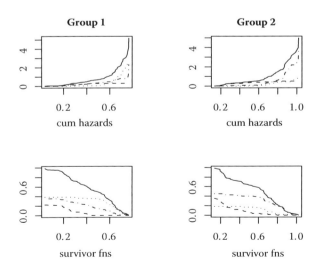

FIGURE 15.1
Cumulative hazards and survivor functions.

plots, which also prints out the figures used to construct Table 15.1, is

```
#mice data (Sec 15.1)
dx1=read.table('hoel.dat',header=T); attach(dx1); #dx1;
dx2=as.numeric(unlist(dx1)); nd=181; dim(dx2)=c(nd,3);
dx2[1:nd,1]=dx2[1:nd,1]/1000; #dx2
#Kaplan Meier
par(mfcol=c(2,2));
n1=99; n2=82; mc=3; kt=1; kc=2;
tvec1=dx2[1:n1,kt]; cvec1=dx2[1:n1,kc];
rtn1=kmplot1(n1,tvec1,cvec1,mc,'Group 1');
tvec2=dx2[(n1+1):(n1+n2),kt]; cvec2=dx2[(n1+1):(n1+n2),kc];
rtn2=kmplot1(n2,tvec2,cvec2,mc,'Group 2');
```

In each case the overall log-survivor curve is at the top since the overall survivor function is larger than each sub-survivor function, being their sum. Also in each case, the Cause-1 curve is at the bottom, though only for times beyond about 0.25 for Group 2. We might be tempted to conclude from this that the Cause-1 events tend to occur earlier than the others. If so, we would have fallen nicely into a trap that others would be keen to point out, purely in the interests of scientific truth, of course. Recall that $\bar{F}(c, t)$ is $P(C = c, T > t)$, which is not the survivor function for type-C failures; the latter is a conditional probability, $\bar{F}(c, t)/p_c$, where $p_c = P(C = c) = \bar{F}(c, 0)$ (Section 12.2). Thus, to arrive at the conclusion stated, we first need to standardise the survivor functions by dividing by the corresponding p_c, so that the standardised versions all start at value 1 at time 0. In Figure 15.1 this just means scaling them vertically to start at 1. It is clear that when this is done, the stated conclusion does appear to stand. Regarding failure-types 2 and 3, these are represented by the middle curves, with type-2 above type-3 in Group 1 and vice versa in Group 2. After scaling to standardise them, the relative positions are unchanged in Group 1 but the curves converge in Group 2 for times up to about 0.7, after which the type-2 curve drops more steeply. The picture does not admit a simple, one-line description.

The Weibull mixture model, fitted to these data in Section 13.4, can accommodate sub-survivor functions at different initial levels and of different shapes, as here, provided that the general form of each curve is a constant times a power of t. However, it was not found possible there either to achieve a simpler picture by reducing the full parameter set.

Dinse (1982) extended the likelihood form to accommodate cases with known failure times and unknown failure causes and cases with known failure causes but non-informatively right-censored failure times; the corresponding likelihood contributions are, respectively, $\nabla \bar{F}(t)$ and $\bar{F}(c, t)$. Dinse presented an iterative maximum likelihood method for computing the nonparametric estimates in this situation.

Pepe (1991) and Pepe and Mori (1993) proposed summary curves for describing competing risks data: they recommended the use of functions associated with cumulative incidence, prevalence, and marginal and conditional probabilties and gave methodology for estimating and testing these functions.

15.2 Actuarial Approach

The h_{jl} and h_l are probabilities of observable events and are thus estimable without too much fuss. In actuarial and demographic work, calculations have routinely been made of certain unobservable aspects from observed ones. Thus, estimates have been derived of the probabilities of death from one set of causes with another set of risks having been eliminated somehow. Performing the impossible in this way might explain why actuaries are so highly paid. Putting aside the question of the real-life relevance of such calculations (Section 14.3), it is inevitable that some assumption connecting the observable and unobservable aspects has to be made. In different contexts such assumptions may have greater or lesser credibility. The Makeham assumption is an example: it says that the *net* (unobservable) hazard, $h_j(t)$, is equal to the *crude* (observable) one, $h(j, t)$. We know from Gail's theorem (Section 14.4.1) that under this assumption the marginals $\bar{G}_j(t)$ are in principle determined by the sub-survivor functions $\bar{F}(j, t)$. A variety of such assumptions is recorded in the actuarial and demograph literature. In the remainder of this section we just give a brief glimpse of the field, more extensive treatments being available in Chiang (1968), Gail (1975), David and Moeschberger (1978), and the references therein.

Let us consider some typical assumptions and the estimation of some unobservable aspects. The assumptions, involving the continuous-time hazard functions, are:

1. that of Makeham, $h_j(t) = h(j, t)$ for each j and all $t > 0$;
2. that of Chiang (1961b), the proportionality $h_j(t) = \omega_{jl}h(t)$ for $t \in I_l = (\tau_{l-1}, \tau_l]$;
3. that of Kimball (1969), that the effect of eliminating risk j is to modify h_{kl} to $h_{kl}/(1 - h_{jl})$ for $k \neq j$;
4. that eliminating risk j has no effect on the remaining hazards.

Assumptions (1) and (2) together imply proportional hazards. In fact, we have then $h(j, t) = \omega_{jl}h(t)$ on I_l and so $f(j, t) = \omega_{jl}f(t)$, $\bar{F}(j, t) = \omega_{jl}\bar{F}(t)$ and $\omega_{+l} = \omega_{1l} + \cdots + \omega_{pl} = 1$.

Example (David and Moeschberger, 1978, Section 4.3)

(a) Let

$$P_{(j)l} = P(\text{survive } I_l \text{ under risk } j \text{ acting alone} \mid \text{enter } I_l)$$

$$= P(T_j \geq \tau_l \mid T_j \geq \tau_{l-1}) = \bar{G}_j(\tau_l-)/\bar{G}_j(\tau_{l-1}-) = \exp\left\{\int_{\tau_{l-1}}^{\tau_l} h_j(t)dt\right\}.$$

Then, under Chiang's proportionality assumption,

$$P_{(j)l} = \exp\left\{-\omega_{jl}\int_{\tau_{l-1}}^{\tau_l} h(t)dt\right\} = \{\bar{F}(\tau_l)/\bar{F}(\tau_{l-1})\}^{\omega_{jl}} = (1 - h_l)^{\omega_{jl}}.$$

We need an estimator for ω_{jl} since we already have $\hat{h}_l$. Under assumptions (1) and (2),

$$h_{jl} = \int_{\tau_{l-1}}^{\tau_l} f(j,t)dt / \bar{F}(\tau_{l-1}) = \int_{\tau_{l-1}}^{\tau_l} h(j,t)\bar{F}(t)dt / \bar{F}(\tau_{l-1})$$

$$= \omega_{jl} \int_{\tau_{l-1}}^{\tau_l} f(t)dt / \bar{F}(\tau_{l-1}) = \omega_{jl}h_l.$$

Thus, we have the estimate

$$\hat{\omega}_{jl} = \hat{h}_{jl}/\hat{h}_l = r_{jl}/r_l,$$

from which follows

$$\hat{P}_{(j)l} = (1 - r_{jl}/q_l)^{r_{jl}/r_l}.$$

Under a natural extension of assumption (3) h_{jl} would be replaced, in the absence of other risks, with

$$h_{jl} / \left(1 - \sum_{k \neq j} h_{kl}\right) = h_{jl}/(1 - h_l + h_{jl}),$$

which can be estimated by $r_{jl}/(q_l - r_l + r_{jl})$. This yields for $P_{(j)l}$ the estimate

$$\tilde{P}_{(j)l} = 1 - r_{jl}/(q_l - r_l + r_{jl}) = (q_l - r_l)/(q_l - r_l + r_{jl})$$
$$= \{1 + r_{jl}/(q_l - r_l)\}^{-1}.$$

(b) Let

$$P_{(-j)l} = P(\text{survive } I_l \text{ with risk } j \text{ eliminated} \mid \text{enter } I_l).$$

Under the assumptions (1), (2), and (4),

$$P_{(-j)l} = \exp\left[-\int_{\tau_{l-1}}^{\tau_l} \{h(s) - h(j,s)\}ds\right] = \exp\left\{-(1-\omega_{jl})\int_{\tau_{l-1}}^{\tau_l} h(s)ds\right\}$$
$$= (1 - h_l)^{1-\omega_{jl}}.$$

Hence,

$$\hat{P}_{(-j)l} = (1 - r_l/q_l)^{1-r_{jl}/r_l}.$$

Alternatively, under assumption (3),

$$P_{(-j)l} = 1 - \sum_{k \neq j} h_{kl}/(1 - h_{jl}) = (1 - h_l)/(1 - h_{jl}),$$

which can be estimated by $(1 - r_l/q_l)/(1 - r_{jl}/q_l)$.

15.3 Proportional Hazards and Partial Likelihood

This section gives a development of the methodology outlined in Section 4.2 to the competing risks context.

15.3.1 The Proportional Hazards (PH) Model

By analogy with the PH model for univariate failure times, the sub-hazard functions for the ith case are specified as

$$h(j, t; \mathbf{x}_i) = \psi_{ij} h_0(j, t),$$

where the $h_0(j, t)$ ($j = 1, \ldots, p$) form a set of baseline sub-hazards, and $\psi_{ij} = \psi(\mathbf{x}_i; \beta_j)$ is some positive function of $\mathbf{x}_i$ and β_j, a vector of explanatory variables and the associated vector of regression coefficients. A very common choice is $\psi_{ij} = \exp(\mathbf{x}_i^T \beta_j)$. This type of model was mentioned in Section 12.5: the difference here is that the $h_0(j, t)$ are left unspecified. In practical applications one would normally seek to limit the number of parameters by testing for restrictions on the βs, for example, that $\beta_1 = \beta_2$, or that particular components of β_3 are zero.

15.3.2 The Partial Likelihood

The observational setup is competing risks with covariates. The observed failure times from all causes are $t_1 < t_2 < \cdots < t_m$, with $t_0 = 0$ and $t_{m+1} = \infty$; R_{jl} is the set of r_{jl} individuals who fail from cause j at time t_l; S_l is the set of s_l individuals whose failure times t_{ls} ($s = 1, \ldots, s_l$) are right-censored during $[t_l, t_{l+1})$; $R(t_l)$ is the risk set at time t_l (Section 4.2).

The probability that individual $i \in R(t_l)$ fails from cause j in the time interval $(t_l, t_l + dt]$ is $h(j, t_l; \mathbf{x}_i)dt$. Let $r_l = r_{1l} + \cdots + r_{pl}$ and suppose for the moment that $r_l = 1$, that is, that there are no ties at time t_l. Suppose that t_l is the failure time of individual i_l, the cause being c_l. Given the events up to time t_l-, and given that there is a failure of type c_l at time t_l, the conditional probability that, among the members of $R(t_l)$, it is individual i_l who steps forward for the chop is

$$h(c_l, t_l; \mathbf{x}_{i_l})dt \div \sum_l h(c_l, t_l; \mathbf{x}_a)dt = \psi_{i_l c_l} \left/ \sum^l \psi_{i c_l} \right.,$$

where $\sum^l$ denotes summation over individuals $i \in R(t_l)$. The corresponding *partial likelihood function* is

$$P(\beta) = \prod_{l=1}^m \left(\psi_{i_l c_l} \left/ \sum^l \psi_{i c_l} \right. \right).$$

The maximum partial likelihood estimator $\hat{\beta}$ is found by maximizing $P(\beta)$ over β. Large-sample inference can be conducted by treating $\log P(\beta)$ as a log-likelihood function in the usual way. Thus,

$$\log P(\beta) = \sum_{l=1}^{m} \left(\log \psi_{i_l c_l} - \log \sum^{l} \psi_{i c_l} \right)$$

with first and second derivatives, respectively,

$$\mathbf{U}(\beta) = \partial \log P(\beta)/\partial \beta = \sum_{l=1}^{m} \mathbf{U}_l(\beta), \quad \mathbf{V}(\beta) = -\partial^2 \log P(\beta)/\partial \beta^2 = \sum_{l=1}^{m} \mathbf{V}_l(\beta),$$

where the vector $\mathbf{U}_l(\beta)$ has kth component $\partial \log(\psi_{i_l c_l}/\sum^{l} \psi_{i c_l})/\partial \beta_k$ and the matrix $\mathbf{V}_l(\beta)$ has (k, k')th entry $-\partial^2 \log(\psi_{i_l c_l}/\sum^{l} \psi_{i c_l})/\partial \beta_k \partial \beta_{k'}$. With the usual choice, $\psi_{ij} = \exp(\mathbf{x}_i^T \beta_j)$, we have $\mathbf{U}_l(\beta) = \mathbf{x}_{i_l} - \mathbf{v}_l$ and $\mathbf{V}_l(\beta) = \mathbf{X}_l - \mathbf{v}_l \mathbf{v}_l^T$, where

$$\mathbf{v}_l = \left(\sum^{l} \psi_{i c_l} \right)^{-1} \sum^{l} \psi_{i c_l} \mathbf{x}_i \text{ and } \mathbf{X}_l = \left(\sum^{l} \psi_{i c_l} \right)^{-1} \sum^{l} \psi_{i c_l} \mathbf{x}_i \mathbf{x}_i^T .$$

Under some standard regularity conditions $\mathbf{V}(\hat{\beta})^{-1}$ provides an estimate for the variance matrix of $\hat{\beta}$, and so follow hypothesis tests and confidence intervals for β.

When there are tied values among the members of $R(t_l)$, the expression for $P(\beta)$ is more complicated and the discussion of Section 4.3 applies. Thus, if there are $r_{jl} > 1$ observed failures from cause j at time t_l $(j = 1, \ldots, p)$ the lth term in the product defining $P(\beta)$ should allow for all such distinct subsets of size r_{jl} selected from the individuals in the risk set $R(t_l)$. This can lead to an unwieldy computation, and a useful approximation is to replace this term by $\prod(\psi_{i_l j}/\sum^{l} \psi_{i c_l})$, where the product is taken over the r_{+l} tied cases. The quality of this approximation deteriorates as the proportion of tied values increases.

15.3.3 A Clinical Trial

We revisit the data of Lagakos (1978), previously fitted with a Weibull sub-hazards model in Section 13.5. Note that in $\psi_{ij} = \exp(\mathbf{x}_i^T \beta_j)$ we do not need an intercept term now because the corresponding factor, $\exp(\beta_{0j})$, is absorbed into $h_0(t)$. The following R-code has been used:

```
#Lagakos data (Sec 15.3)
dx1=read.table('lagakos1.dat',header=T); attach(dx1);
dx2=as.numeric(unlist(dx1)); nd=194; dim(dx2)=c(nd,5);
dx2[1:nd,1]=time/100; dx2[1:nd,5]=age/100; #rescale
#Cox PH fit x=(x1,x2,age)
mc=2; kt=1; kc=2; iwr=0; opt=0; tv=dx2[1:nd,kt]; cv=dx2[1:nd,kc];
rtn01=km01(nd,tv,cv,mc,1); m=rtn01[1]; iv=rtn01[2:(nd+1)];
ia=1+nd+m; mt=rtn01[(ia+1):(ia+m)]; xml=dx2[iv,1:5]; #order rows
```

```
#full model x=(x1,x2,x3)
mx=3; kx=c(3,4,5); np=mc*mx; par0=runif(np);
adt=c(nd,mc,kt,kc,opt,0,m,mt,mx,kx);
par1=fitmodel(plikd,np,par0,xml,adt);
#null model (betas all zero)
mx=1; kx=c(4); np=mc*mx; par0=c(0,0);
adt=c(nd,mc,kt,kc,opt,1,m,mt,mx,kx);
plkd=plikd(par0,xml,adt);
#test for b21,b22,b31,b32 all zero
mx=2; kx=c(4,5); np=mc*mx; par0=runif(np);
adt=c(nd,mc,kt,kc,opt,0,m,mt,mx,kx);
par1=fitmodel(plikd,np,par0,xml,adt);
```

The full model fit, applying maximum partial likelihood, yields estimates $(0.61, 0.36, -2.04)$ for β_1 and $(0.56, 0.43, 0.85)$ for β_2; the corresponding standard errors are $(0.23, 0.28, 1.14)$ and $(0.32, 0.39, 1.64)$. The previous estimates given in Section 13.5 were $(-0.47, -0.32, 1.26)$ for β_1 and $(-0.43, -0.32, -0.96)$. Apart from the signs, due to the previous parametrisation in terms of ξ_j^{-1}, the comparison reveals a certain qualitative resemblance.

Regarding the effects of the covariates, the first assessment is whether all the βs are zero. For this, the log-partial likelihoods for the zero-constrained model, -542.7321, is compared with that of the full model, -535.8724, giving $\chi_6^2 = 13.72$ ($p = 0.03$); so it looks as though at least some of the βs are not zero. Incidentally, a similar test based on the Weibull sub-hazards model gives $\chi_6^2 = 14.33$. The next hypothesis to be examined is that $\beta_{21} = \beta_{22} = \beta_{31} = \beta_{32} = 0$. The constrained fit gives log-partial likelihood 540.5731 and then $\chi_4^2 = 9.40$ ($p = 0.05$); the corresponding value for the Weibull sub-hazards model was $\chi_4^2 = 6.70$ ($p = 0.15$).

15.4 The Baseline Survivor Functions

The likelihood function, like that in Section 15.1 but now with explanatory variables, is

$$L = \prod_{l=1}^{m} \prod_{i \in R_l} \nabla \bar{F}(c_l, t_l; \mathbf{x}_i) \times \prod_{l=0}^{m} \prod_{i \in S_l} \bar{F}(t_{ls}; \mathbf{x}_i)$$

To maximise L we must take $\hat{\bar{F}}(t_{ls}; \mathbf{x}_i)$ equal to $\hat{\bar{F}}(t_l; \mathbf{x}_i)$ for $i \in S_l$, for which we must take $\hat{\bar{F}}(j, t_{ls}; \mathbf{x}_i) = \hat{\bar{F}}(j, t_l; \mathbf{x}_i)$ for each j. In particular, $\hat{\bar{F}}(t_{0s}; \mathbf{x}_i) = 1$, which, in effect, removes the factor for $l = 0$ from the second term in L. Also, $\hat{\bar{F}}(c_i, t; \mathbf{x}_i)$ must have a discontinuity at $t = t_l$ for $i \in R_l$, otherwise $L = 0$. Therefore, $\hat{\bar{F}}(t; \mathbf{x}_i)$, which equals $\sum_{j=1}^{p} \hat{\bar{F}}(j, t; \mathbf{x}_i)$, also has such a discontinuity. Take

$$\hat{\bar{F}}(t_l; \mathbf{x}_i) = \prod_{s=1}^{l} \{1 - \hat{h}(t_s; \mathbf{x}_i)\},$$

which allows for discontinuities at each t_s, and use

$$h(t_s; \mathbf{x}_i) = \sum_{j=1}^{p} h(j, t_s; \mathbf{x}_i) = \sum_{j=1}^{p} \psi_{ij} h_0(j, t_s).$$

Then

$$\hat{F}(t_l; \mathbf{x}_i) = \prod_{s=1}^{l} \left\{ 1 - \sum_{j=1}^{p} \psi_{ij} \hat{h}_0(j, t_s) \right\},$$

and

$$\nabla \hat{F}(j, t_l; \mathbf{x}_i) = \hat{h}(j, t_l; \mathbf{x}_i) \hat{F}(t_l - ; \mathbf{x}_i) = \psi_{ij} \hat{h}_0(j, t_l) \prod_{s=1}^{l-1} \left\{ 1 - \sum_{j=1}^{p} \psi_{ij} \hat{h}_0(j, t_s) \right\}.$$

Hence, as in Section 4.3 (The baseline survivor function), and with Q_l defined as there, we have to maximize

$$L = \prod_{l=1}^{m} \prod_{i \in R_l} \left[\psi_{ic_l} h_0(c_i, t_l) \prod_{s=1}^{l-1} \left\{ 1 - \sum_{j=1}^{p} \psi_{ij} h_0(j, t_s) \right\} \right]$$

$$\times \prod_{l=1}^{m} \prod_{i \in S_l} \prod_{s=1}^{l} \left\{ 1 - \sum_{j=1}^{p} \psi_{ij} h_0(j, t_s) \right\}$$

$$= \prod_{l=1}^{m} \left[\prod_{i \in R_l} \psi_{ic_l} h_0(c_i, t_l) \times \prod_{i \in Q_l} \left\{ 1 - \sum_{j=1}^{p} \psi_{ij} h_0(j, t_l) \right\} \right]$$

over the "parameters" $h_0(j, t_s)$ $(j = 1, \ldots, p; \ s = 1, \ldots, m)$, having replaced ψ_{ij} with $\psi(\mathbf{x}_i; \hat{\beta}_j)$ from the partial likelihood estimation.

The function is orthogonal in the parameter sets $\{h_0(j, t_l) : \ j = 1, \ldots, p\}$ so the numerical maximisation is simplified (Appendix D). Armed with the estimates $\hat{h}_0(j, t_s)$, those for the baseline survivor functions follow from

$$\hat{F}_0(t_l) = \prod_{s=1}^{l} \{1 - \hat{h}_0(t_s)\}, \quad \nabla \hat{F}_0(j, t_l) = \hat{h}_0(j, t_l) \hat{F}_0(t_{l-1}),$$

where $\hat{h}_0(t_s) = \sum_{j=1}^{p} \hat{h}_0(j, t_s)$.

15.5 Other Methods and Applications

In this subsection a short round-up of articles is corralled without trying to be definitive or comprehensive. There are a number of good reviews in the literature, among which we mention here Moeschberger and Klein (1995).

Many applications of competing risks have appeared in the literature. Kay (1986) gives an early example of regression modelling. Other examples are

Gaynor et al. (1993), Serio (1997), Beyersmann et al. (2007), and Grambauer et al. (2010).

The case where causes of failure may be missing in the data has been addressed variously by Goetghebeur and Ryan (1990, 1995), Dewanji (1992), Andersen et al. (1996), Lu and Tsiatis (2001, 2005), Tsiatis et al. (2002), Dewanji and Sengupta (2003).

Much work has been produced on the question of tests for survival distributions. Tests for equality of survivor and sub-survivor functions have been proposed by Lindkvist and Belyaev (1998), Carriere and Kochar (2000), Kochar et al. (2002), Kulathinal and Gasbarra (2002), DiRienzio (2003), and Solari et al. (2008).

Some authors have pointed out that, since the covariate effects act directly on the sub-hazards in the proportional-hazards model, they do not act directly on the sub-survivor functions, the latter being non-linearly related to the former. So, adopting a proportional-hazards model makes it harder to interpret the covariate effects if your focus is on the subsurvivor functions. And this is the focus advocated by some for practical applications. The discussion is often presented in terms of *cumulative incidence functions*, or sub-distribution functions, rather than sub-survivor functions. The problem has been addressed in a variety of ways: see, for example, Shen and Cheng (1999), Fine and Gray (1999), Klein and Andersen (2005), Sun et al. (2006), Jeong and Fine (2006), Lu and Peng (2008), Scheike and Zhang (2008), and Dixon et al. (2011).

A fairly recent development concerns a combination of longitudinal and survival data. For a competing-risks version see Huang et al.; the reference list given in this paper can serve as a guide to the field.

16

Discrete Lifetimes

16.1 Basic Probability Functions

In this first section we will set out the basic probability functions. We consider the general case where the set of possible failure times is $0 = \tau_0 < \tau_1 < \cdots < \tau_m$; m and τ_m may each be finite or infinite.

The *sub-distribution* and *sub-survivor functions* are defined as in the continuous-time case (Section 12.2):

$$F(c, t) = P(C = c, T \leq t), \quad \bar{F}(c, t) = P(C = c, T > t);$$

C is the failure cause and T is the discrete failure time. Thus,

$$F(c, t) + \bar{F}(c, t) = P(C = c) = p_c.$$

Zero lifetimes will be discounted by taking $\bar{F}(c, 0) = p_c$ for each c; this is not mandatory—what follows needs little adaptation to accommodate zeros. The overall system distribution and survivor functions are given by

$$F(t) = P(T \leq t) = \sum_c F(c, t), \quad \bar{F}(t) = P(T > t) = \sum_c \bar{F}(c, t).$$

The *discrete sub-density* and overall system density functions are defined for $l = 1, \ldots, m$ by

$$f(c, \tau_l) = P(C = c, T = \tau_l) = \bar{F}(c, \tau_{l-1}) - \bar{F}(c, \tau_l),$$
$$f(\tau_l) = P(T = \tau_l) = \bar{F}(\tau_{l-1}) - \bar{F}(\tau_l) = \sum_c f(c, \tau_l);$$

if t is not equal to one of the τ_l then $f(c, t) = 0$. Also, $f(c, 0) = 0$ and, for $l > 1$,

$$\bar{F}(c, \tau_l) = \sum_{s=l+1}^{m} f(c, \tau_s) = p_c - \sum_{s=1}^{l} f(c, \tau_s).$$

The *sub-hazard* and overall hazard functions are defined for $l = 1, \ldots, m$ as

$$h(c, \tau_l) = f(c, \tau_l)/\bar{F}(\tau_{l-1}), \ h(\tau_l) = f(\tau_l)/\bar{F}(\tau_{l-1}) = \sum_c h(c, \tau_l);$$

for $l = 0, h(c, \tau_0) = 0$. At the last possible failure time, τ_m, we have $\bar{F}(\tau_m) = 0$, $\bar{F}(\tau_{m-1}) = f(\tau_m)$, and $h(\tau_m) = 1$. The following relations hold for $l = 1, \ldots, m$:

$$\bar{F}(\tau_l) = \prod_{s=0}^{l}\{1 - h(\tau_s)\}, \ \ f(\tau_l) = h(\tau_l)\prod_{s=0}^{l-1}\{1 - h(\tau_s)\}.$$

Proportional hazards obtains when $h(c, \tau_l)/h(\tau_l)$ is independent of l for each c, and the theorem of Section 12.4 goes through as before: thus, proportional hazards is equivalent to independence of T and C, the time and cause of failure.

16.1.1 Geometric Mixture

Consider the sub-survivor function

$$\bar{F}(j, t) = \pi_j \rho_j^t$$

$$(j = 1, \ldots, p; \ t = 1, \ldots, m = \infty; 0 < \rho_j < 1; \ 0 < \pi_j < 1; \ \sum_{j=1}^{p}\pi_j = 1).$$

This represents a mixture of geometric distributions: the setup can be interpreted as first choosing a component, the jth with probability π_j, and then observing that component's geometric failure time. This is the discrete-time analogue of the exponential mixture model. The discrete sub-densities are $f(j, t) = \pi_j \rho_j^{t-1}(1 - \rho_j)$, the system survivor function is $\bar{F}(t) = \sum_{j=1}^{p} \pi_j \rho_j^t$ with discrete density $f(t) = \sum_{j=1}^{p} \pi_j \rho_j^{t-1}(1 - \rho_j)$, and the marginal probabilities for C are the π_j. The conditional distributions are defined by $P(T > t \mid C = j) = \rho_j^t$ (geometric) and

$$P(C = j \mid T = t) = \pi_j \rho_j^{t-1}(1 - \rho_j) \left/ \sum_{l=1}^{p} \pi_l \rho_l^{t-1}(1 - \rho_l) \right.$$

Thus, C and T are independent if and only if the ρ_j are all equal. The sub-hazards are

$$h(j, t) = f(j, t)/\bar{F}(t - 1) = \pi_j(1 - \rho_j)\rho_j^{t-1} \left/ \sum_{l=1}^{p} \pi_l \rho_l^{t-1} \right.$$

and the system hazard is

$$h(t) = \sum_{j=1}^{p} \pi_j \rho_j^{t-1}(1 - \rho_j) \left/ \sum_{l=1}^{p} \pi_l \rho_l^{t-1}. \right.$$

The condition for proportional hazards is that the ρ_j are all equal, the same as for independence of C and T. In that case, when $\rho_j = \rho$ for all j, $h(j, t) = \pi_j(1 - \rho)$ and $h(t) = 1 - \rho$ are independent of t.

16.2 Latent Lifetimes and Sub-Odds Functions

As a basis for developing the theory we now assume an underlying vector $\mathbf{T} = (T_1, \ldots, T_p)$ of discrete latent failure times with joint survivor function $\bar{G}(\mathbf{t})$. The lifetime is determined as $T = \min(T_1, \ldots, T_p)$, in which each T_j can take values from the set $0 = \tau_0 < \tau_1 < \cdots < \tau_m$. The marginal survivor and density functions of T_j are $\bar{G}_j(t)$ and $g_j(t)$, and its marginal hazard function is $h_j(t) = g_j(t)/\bar{G}_j(t-)$. Ties between the T_j can occur in the discrete-time case, so we extend the definition of C from a single cause to a multiple cause, or *configuration* of causes. For example, if components j, k, and l fail simultaneously, and before any others, we define $T = T_j = T_k = T_l$ and $C = \{j, k, l\}$. Thus, C can now be any non-empty subset of $\{1, \ldots, p\}$, of which there are $2^p - 1$ in number. The functions $F(c, t)$ and $\bar{F}(c, t)$ are thus augmented, with c representing any failure configuration, not necessarily just a single index.

A version of Tsiatis's lemma (Section 14.1) can be established for discrete failure times. We will need the following difference operator. Let $w(\mathbf{t})$ be a scalar function of $\mathbf{t} = (t_1, \ldots, t_p)$ and define the operator ∇_1 by

$$\nabla_1(j)w(\mathbf{t}) = w(t_1, \ldots, t_j, \ldots, t_p) - w(t_1, \ldots, t_j-, \ldots, t_p);$$

thus, $\nabla_1(j)$ performs backward differencing on the jth component of $\mathbf{t}$. The general s-dimensional version can now be defined for $s \leq p$ by

$$\nabla_s(\mathbf{c})w(\mathbf{t}) = \prod_{j=1}^{s} \nabla_1(c_j)w(\mathbf{t}),$$

where the components of the vector $\mathbf{c} = (c_1, \ldots, c_s)$ are distinct elements of the set $\{1, \ldots, p\}$. Thus, ∇_s differences $w(\mathbf{t})$ on components $c_1, \ldots, c_s$, the order being immaterial. For example, if $p = 4$ and $\mathbf{c} = (1, 3)$, then $s = 2$ and

$$\nabla_s(\mathbf{c})w(\mathbf{t}) = w(t_1-, t_2, t_3-, t_4) - w(t_1-, t_2, t_3, t_4) - w(t_1, t_2, t_3-, t_4)$$
$$+ w(t_1, t_2, t_3, t_4).$$

We have $\bar{F}(t) = \bar{G}(t\mathbf{1}_p)$ and the modified form of Tsiatis's lemma states that

$$f(c, t) = (-1)^s \nabla_s(\mathbf{c})\bar{G}(\mathbf{t}) \mid_{\mathbf{t}=t\mathbf{1}_p},$$

where s is the number of components in configuration $\mathbf{c}$. Note that if t is not one of the τ_l then $f(c, t) = 0$.

Peterson's bounds (Section 14.3) hold as before, but with $\sum_{j=1}^{p}$ now replaced by $\sum_c$, summation over all configurations.

Much of the subsequent material in this section was first presented in Crowder (1996). We now define a set of $\bar{h}$-functions by $\bar{h}(c, t) = f(c, t)/\bar{F}(t)$. They not proper hazard functions but are related to the real ones $h(c, t)$ via

$$\bar{h}(c, t) = h(c, t)\bar{F}(t-)/\bar{F}(t) = h(c, t)/\{1 - h(t)\}.$$

Further,

$$h(c, \tau_l) = f(c, \tau_l) \left/ \sum_{s=l}^{m} f(\tau_s), \quad \bar{h}(c, \tau_l) = f(c, \tau_l) \right/ \sum_{s=l+1}^{m} f(\tau_s),$$

that is, $\bar{h}(c, \tau_l)$ is like $h(c, \tau_l)$ but with the term $f(\tau_l)$ missing in the denominator. Note that $\bar{h}(c, \tau_l)$ can exceed unity, unlike $h(c, \tau_l)$, the latter being a conditional probability. We will also make use of marginal $\bar{h}$-functions, $\bar{h}_j(t) = g_j(t)/\bar{G}_j(t)$, and of

$$\bar{h}(t) = f(t)/\bar{F}(t) = \sum_c \bar{h}(c, t) = h(t)/\{1 - h(t)\}$$

$$= \mathrm{P}(T = t \mid T \geq t)/\mathrm{P}(T \neq t \mid T \geq t).$$

The last expression shows that $\bar{h}(t)$ is the odds for failure at time t, given survival to time t. Thus we are led inevitably to refer to the $\bar{h}(c, t)$ as *sub-odds functions*. At the upper end point τ_m of the T-distribution $h(t)$ and $\bar{h}(t)$ respectively take the values 1 and ∞, both for the first time.

The case of independent risks is defined by $\bar{G}(\mathbf{t}) = \prod_{j=1}^{p} \bar{G}_j(t_j)$, as for continuous failure times. The convenient notation $f(j, t)$ will be used for $f(\{j\}, t)$ whenever the configuration is a simple index; likewise, $\bar{F}(j, t)$, $h(j, t)$ and p_j respectively stand for $\bar{F}(\{j\}, t)$, $h(\{j\}, t)$ and $p_{\{j\}}$. The following theorem is the discrete-time analogue of Gail's theorem (Section 14.4.1).

16.2.1 Sub-Odds Theorem (Crowder, 1996)

The following implications hold for the statements given below: (1) $\Rightarrow$ (2), and (2) $\Rightarrow$ the rest.

1. independent risks obtains;
2. $\bar{h}(c, t) = \prod_{j \in c} \bar{h}_j(t)$ for all (c, t), in particular, $\bar{h}(j, t) = \bar{h}_j(t)$ for $j = 1, \ldots, p$;
3. $\bar{h}(c, t) = \prod_{j \in c} \bar{h}(j, t)$ for all (c, t);
4. $\bar{G}_j(t) = \prod_{s=1}^{l(t)} \{1 + \bar{h}(j, \tau_s)\}^{-1}$, where $l(t) = \max\{l : \tau_l \leq t\}$;
5. $\bar{F}(t) = \prod_{s=1}^{l(t)} \prod_{j=1}^{p} \{1 + \bar{h}_j(\tau_s)\}^{-1}$;
6. $\bar{G}(t\mathbf{1}_p) = \prod_{j=1}^{p} \bar{G}_j(t)$, i.e., $\bar{R}(t\mathbf{1}_p) = 1$ (for $\bar{R}$ see Section 6.2);
7. $f(c, t) = \{\prod_{j \in c} g_j(t)\} \{\prod_{j \notin c} \bar{G}_j(t)\}$.

Proof According to (1), there is no interaction among the T_j, so ties can only occur through chance coincidence. In that case,

$$f(c,t) = P\left[\{\cap_{j\in c}(T_j = t)\} \cap \{\cap_{j\notin c}(T_j > t)\}\right] = \left\{\prod_{j\in c} g_j(t)\right\} \times \left\{\prod_{j\notin c} \bar{G}_j(t)\right\}$$

$$= \left[\prod_{j\in c}\{g_j(t)/\bar{G}_j(t)\}\right] \times \left[\left\{\prod_{j=1}^{p} \bar{G}_j(t)\right\}\right] = \left\{\prod_{j\in c} \bar{h}_j(t)\right\} \bar{F}(t),$$

and (2) follows. Statement (3) follows immediately from (2). For (4) substitute

$$\bar{h}_j(t) = \{\bar{G}_j(t-)/\bar{G}_j(t)\} - 1$$

into the particular case of (2). This yields the recurrence relation

$$\bar{G}_j(t) = \bar{G}_j(t-)\{1 + \bar{h}(j,t)\}^{-1}$$

from which (4) follows, after setting $t = \tau_l$ and noting that $\bar{G}_j(\tau_l-) = \bar{G}_j(\tau_{l-1})$. For (5), sum (2) over c: the left-hand side gives

$$\sum_c \bar{h}(c,t) = \bar{h}(t) = \{\bar{F}(t-)/\bar{F}(t)\} - 1,$$

and the right-hand side gives

$$\sum_c \prod_{j\in c} \bar{h}_j(t) = \prod_{j=1}^{p}\{1 + \bar{h}_j(t)\} - 1.$$

Hence,

$$\bar{F}(t) = \bar{F}(t-) \div \prod_{j=1}^{p}\{1 + \bar{h}_j(t)\}$$

and the result follows as for (4). For (6) note that (2), (4) and (5) together imply that $\bar{F}(t) = \prod_{j=1}^{p} \bar{G}_j(t)$, and that $\bar{F}(t) = \bar{G}(t\mathbf{1}_p)$. For (7) we have, using (2),

$$f(c,t) = \bar{F}(t)\bar{h}(c,t) = \bar{F}(t)\prod_{j\in c}\{g_j(t)/\bar{G}_j(t)\} = \bar{R}(t\mathbf{1}_p)\left\{\prod_{j\in c} g_j(t)\right\}\left\{\prod_{j\notin c}\bar{G}_j(t)\right\},$$

and the result now follows from (6).

Some interpretations can be made of the statements in the theorem.

1. The equality $\bar{h}(j,t) = \bar{h}_j(t)$ looks suspiciously like the Makeham assumption (Section 14.5), but is not—it is its discrete replacement. Proportional hazards cannot hold under

2. because setting $h(c, t) = p_c h(t)$ gives

$$p_c \bigg/ \prod_{j \in c} p_j = \bar{h}(t)^{c_1 - 1},$$

where c_1 is the number of single indices in configuration c; unless $c_1 = 1$, one side varies with t and the other does not.

4. The set of $\bar{h}(j, t)$, or equivalently the set $\{\bar{F}(j, t) : j = 1, \ldots, p\}$ of sub-survivor functions, determines the set of marginals $\bar{G}_j(t)$, that is, the crude risks determine the net ones.

6. *Independence*, that is, $\bar{G}(\mathbf{t})$ equal to $\prod_j \bar{G}_j(t_j)$, holds along the diagonal line $\mathbf{t} = t\mathbf{1}_p$.

7. $f(c, t)$ is deceptively expressible in the same form that it would take if the risks acted independently. A similar point was made after Gail's theorem, which deals with proper hazard functions and simple configurations.

The theorem shows that (2) is satisfied by any independent-risks model. Examples to be given in the following section show that the condition holds in some dependent-risks cases but not others. Such systems can be constructed in general by starting with an independent-risks model and then shifting probability mass around in the style of Theorem 4 of Crowder (1991). In this way, one can change the joint $\bar{G}(\mathbf{t})$ without disturbing the sub-survivor functions $\bar{G}(j, t)$ nor the marginals $\bar{G}_j(t)$. By this process the ratio $\bar{h}(c, t)/\prod_{j \in c} \bar{h}_j(t)$ retains its original value, 1, whereas the dependence measure $\bar{R}(\mathbf{t})$ is changed from 1 at one or more points off the diagonal $\mathbf{t} = t\mathbf{1}_p$. This also shows that (3) does not imply (2): probability mass can be shifted so that the $\bar{h}(c, t)$ are unaltered, so (3) is retained, but any particular marginal is altered, so (2) is not retained.

Lemma (Crowder, 1996)

Under the condition $\bar{h}(j, t) = \bar{h}_j(t)$ $(j = 1, \ldots, p)$, proportional hazards obtains for single-index configurations, that is, $h(j, t) = p_j h(t)$, if and only if

$$\bar{G}_j(t) = \prod_{s=1}^{l(t)} \{1 + p_j \bar{h}(\tau_s)\}^{-1},$$

where $l(t) = \max\{l : \tau_l < t\}$.

Proof The limited proportional hazards condition is equivalent to $\bar{h}(j, t) = p_j \bar{h}(t)$. Substituting this into (4) of the sub-odds theorem gives the result. Conversely, given the stated form for $\bar{G}_j(t)$,

$$g_j(t) = \bar{G}_j(t-) - \bar{G}_j(t) = \bar{G}_j(t)[\{1 + p_j \bar{h}(t)\} - 1],$$

from which $\bar{h}_j(t) = p_j \bar{h}(t)$ follows immediately.

In the case covered by the lemma, the $\bar{G}_j(t)$, and hence $\bar{G}(\mathbf{t})$, are expressed in terms of the single function $\bar{h}(t)$, or equivalently $\bar{F}(t)$, alone. This is the discrete counterpart of the lemma in Section 14.5.

16.3 Some Examples

In this section some discrete counterparts of certain joint continuous failure time distributions will be examined. The basic construction is via replacement of the exponential distribution by the geometric, as illustrated in the geometric mixture (Section 16.1).

16.3.1 Discrete Version of Gumbel

Let $\rho_j = e^{-\lambda_j}$ $(j = 1, 2)$ and $\sigma = e^{-\nu}$ in Gumbel's bivariate exponential (Section 6.2). Then we have a discrete bivariate survivor function

$$\bar{G}(t_1, t_2) = \rho_1^{t_1} \rho_2^{t_2} \sigma^{t_1 t_2} \text{ for } t_1, t_2 = 0, 1, 2, \ldots.$$

The marginal distributions are geometric:

$$\bar{G}_j(t) = \rho_j^t, \quad f_j(t) = (1 - \rho_j)\rho_j^{t-1}, \quad h_j(t) = (1 - \rho_j) \text{ and } \bar{h}_j(t) = (1 - \rho_j)/\rho_j.$$

The overall survivor function is $\bar{F}(t) = (\rho_1 \rho_2)^t \sigma^{t^2}$, the discrete density is $f(t) = (\rho_1 \rho_2)^{t-1} \sigma^{(t-1)^2} \{1 - \rho_1 \rho_2 \sigma^{2t-1}\}$, and the hazard function is $h(t) = 1 - \rho_1 \rho_2 \sigma^{2t-1}$.

For the sub-distributions we have sub-density

$$
\begin{aligned}
f(1, t) &= -\nabla_1(1)\bar{G}(t_1, t_2)\,|_{t_1 = t_2 = t} = \bar{G}(t - 1, t) - \bar{G}(t, t) \\
&= \rho_1^{t-1} \rho_2^t \sigma^{(t-1)t} - \rho_1^t \rho_2^t \sigma^{t^2} = (\rho_1 \rho_2)^t \sigma^{t^2} (\rho_1^{-1} \sigma^{-t} - 1),
\end{aligned}
$$

with a similar form for $f(2, t)$, and

$$
\begin{aligned}
f(\{1, 2\}, t) &= -\nabla_2(\{1, 2\})\bar{G}(t_1, t_2)\,|_{t_1 = t_2 = t} \\
&= \bar{G}(t - 1, t - 1) - \bar{G}(t - 1, t) - \bar{G}(t, t - 1) + \bar{G}(t, t) \\
&= \cdots = (\rho_1 \rho_2)^{t-1} \sigma^{(t-1)^2} \{1 - (\rho_1 + \rho_2)\sigma^{t-1} + \rho_1 \rho_2 \sigma^{2t-1}\}.
\end{aligned}
$$

The sub-hazards are

$$h(1, t) = \rho_2 \sigma^{t-1}(1 - \rho_1 \sigma^t), \quad h(2, t) = \rho_1 \sigma^{t-1}(1 - \rho_2 \sigma^t)$$

and

$$h(\{1, 2\}) = 1 - (\rho_1 + \rho_2)\sigma^{t-1} + \rho_1 \rho_2 \sigma^{2t-1};$$

the sub-odds are

$$\bar{h}(1, t) = \rho_1^{-1}\sigma^{-t} - 1, \quad \bar{h}(2, t) = \rho_2^{-1}\sigma^{-t} - 1$$

and

$$\bar{h}(\{1, 2\}) = (\rho_1\rho_2)^{-1}\sigma^{1-2t}\{1 - (\rho_1 + \rho_2)\sigma^{t-1} + \rho_1\rho_2\sigma^{2t-1}\}.$$

Independent-risks obtains if $\sigma = 1$. Also, $\bar{h}(1, t) = \bar{h}_1(t)$ if and only if $\sigma = 1$, and then $\bar{h}(\{1, 2\}, t) = \bar{h}_{\{1,2\}}(t)$ in which case condition (2) of the sub-odds theorem is met.

16.3.2   Discrete Version of Freund

Referring back to Section 7.3, the marginal densities and survivor functions are $g_j(t) = \pi_j^{t-1}(1 - \pi_j)$ and $\bar{G}_j(t) = \pi_j^t$ ($j = 1, 2$; $t = 1, \ldots, m = \infty$). The sub-densities and sub-survivor functions are

$$f(j, t) = \pi_j^{t-1}(1 - \pi_j)\pi_k^t \; (k = 3 - j), \quad f(\{1, 2\}, t) = \pi_1^{t-1}(1 - \pi_1)\pi_2^{t-1}(1 - \pi_2)$$

and

$$\bar{F}(j, t) = \sum_{s=t+1}^{\infty} f(j, s) = \pi_j^t(1 - \pi_j)\pi_k^{t+1}/(1 - \pi_1\pi_2),$$

$$\bar{F}(\{1, 2\}, t) = \pi_1^t(1 - \pi_1)\pi_2^t(1 - \pi_2)/(1 - \pi_1\pi_2).$$

The marginal distribution of T has survivor function $\bar{F}(t) = \pi_1^t\pi_2^t$ and discrete density $f(t) = \pi_1^{t-1}\pi_2^{t-1}(1 - \pi_1\pi_2)$. The probability of the first failure's occurring in sequence j is

$$p_j = \bar{F}(j, 0) = (1 - \pi_j)\pi_k/(1 - \pi_1\pi_2),$$

and that of simultaneous failure is

$$p_{\{1,2\}} = \bar{F}(\{1, 2\}, 0) = (1 - \pi_1)(1 - \pi_2)/(1 - \pi_1\pi_2);$$

one can check that $p_1 + p_2 + p_{\{12\}} = 1$.

The hazard functions for this system are are as follows:

marginal hazards $h_j(t) = g_j(t)/\bar{G}_j(t - 1) = 1 - \pi_j$;

sub-hazards $h(j, t) = (1 - \pi_j)\pi_k$ and $h(\{1, 2\}, t) = (1 - \pi_1)(1 - \pi_2)$;

overall hazard $h(t) = 1 - \pi_1\pi_2$.

The sub-odds functions are

$$\bar{h}(j, t) = (1 - \pi_j)/\pi_j, \quad \bar{h}(\{1, 2\}, t) = (1 - \pi_1)(1 - \pi_2)/(\pi_1\pi_2),$$
$$\bar{h}_j(t) = (1 - \pi_j)/\pi_j.$$

Condition (2) of the sub-odds theorem, which reduces to $\bar{h}(\{1,2\},t) = \bar{h}_1(t)\bar{h}_2(t)$ for the case $p = 2$, holds here. This shows that the condition can hold for models with dependent risks, a fact that will be of some significance in Section 17.4. That it is satisfied for this example is predictable from the construction of the model in which the risks act independently up to the first failure.

16.3.3 Discrete Version of Marshall–Olkin

This system has marginal densities $g_j(t) = (\pi_j\pi_{12})^{t-1}(1-\pi_j\pi_{12})$ and marginal hazards $h_j(t) = 1 - \pi_j\pi_{12}$ (Section 7.4). The sub-densities are

$$f(j,t) = \pi_j^{t-1}(1-\pi_j)(\pi_{3-j}\pi_{12})^t, \quad f(\{1,2\},t) = \rho(\pi_1\pi_2\pi_{12})^{t-1},$$

where

$$\rho = (1 - \pi_{12}) + \pi_{12}(1 - \pi_1)(1 - \pi_2)$$

is the probability of failure configuration $\{1,2\}$ at any given trial. The sub-survivor functions are

$$\bar{F}(j,t) = \pi_j^{-1}(1-\pi_j)(\pi_1\pi_2\pi_{12})^{t+1}/(1 - \pi_1\pi_2\pi_{12}),$$
$$\bar{F}(\{1,2\},t) = \rho(\pi_1\pi_2\pi_{12})^t/(1 - \pi_1\pi_2\pi_{12}).$$

The overall system failure time has survivor function $\bar{F}(t) = (\pi_1\pi_2\pi_{12})^t$ with density $f(t) = (\pi_1\pi_2\pi_{12})^{t-1}(1-\pi_1\pi_2\pi_{12})$. The probabilities of first failure type are given by

$$p_j = P(C = j) = \bar{F}(j,0) = \pi_j^{-1}(1-\pi_j)(\pi_1\pi_2\pi_{12})/(1 - \pi_1\pi_2\pi_{12}),$$
$$p_{\{1,2\}} = P(C = \{1,2\}) = \bar{F}(\{1,2\},0) = \rho/(1 - \pi_1\pi_2\pi_{12}).$$

The sub-hazards are $h(j,t) = \pi_j^{-1}(1-\pi_j)(\pi_1\pi_2\pi_{12})$ and $h(\{1,2\},t) = \rho$, and the overall hazard rate is $h(t) = 1 - \pi_1\pi_2\pi_{12}$. The $\bar{h}$-functions are

$$\bar{h}(j,t) = (1 - \pi_j)/\pi_j, \quad \bar{h}(\{1,2\},t) = \rho/(\pi_1\pi_2\pi_{12}),$$
$$\bar{h}_j(t) = (1 - \pi_j\pi_{12})/(\pi_j\pi_{12}).$$

Condition (2) of the sub-odds theorem, which boils down to $\bar{h}(\{1,2\},t) = \bar{h}_1(t)\bar{h}_2(t)$ here, evidently fails unless $\pi_{12} = 1$.

16.3.4 Mixture Models

Consider p independent sequences of Bernoulli trials with success probabilities π_j ($j = 1, \ldots, p$). The joint survivor function of the p failure times is $\bar{G}(\mathbf{t}) = \prod_{j=1}^{p} \pi_j^{t_j}$. Now suppose that the sequences are linked through some common *random effect* $z \in (0,1)$ such that π_j is replaced by $z\pi_j$. Then the formula just given becomes conditional on z, and the unconditional version is

$$\bar{G}(\mathbf{t}) = \int_0^1 \left\{ \prod_{j=1}^p (z\pi_j)^{t_j} \right\} dK(z) = \prod_{j=1}^p \pi_j^{t_j} \int_0^1 z^{s_t} dK(z),$$

where $s_t = t_1 + \cdots + t_p$ and K is the distribution function of z. The original formula thus acquires an extra factor, this being the s_tth moment of K.

For example, taking K to be a beta distribution,

$$\bar{G}(\mathbf{t}) = \left\{ \prod_{j=1}^p \pi_j^{t_j} \right\} \int_0^1 z^{s_t} \{z^{\nu-1}(1-z)^{\tau-1} / B(\nu, \tau)\} dz$$

$$= \left\{ \prod_{j=1}^p \pi_j^{t_j} \right\} B(\nu + s_t, \tau) / B(\nu, \tau)$$

$$= \left\{ \prod_{j=1}^p \pi_j^{t_j} \right\} \prod_{l=1}^{s_t} \{(\nu + s_t - l)/(\nu + \tau + s_t - l)\};$$

B here denotes the beta function, $B(a, b) = \Gamma(a)\Gamma(b)/\Gamma(a + b)$. The subdensity functions for single-index causes of failure are given by (Section 15.2)

$$f(j, t) = -\nabla_1(j)\bar{G}(\mathbf{t}) = \bar{G}(t, \ldots, t-1, \ldots, t) - \bar{G}(t, \ldots, t)$$

$$= \left\{ \prod_{j=1}^p \pi_j \right\}^t \{\pi_j^{-1} - (\nu + pt - 1)/(\nu + \tau + pt - 1)\}$$

$$\times \prod_{l=2}^{pt} \{(\nu + pt - l)/(\nu + \tau + pt - l)\},$$

and those for more complicated configurations can be derived similarly, using $f(c, t) = (-1)^s \nabla_s(\mathbf{c})\bar{G}(t\mathbf{1}_p)$ (Section 16.2), but with rather more trouble. The calculation of the various other associated functions is algebraically untidy though simple to program for computer evaluation.

16.4 Parametric Estimation

We turn now to estimation. Consider the general case where the possible failure times are $\tau_0 < \tau_1 < \cdots < \tau_m$, with $\tau_0 = 0$. Denote by R_{cl} the set of r_{cl} individuals who fail from cause/configuration c at time τ_l, with $R_l = R_{1l} \cup \ldots \cup R_{pl}$, and R_0 empty, and let $r_l = \sum_c r_{cl}$. Denote by S_l the set of s_l individuals whose failure times are right-censored at time τ_l. For example, in the special but common situation where all individuals are kept under observation up to time τ_a, but none beyond, R_l would be empty for $l > a$ and S_l would be empty for $l \neq a$.

16.4.1 Likelihood Function

The likelihood function is

$$L = \prod_{l=1}^{m} \prod_{i \in R_l} f(c_i, \tau_l; \mathbf{x}_i) \times \prod_{l=0}^{m} \prod_{i \in S_l} \bar{F}(\tau_l; \mathbf{x}_i),$$

$\mathbf{x}_i$ denoting the vector of explanatory variables attached to the ith individual. For an individual whose failure time, but not cause, is observed, a factor $f(\tau_l; \mathbf{x})$ must be attached; likewise, a factor $\bar{F}(c, \tau_l; \mathbf{x})$ is needed when the cause is known but the time is (non-informatively) right-censored. We will not deal with these elaborations here explicitly.

In terms of the discrete sub-hazard functions defined in Section 16.1,

$$\bar{F}(\tau_l; \mathbf{x}) = \prod_{s=1}^{l} \{1 - h(\tau_s; \mathbf{x})\}, \quad f(c, \tau_l; \mathbf{x}) = h(c, \tau_l; \mathbf{x}) \bar{F}(\tau_{l-1}; \mathbf{x}),$$

for $l = 1, \ldots, m$, where $h(\tau; \mathbf{x}) = \sum_c h(c, \tau; \mathbf{x})$; also, $\bar{F}(\tau_0; \mathbf{x}_i) = 1$, so the $l = 0$ term in the second product in L can be dropped. Algebraic reduction yields

$$L = \prod_{l=1}^{m} \prod_{i \in R_l} \left[h(c_i, \tau_l; \mathbf{x}_i) \prod_{s=1}^{l-1} \{1 - h(\tau_s; \mathbf{x}_i)\} \right] \times \prod_{l=1}^{m} \prod_{i \in S_l} \prod_{s=1}^{l} \{1 - h(\tau_s; \mathbf{x}_i)\}$$

$$= \prod_{l=1}^{m} \left[\prod_{i \in R_l} h(c_i, \tau_l; \mathbf{x}_i) \times \prod_{i \in Q_l} \{1 - h(\tau_l; \mathbf{x}_i)\} \right],$$

where $Q_l = R(\tau_l) - R_l$, $R(\tau_l)$ being the *risk set* at time τ_l.

When parametric functions are specified for the discrete sub-densities $f(c, t)$, or equivalently for the sub-hazards $h(c, t)$, L can be used in the standard manner for inference.

16.4.2 Discrete Marshall–Olkin

The likelihood function for random samples is

$$L(\pi_1, \pi_2, \pi_{12}) = \prod_{l=1}^{\infty} \{ f_{1l}^{r_{1l}} f_{2l}^{r_{2l}} f_{3l}^{r_{3l}} \} \times \prod_{l=0}^{\infty} (\pi_1 \pi_2 \pi_{12})^l,$$

where

$$f_{jl} = \pi_j^{-1}(1 - \pi_j)(\pi_1 \pi_2 \pi_{12})^l \ (j = 1, 2), \quad f_{3l} = \rho(\pi_1 \pi_2 \pi_{12})^{l-1},$$
$$\rho = 1 - (\pi_1 + \pi_2 - \pi_1 \pi_2)\pi_{12}.$$

Collecting terms,

$$L = \prod_{l=1}^{\infty} \{ \pi_1^{-r_{1l}}(1 - \pi_1)^{r_{1l}} \pi_2^{-r_{2l}}(1 - \pi_2)^{r_{2l}} \rho^{r_{3l}} (\pi_1 \pi_2 \pi_{12})^{l(r_{1l}+r_{2l})+(l-1)r_{3l}+l} \}$$
$$= \pi_1^{a_+ - r_{1+}}(1 - \pi_1)^{r_{1+}} \pi_2^{a_+ - r_{2+}}(1 - \pi_2)^{r_{2+}} \pi_{12}^{a_+} \rho^{r_{3+}},$$

where $a_l = l(r_l + 1) - r_{3l}$ and a + subscript indicates summation over that index. The log-likelihood derivatives are

$$\partial \log L / \partial \pi_1 = (a_+ - r_{1+})/\pi_1 - r_{1+}/(1 - \pi_1) - r_{3+}(1 - \pi_2)\pi_{12}/\rho,$$
$$\partial \log L / \partial \pi_2 = (a_+ - r_{2+})/\pi_2 - r_{2+}/(1 - \pi_2) - r_{3+}(1 - \pi_1)\pi_{12}/\rho,$$
$$\partial \log L / \partial \pi_{12} = a_+/\pi_{12} - r_{3+}(\pi_1 + \pi_2 - \pi_1\pi_2)/\rho.$$

The likelihood equations do not have explicit solution, so iterative maximisation of L is called for. The information matrix can be calculated by further differentiation to provide an estimated covariance matrix for the maximum likelihood estimates of π_1, π_2, and π_{12}.

16.4.3   Psychiatric Wards

Patients may be admitted to a psychiatric ward on a voluntary basis or through legal enforcement. Records are kept on a daily basis of incidents in the ward. These might be minor infringements, such as refusing to comply with the rules, or more serious misbehaviour, involving verbal or physical abuse. Such records, held by the health authorities, are confidential, but the data in Table 16.1 are typical of the type. For each patient the case identifier (id), time (days) from admission to first incident (t in the table), and what kind of incident (c) are listed. The accompanying covariates are age of patient (age), whether voluntary or compulsory admission (adm), the number of beds in the ward (nb), and the number of staff on duty (ns).

TABLE 16.1

Times to First Incident in a Psychiatric Ward

id	t	c	age	adm	nb	ns	id	t	c	age	adm	nb	ns	id	t	c	age	adm	nb	ns
1	3	1	26	0	15	9	18	5	0	50	0	20	8	35	3	1	25	0	25	9
2	3	1	40	1	15	9	19	4	2	33	1	20	10	36	0	2	38	1	25	10
3	2	2	36	1	15	8	20	1	2	40	0	20	8	37	2	2	24	0	25	8
4	3	1	38	0	15	7	21	3	2	33	1	20	9	38	5	2	46	0	25	12
5	3	2	40	0	15	6	22	1	0	45	1	20	8	39	1	2	32	1	25	11
6	5	1	49	0	15	6	23	2	2	31	1	20	10	40	1	1	39	1	25	11
7	6	0	38	1	15	7	24	4	2	33	0	20	8	41	0	2	32	1	25	9
8	3	1	37	0	15	7	25	3	0	23	0	20	7	42	3	1	40	0	25	9
9	6	0	41	0	15	7	26	2	0	43	0	20	11	43	4	0	23	0	25	12
10	1	1	33	1	15	5	27	5	2	51	0	20	9	44	2	1	31	0	25	12
11	2	1	18	1	15	8	28	1	1	18	1	20	9	45	5	2	43	0	25	10
12	3	0	26	0	15	7	29	3	0	27	0	20	10	46	3	0	26	0	25	8
13	3	2	36	0	15	8	30	5	1	35	1	20	12	47	2	0	48	1	25	11
14	2	2	28	0	15	10	31	1	2	36	1	20	7	48	2	0	49	0	25	10
15	4	0	39	1	15	7	32	3	1	52	1	20	10	49	4	1	45	1	25	11
16	3	2	50	1	15	5	33	4	0	28	0	20	12	50	0	1	21	0	25	11
17	4	1	38	1	15	6	34	4	2	30	0	20	12	51	2	0	37	0	25	8

The data have been generated from a geometric mixture model (Section 16.1), so it seems only fair to fit that model back to them. The details are

$$\bar{F}(j, t) = \pi_j \rho^t \ (j = 1, 2; \ t = 1, 2, \ldots),$$
$$\pi_1 = 1/\{1 + \exp(-\mathbf{x}^T \beta)\}, \text{ and } \pi_2 = 1 - \pi_1.$$

Thus, the type infringement is influenced by the covariates via a logit-linear model; ρ, which governs the time to first infringement of given type, is taken to be independent of $\mathbf{x}$. The R-code for fitting the full model, with $\mathbf{x} = (age, adm, nb, ns)$, is

```
#psychiatric wards (Sec 16.4)
dx1=read.table('pwards.dat',header=T); attach(dx1); #dx1;
nd=51; md=7; dx2=as.numeric(unlist(dx1)); dim(dx2)=c(nd,md);
dx2[1:nd,2]=dx2[1:nd,2]+1; dx2[1:nd,4]=dx2[1:nd,4]/10;
dx2[1:nd,6]=dx2[1:nd,6]/10; dx2[1:nd,7]=dx2[1:nd,7]/10;
#geometric mixture: full model
mc=2; kt=2; kc=3; mx=5; kx=c(0,4,5,6,7);
mdl=0; np=(mc-1)*mx+mc; par0=runif(np,0,1);
opt=0; iwr=0; adt=c(nd,mc,kt,kc,mx,kx,mdl,opt,iwr);
par1=fitmodel(geomix1,np,par0,dx2,adt);
#null model
mdl=0; mx=1; kx=c(0); np=(mc-1)*mx+mc; par0=runif(np,0,1);
opt=0; iwr=0; adt=c(nd,mc,kt,kc,mx,kx,mdl,opt,iwr);
par1=fitmodel(geomix1,np,par0,dx2,adt);
```

The meanings of the various quantities can be found in function `geomix1`, which itself can be found on the CRC Press Web site noted in the Preface. The fit gives maximised log-likelihood $\hat{l} = -173.5817$. A null model, with no covariates, gives $\hat{l} = -176.127$; comparing this with the full model, $\chi_4^2 = 5.09$ with an inoffensive p-value 0.72. So, the explanatory variables look short of an explanation. The ratios of β-estimates to standard errors in the full model bear this out—altogether a disappointing result for the earnest seeker after enlightenment. Maybe a contributing factor to non-significance is that the geometric distribution does give relatively large variability.

16.5 Non-Parametric Estimation from Random Samples

For non-parametric inference we follow here Davis and Lawrance (1989). In the absence of explanatory variables the likelihood function is, writing h_{cl} for $h(c, \tau_l)$ and h_l for $h(\tau_l)$,

$$L = \prod_{l=1}^{m} \prod_{c} \left\{ h_{cl} \prod_{s=1}^{l-1} (1 - h_s) \right\}^{r_{cl}} \times \prod_{l=1}^{m} \left\{ \prod_{s=1}^{l} (1 - h_s) \right\}^{s_l}$$
$$= \prod_{l=1}^{m} \left\{ \prod_{c} h_{cl}^{r_{cl}} \times (1 - h_l)^{q_l - r_l} \right\},$$

where $\prod_{s=1}^{l-1}(1 - h_s)$ is interpreted as 1 for $l = 1$, $r_l = \sum_c r_{cl}$, and

$$q_l = (r_l + s_l) + \cdots + (r_m + s_m)$$

is the size of the risk set $R(\tau_l)$. The estimates $\hat{h}_{cl}$ that maximise L can be calculated by equating

$$\partial \log L / \partial h_{cl} = r_{cl}/h_{cl} - (q_l - r_l)/(1 - h_l)$$

to zero, where $h_l = \sum_c h_{cl}$ has been used. This gives

$$\hat{h}_{cl} = r_{cl}(1 - \hat{h}_l)/(q_l - r_l)$$

which may be summed over c to yield $\hat{h}_l = r_l/q_l$, and this then leads to

$$\hat{h}_{cl} = r_{cl}/q_l \ (l = 1, \ldots, m).$$

This is the intuitive estimate of h_{cl}, that is, the observed number of failures divided by the number at risk. If there are no observed failures in the data of type c at time τ_l, the hazard h_{cl} is estimated as zero; this could be true for all c, in which case $r_l = 0$ and $\hat{h}_l = 0$. The *mle* for $\bar{F}(\tau_l)$ and $f(c, \tau_l)$ now follow as

$$\hat{\bar{F}}(\tau_l) = \prod_{s=0}^{l}(1 - \hat{h}_s) \ \text{ and } \ \hat{f}(c, \tau_l) = \hat{h}_{cl}\hat{\bar{F}}(\tau_{l-1}),$$

where $\hat{h}_s = \sum_c \hat{h}_{cs}$, and those for the p_c and $\bar{F}(c, \tau)$ are

$$\hat{p}_c = \sum_{s=1}^{m} \hat{f}(c, \tau_s), \ \hat{\bar{F}}(c, \tau) = \sum \hat{f}(c, \tau_s),$$

where the second summation is over $\{s : \tau_s > \tau\}$.

There is a slight qualification to the maximum likelihood solution here. In the likelihood function, L, the mth factor should be just $\prod_c h_{cm}^{r_{cm}}$, since $s_m = 0$ implies that $q_m - r_m = 0$ and so $(1 - h_m)^{q_m - r_m} = 1$. As far as the h_{cm} are concerned, then, maximisation of L just entails maximisation of $\prod_c h_{cm}^{r_{cm}}$ subject to the constraint $\sum_c h_{cm} = h_m = 1$. Luckily, this just yields $\hat{h}_{cm} = r_{cm}/q_m$ as before. However, a real hiccup occurs if q_l becomes zero first at $l = m' \leq m$, which entails $q_l = 0$ for $l = m', \ldots, m$; this will certainly occur if $m = \infty$. The estimates $\hat{h}_{cl} = r_{cl}/q_l$ are then 0/0, and thus undefined for $l \geq m'$, as are the $\hat{h}_l$. An interpretation of this is that, because the conditioning event $\{T > \tau_{l-1}\}$ in the definitions of h_{cl} and h_l is not observed for $l \geq m'$, they are not estimable. On the face of it, then, the estimates of all the other quantities that depend on the $\hat{h}_{cl}$ will also be unobtainable for $l > m'$. But all is not lost if $s_{m'-1} = 0$, that is, if there are no censored times outstanding. In that case, $r_{m'-1} = q_{m'-1} > 0$, so $\hat{h}_{m'-1} = 1$ which implies $\hat{\bar{F}}(\tau_{m'-1}) = 0$; then $\hat{\bar{F}}(\tau_l) = 0$ and $\hat{\bar{F}}(c, \tau_l) = 0$ for $l \geq m' - 1$, $\hat{f}(\tau_l) = 0$ and $\hat{f}(c, \tau_l) = 0$ for $l \geq m'$, and $\hat{p}_c = \sum_{s=1}^{m'-1} \hat{f}(c, \tau_l)$. When there are censored times outstanding, beyond the last observed failure time, the sum $\hat{p}_+$ of the estimates $\hat{p}_c$ will fall short of 1. In that case a pragmatic solution is to adjust them by dividing them by $\hat{p}_+$, so that the adjusted components do sum to 1.

When there are no censored observations, the situation is much simpler. In that case, $q_l = r_l + \cdots + r_m$ and $\hat{F}(\tau_l) = q_{l+1}/q_1$, the overall empirical survivor function, q_1 being the sample size. Then, $\hat{f}(c, \tau_l) = r_{cl}/q_1$. It follows that

$$\hat{F}(c, \tau_l) = (r_{c,l+1} + \cdots + r_{cm})/q_1,$$

the sample proportion of type-c failures beyond time τ_l, and then

$$\hat{p}_c = (r_{c1} + \cdots + r_{cm})/q_1,$$

the overall proportion of type-c failures.

Useful plots can be constructed from the estimates. For example, $\hat{F}(\tau_l)$ and $\hat{F}(c, \tau_l)$ can be plotted against τ_l with either or both scales transformed, for example, logarithmically, for convenience; plots of the standardised versions, $\hat{F}(c, \tau_l)/\hat{p}_c$, are probably better for comparing the shapes of the sub-survivor functions. The hazard functions can be represented in *cusum* form to provide smoother plots, that is, as $\sum_{s=1}^{l} \hat{h}_s$ and $\sum_{s=1}^{l} \hat{h}_{cs}$ versus τ_l. The plots will often suggest behaviour such as increasing hazard or failure earlier from one cause than another. Given sufficient data, such hypotheses can be appraised as described below.

16.5.1 Gubbins

Let us apply the procedure to these data (Section 12.1); the breaking strengths are essentially discrete because of the numerical rounding, but there are only failure configurations {1} and {2} present, none of type {1, 2} occurring. The estimates, given in Table 16.2, are based on the assumption, almost certainly

TABLE 16.2

Estimates for Gubbins Data

l	τ_l	q_l	r_{1l}	r_{2l}	s_l	$\hat{h}_{1l}$	$\hat{h}_{2l}$	$\hat{h}_l$	$\hat{F}(\tau_l)$	$\hat{F}(1, \tau_l)$	$\hat{F}(2, \tau_l)$
1	1.8	55	0	5	0	0.000	0.091	0.091	0.447	0.462	0.909
2	1.9	50	0	7	0	0.000	0.140	0.140	0.447	0.335	0.782
3	2.0	43	0	2	0	0.000	0.047	0.047	0.447	0.299	0.745
4	2.1	41	4	0	0	0.098	0.000	0.098	0.374	0.299	0.673
5	2.2	37	2	1	0	0.054	0.027	0.081	0.338	0.280	0.618
6	2.3	34	2	5	0	0.059	0.147	0.206	0.301	0.190	0.491
7	2.4	27	3	6	0	0.110	0.220	0.330	0.247	0.080	0.327
8	2.5	18	2	0	0	0.110	0.000	0.110	0.210	0.080	0.291
9	2.6	16	3	0	0	0.190	0.000	0.190	0.156	0.080	0.236
10	2.7	13	2	0	0	0.154	0.000	0.154	0.120	0.080	0.200
11	2.8	11	3	0	0	0.273	0.000	0.273	0.065	0.080	0.145
12	3.0	8	0	0	8	0.000	0.000	0.000	0.065	0.080	0.145

false, that $\tau_m = 3.0$, that is, that this is the maximum possible breaking strength. In reality, because of the censoring, the value τ_m has not been observed, so we are in danger of the real hiccup mentioned above.

The layout in Table 16.2 follows the computations, left to right. Thus, the columns for q_l, r_{1l}, and r_{2l} come directly from inspecting the data; the three columns for the hazard functions are computed as $\hat{h}_{cl} = r_{cl}/q_l$ ($c = 1, 2$) and $\hat{h}_l = \hat{h}_{1l} + \hat{h}_{2l}$; the next column, for $\hat{\bar{F}}(\tau_l)$, is computed as $\prod_{s=1}^{l}(1 - \hat{h}_s)$; the last two columns are computed using $f(c, \tau_s) = h(c, \tau_s)\bar{F}(\tau_{s-1})$ (carefully noting the subscript $s - 1$) and then summing over s. The estimates for the overall probabilities of the two types of failure, p_1 and p_2, come out as 0.45 and 0.55 for these data.

The computations can be performed using the R-function `survfit` from the `survival` package; here I have used the homegrown R-function `km10`, which makes the various intermediate quantities more accessible for printing out. The R-code for the table and the plot described below is

```
#gubbins data (Sec 16.5)
n0=8; n1=21; n2=26; gb0=rep(3.0,n0);
gb1=c(2.7,2.4,2.2,2.1,2.8,2.1,2.8,2.3,2.1,2.6,2.4,2.7,2.6,2.5,2.3,
    2.8,2.1,2.5,2.6,2.4,2.2);
gb2=c(2.4,1.9,1.8,2.3,2.0,2.4,2.3,1.9,1.8,1.9,2.2,1.9,1.9,2.3,2.4,
    1.9,1.8,2.4,1.8,2.3,2.4,1.9,1.8,2.0,2.4,2.3);
cvec=c(rep(0,n0),rep(1,n1),rep(2,n2)); tvec=c(gb0,gb1,gb2);
nd=n0+n1+n2; mc=2; rtn10=km10(nd,tvec,cvec,mc,1);
par(mfrow=c(1,2)); kmplot1(nd,tvec,cvec,mc,'');
```

Plots of the cumulative hazard functions, $\sum_{s=1}^{l} \hat{h}_s$ and $\sum_{s=1}^{l} \hat{h}_{js}$ ($j = 1, 2$) versus τ_l, are given in the left-side panel of Figure 16.1. Corresponding plots

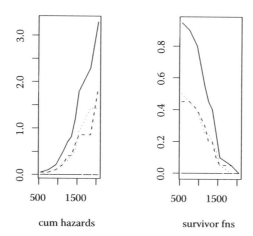

cum hazards   survivor fns

FIGURE 16.1
Cumulative hazard and survivor functions.

of the survivor functions, $\hat{\bar{F}}(\tau_l)$ and $\hat{\bar{F}}(j, \tau_l)$ ($j = 1, 2$) versus τ_l, are given in the right-side panel; the continuous line is for $\hat{\bar{F}}(\tau_l)$, the short-dashed line is for $\hat{\bar{F}}(1, \tau_l)$, and the long-dashed line is for $\hat{\bar{F}}(2, \tau_l)$. The points are joined by straight lines just to help the eye, though strictly, these are step functions.

The procedure described in this section gives estimates of the sub-hazard functions and thence the sub-densities and sub-survivor functions. In the case of independent risks, more strictly, when the Makeham assumption (Section 14.5) holds, these are estimates for the marginal distributions of the components T_j in the latent failure time representation. Zheng and Klein (1995) presented an extended version of this. Independence is replaced by known dependence structure, in the form of a specified *copula*. Then, a procedure analogous to the Kaplan–Meier method is employed to produce estimates of the marginal distributions. Rivest and Wells (2001) investigated the proposal further.

16.5.2 Interval-Censored Data

The foregoing development, though strictly for discrete failure times, applies almost immediately to interval-censored, or grouped, continuous failure times. Suppose that the grouping intervals are $I_l = (\tau_{l-1}, \tau_l]$ for $l = 1, \ldots, m$. The appropriate likelihood function is as given in Section 16.4, with $f(c_i, \tau_l; x_i)$ replaced by $\bar{F}(c_i, \tau_{l-1}; x_i) - \bar{F}(c_i, \tau_l; x_i)$. The equivalent form in terms of hazard contributions, follows.

In the case of random samples, without explanatory variables, we have the hazard estimates $\hat{h}_{jl} = r_{jl}/q_l$ and $\hat{h}_l = r_l/q_l$ for

$$h_{jl} = \text{P(}failure\ from\ cause\ j\ in\ I_l\ |\ enter\ I_l\text{)}$$

and

$$h_l = \text{P(}failure\ in\ I_l\ |\ enter\ I_l\text{)}.$$

Then P(*survive to enter* I_l) and P(*failure from cause j in* I_l) are estimated as

$$\hat{\bar{F}}(\tau_{l-1}) = \prod_{s=0}^{l-1}(1 - r_s/q_s) \quad \text{and} \quad \hat{f}(j, \tau_l) = \hat{h}_{jl}\hat{\bar{F}}(\tau_{l-1}),$$

and then P(*eventual failure from cause j*) as $\hat{p}_j = \sum_{l=1}^{m} \hat{f}(j, \tau_l)$.

16.5.3 Superalloy Testing

These data (Section 12.1) provide an example where the failure times, numbers of cycles to failure in fatigue testing of a superalloy, have been grouped on a log scale. Table 16.3 shows the first half-dozen τ_l-values, together with the estimates of the hazard and survivor functions. The R-code for this and the

TABLE 16.3

Estimates for Superalloy Data

l	τ_l	$\hat{h}_{1l}$	$\hat{h}_{2l}$	$\hat{h}_l$	$\hat{F}(1, \tau_l)$	$\hat{F}(2, \tau_l)$	$\hat{F}(\tau_l)$
1	3.55	0.000	0.008	0.008	0.156	0.836	0.992
2	3.60	0.000	0.016	0.016	0.156	0.820	0.977
3	3.65	0.000	0.048	0.048	0.156	0.773	0.930
4	3.70	0.003	0.064	0.067	0.154	0.714	0.867
5	3.75	0.009	0.141	0.150	0.146	0.591	0.737
6	3.80	0.004	0.155	0.159	0.143	0.477	0.620
...	...	...	...	...	...	...	...

following plots is

```
#Nelson data (Sec 16.5)
dx1=read.table('nelson2.dat',header=T); attach(dx1); dx1;
n1=27; m1=4; dx2=as.numeric(unlist(dx1)); dim(dx2)=c(n1,m1);
mc=2; jt=1; jc=2;
#KM
rtn1=nelsonxpand(n1,m1,dx2); n2=rtn1[1];
dx3=rtn1[2:(2*n2+1)]; dim(dx3)=c(n2,2);
tvec=dx3[1:n2,1]; cvec=dx3[1:n2,2];
par(mfrow=c(1,2));
mc=2; km=kmplot1(n2,tvec,cvec,mc,'');
```

Plots of the survivor functions and cumulative hazards are given in Figure 16.2. The latter show a fairly clear pattern: the overall cumulative hazard is the top curve, being the sum of the other two, and the cumulative sub-hazard for defect-type 1 does not really get off the ground until about

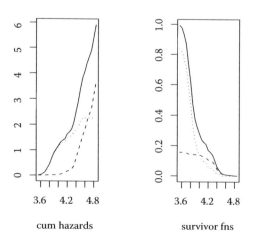

cum hazards   survivor fns

FIGURE 16.2

Cumulative hazard and survivor functions.

$\tau = 4.05$, after which it rises more sharply, eventually overtaking the sub-hazard for defect type 2, which has flattened out. It seems, then, that defects of type 2 are predominant at lower numbers of cycles and then defects of type 1 catch up and overtake later on, at which stage defects of type 2 have all but disappeared. The survivor functions tell a similar tale (all right, the same tale): the overall curve is at the top with the defect-2 curve in close attendance at the lower numbers of cycles; the defect-1 curve starts off lower but is flatter and crosses its rival at about $\tau = 4.05$.

16.6 Asymptotic Distribution of Non-Parametric Estimators

When m is finite, we can apply standard likelihood theory (Appendix B) to argue that the *mle* $\hat{h}_{cl} = r_{cl}/q_l$ are jointly asymptotically normal with means h_{cl} and covariances $V_{cl,bk} = acov(\hat{h}_{cl}, \hat{h}_{bk})$ obtained from the information matrix, where *acov* means *asymptotic covariance*. Let $\mathbf{h}_l = (h_{1l}, \ldots, h_{dl})^{\mathsf{T}}$ be the $d \times 1$ vector of discrete sub-hazards at time τ_l, where $d = 2^p - 1$, and stack them all into the single $md \times 1$ vector $\mathbf{h} = (\mathbf{h}_1^{\mathsf{T}}, \ldots, \mathbf{h}_m^{\mathsf{T}})^{\mathsf{T}}$. The $V_{cl,bk}$ are then found as the elements of the inverse of the $md \times md$ matrix $\mathrm{E}(-\partial^2 \log L/\partial \mathbf{h}^2)$. Recall, from Section 16.5,

$$\partial \log L/\partial h_{cl} = r_{cl}/h_{cl} - (q_l - r_l)/(1 - h_l).$$

Differentiating again, now with respect to h_{bk}, and using the Kronecker delta symbol,

$$-\partial^2 \log L/\partial h_{cl}\partial h_{bk} = \delta_{lk}\{\delta_{cb}r_{cl}/h_{cl}^2 + (q_l - r_l)/(1 - h_l)^2\};$$

recall that the second term is absent for $l = m$. The presence of δ_{lk} implies that $\hat{h}_l$ and $\hat{h}_k$ are asymptotically independent for $l \neq k$, so the asymptotic covariance matrix $\mathbf{V}_h$ of $\hat{\mathbf{h}}$ is block diagonal with m $d \times d$ blocks $\mathbf{V}_l = acov(\hat{\mathbf{h}}_l)$ whose elements can be estimated from

$$\begin{aligned}(\hat{\mathbf{V}}_l^{-1})_{cb} &= \delta_{cb}r_{cl}/\hat{h}_{cl}^2 + (1 - \delta_{lm})(q_l - r_l)/(1 - \hat{h}_l)^2 \\ &= q_l\{\delta_{cb}/\hat{h}_{cl} + (1 - \delta_{lm})/(1 - \hat{h}_l)\}.\end{aligned}$$

For this asymptotic formula to be useful, we need $\hat{h}_{cl} > 0$ for every c and l. This in turn requires $h_{cl} > 0$ and a sample size large enough to ensure that $r_{cl} > 0$. In cases where it is known that $h_{cl} = 0$, this parameter would simply be omitted from $\mathbf{h}$. In matrix terms,

$$\mathbf{V}_l^{-1} = q_l\{diag(\hat{\mathbf{h}}_l)^{-1} + (1 - \delta_{lm})(1 - \hat{h}_l)^{-1}\mathbf{J}_d\},$$

where $diag(\mathbf{h}_l)$ is the diagonal matrix with non-zero elements $\mathbf{h}_l = (h_{1l}, \ldots, h_{dl})$ and $\mathbf{J}_d$ is the $d \times d$ matrix of 1s. Inverting the matrix (see the Exercises, Section 16.8),

$$\mathbf{V}_l = q_l^{-1}\{diag(\hat{\mathbf{h}}_l) - v\hat{\mathbf{h}}_l\hat{\mathbf{h}}_l^{T}\},$$

with

$$v = (1 - \delta_{lm})(1 - \hat{h}_l)^{-1}/\{1 + (1 - \delta_{lm})(1 - \hat{h}_l)^{-1}\hat{h}_l\} = (1 - \delta_{lm}).$$

Hence,

$$\hat{V}_{cl,bk} = \delta_{lk}(\hat{\mathbf{V}}_l)_{cb} = \delta_{lk}q_l^{-1}\hat{h}_{cl}\{\delta_{bc} - (1 - \delta_{lm})\hat{h}_{bl}\}.$$

For example, it follows from $h_l = \mathbf{1}_d^{\mathsf{T}}\mathbf{h}_l$ (where $\mathbf{1}_d$ is the $d \times 1$ vector of 1s) that the $\hat{h}_l$ ($l = 1, \ldots, m$) are asymptotically independent with asymptotic variances estimated as

$$\mathbf{1}_d^{\mathsf{T}}\hat{\mathbf{V}}_l\mathbf{1}_d = q_l^{-1}\hat{h}_l\{1 - (1 - \delta_{lm})\hat{h}_l\}.$$

Likewise, the $\hat{p}_c$ are jointly asymptotically normal with means p_c and covariance matrix calculated as follows. We have

$$p_c = \sum_{l=1}^{m} f(c, \tau_l) = \sum_{l=1}^{m} h_{cl}\bar{F}(\tau_{l-1}) = h_{cl} + \sum_{l=2}^{m}\left\{h_{cl}\prod_{s=1}^{l-1}(1 - h_s)\right\}.$$

Thus, taking differentials,

$$\partial p_c = \partial h_{cl} + \sum_{l=2}^{m}\left[\left\{\partial h_{cl} - h_{cl}\sum_{s=1}^{l-1}\partial h_s/(1 - h_s)\right\}\prod_{s=1}^{l-1}(1 - h_s)\right].$$

Now applying the delta method (Appendix B), the asymptotic covariance between estimates $\hat{p}_c$ and $\hat{p}_b$ may be evaluated as

$$\mathrm{E}\{(\partial\hat{p}_c)(\partial\hat{p}_b)\} \sim V_{cl,bl} - \sum_{l=2}^{m}h_{bl}S_{l-1}V_{cl,+l}/(1 - h_1) + \sum_{l=2}^{m}S_{l-1}^2 V_{cl,bl}$$

$$- \sum_{l=2}^{m}\sum_{l'>l}(h_{bl'}V_{cl,+l} + h_{cl'}V_{bl,+l})S_{l-1}S_{l'-1}/(1 - h_l)$$

$$- \sum_{l=2}^{m}\sum_{l'=2}^{m}h_{cl}h_{bl'}S_{l-1}S_{l'-1}\sum_{s=2}^{l''-1}\left(\sum_{j=1}^{p}V_{js,+s}\right)\bigg/(1 - h_s)^2,$$

where we have used $\mathrm{E}\{(\partial\hat{h}_{cl})(\partial\hat{h}_{bk})\} \approx V_{cl,bk}$, which is zero for $l \neq k$, and adopted the notations S_l for $\prod_{s=1}^{l}(1 - h_s)$, $V_{cl,+l}$ for $\sum_{j=1}^{p}V_{cl,jl}$ and l'' for $\min(l, l')$. This expression for $acov(\hat{p}_c, \hat{p}_b)$ is not a pretty sight but is easily computed.

16.7 Proportional Odds and Partial Likelihood

The extension of Section 5.3 to Competing Risks can be outlined as follows: The argument for constructing the partial likelihood is the same as before. Suppose for the moment that individual i_l is the only one to fail at time τ_l and that the cause of failure is c_l. Then the lth contribution to the partial likelihood function is

$$[h_l(c_l; \mathbf{x}_{i_l})/\{1 - h_l(c_l; \mathbf{x}_{i_l})\}] \div \sum_{i \in R(\tau_l)} [h_l(c_l; \mathbf{x}_i)/\{1 - h_l(c_l; \mathbf{x}_i)\}],$$

writing $h_l(j; \mathbf{x})$ for $h(j, \tau_l; \mathbf{x})$. The assumption of a proportional-odds model

$$[h_l(j; \mathbf{x}_i)/\{1 - h_l(j; \mathbf{x}_i)\}] = \psi_{ij}[h_0(j, l)/\{1 - h_0(j, l)\}],$$

where $\psi_{ij} = \psi(\mathbf{x}_i; \beta_j)$, reduces this contribution to $\psi_{i_l c_l} / \sum^l \psi_{i c_l}$, where $\sum^l \psi_{i c_l}$ denotes $\sum_{i \in R(\tau_l)} \psi_{i c_l}$. Hence, the *partial likelihood* is

$$P(\beta) = \prod_{l=1}^{m} \left(\psi_{i_l c_l} \Big/ \sum\nolimits^l \psi_{i c_l} \right)$$

and it is put to use as before. Ties are a nuisance and must be dealt with along the lines set out in Section 5.3.

For estimation of the baseline hazards $h_0(j, l)$ the relevant likelihood function is derived in Section 16.4. Thus, we take L as there with $h(c_i, \tau_l; \mathbf{x}_i)$ replaced by

$$\psi_{ij} h_0(c_i, l)/\{1 + (\psi_{ij} - 1)h_0(c_i, l)\},$$

according to the proportional-odds model; in the expression for L, $h(\tau_l; \mathbf{x}_i)$ is replaced by $\sum_j h(j, \tau_l; \mathbf{x}_i)$. In ψ_{ij}, β_j can be replaced by $\hat{\beta}_j$, the maximum partial-likelihood estimator, and then L is to be maximised over the remaining parameter set $\{h_0(j, l) : j = 1, \ldots, p; l = 1, \ldots, m\}$.

16.7.1 Psychiatric Wards

Partial likelihood was applied to the data using the following R-code, which follows on from that listed in Section 16.4:

```
#psychiatric wards (Sec 16.7)
dx1=read.table('pwards.dat',header=T); attach(dx1); #dx1;
nd=51; md=7; dx2=as.numeric(unlist(dx1)); dim(dx2)=c(nd,md);
dx2[1:nd,2]=dx2[1:nd,2]+1; dx2[1:nd,4]=dx2[1:nd,4]/10;
dx2[1:nd,6]=dx2[1:nd,6]/10; dx2[1:nd,7]=dx2[1:nd,7]/10;
#Cox PH fit (Sec 16.7)
nd=51; mc=2; kt=2; kc=3; tv=dx2[1:nd,kt]; cv=dx2[1:nd,kc];
iwr=1; rtn01=km01(nd,tv,cv,mc,iwr); m=rtn01[1]; iv=rtn01[2:(nd+1)];
it=1+nd+m; mt=rtn01[(it+1):(it+m)]; xml=dx2[iv,1:7];
#full model x=(age,adm,nbeds,nstaff)
mx=4; kx=c(4,5,6,7); np=mc*mx; par0=runif(np);
opt=0; iwr=0; adt=c(nd,mc,kt,kc,opt,iwr,m,mt,mx,kx);
par1=fitmodel(plikd,np,par0,xml,adt);
#x=(age,adm)
mx=2; kx=c(4,5); np=mc*mx; par0=runif(np);
adt=c(nd,mc,kt,kc,opt,0,m,mt,mx,kx);
par1=fitmodel(plikd,np,par0,xml,adt);
#x=(adm)
mx=1; kx=c(5); np=mc*mx; par0=runif(np);
```

```
adt=c(nd,mc,kt,kc,opt,0,m,mt,mx,kx);
par1=fitmodel(plikd,np,par0,xm1,adt);
#null model (betas all zero)
mx=1; kx=c(0); np=mc*mx; par0=c(0,0);
adt=c(nd,mc,kt,kc,opt,1,m,mt,mx,kx);
plkd=plikd(par0,xm1,adt);
```

The maximised log-partial likelihood with the full model (all four covariates) is -114.0070, with estimated regression coefficients $(-0.59, 1.01, 0.55, -0.92)$ for incident type 1 and $(-0.29, 0.83, 1.31, -1.75)$ for incident type 2; the corresponding standard errors are $(0.32, 0.54, 0.88, 1.63)$ and $(0.28, 0.52, 0.84, 1.62)$. The last two covariates, *nbeds* and *nstaff*, have unimpressive ratios $\hat{\beta}/se(\hat{\beta})$, so let's try a fit without them. The maximised log-partial likelihood is now -115.4628, and that for the fit with just *adm* alone is 117.4943, and that of a null fit (without any covariates) is -119.3381. First, comparing the null fit with the full fit gives $\chi^2_8 = 10.6622$ ($p = 0.22$): it is a little disappointing that the null fit appears to be not much worse than the full one. However, let's see if this might be just due to some masking. It is likely that *adm* will be a factor in this context, and the ratios $\beta/se(\hat{\beta})$ seem to bear this out to some extent. Comparing the *adm* fit with the full fit gives $\chi^2_8 = 6.97$ ($p = 0.32$), and comparing it with the null fit gives $\chi^2_2 = 3.69$ ($p = 0.16$), which fails to achieve the magic 5% level. So, we must conclude that the evidence for any effect of these covariates, based on these data, is not strong.

16.8 Exercises

1. How can the joint survivor function of an independent-risks model be constructed to conform to a set of discrete sub-distributions?
2. Apply the process of Question 1 to the geometric mixture model (Section 16.1). Specialise to the case of equal ρ_j.
3. Verify the form of matrix inverse $\mathbf{V}_l^{-1}$ quoted in Section 16.6.

16.9 Hints and Solutions

1. Start with the $h(j, t)$, equate these to marginals $h_j(t)$ (Makeham assumption), obtain marginal survivor functions $\bar{G}_j(t)$, and then multiply: $\bar{G}(\mathbf{t}) = \prod_{j=1}^p G_j(t_j)$.
2. Some algebra is needed here.
3. Let $\mathbf{b} = (b_1, \ldots, b_d)^T$ and $\mathbf{B} = diag(\mathbf{b})$: then $(\mathbf{B}^{-1}+u\mathbf{J}_d)(\mathbf{B}-v\mathbf{b}\mathbf{b}^T) = \mathbf{I}_d$ (unit matrix), where $v = u/(1 + ub_+)$. This follows from the product formulae

$$\mathbf{B}^{-1}\mathbf{b}\mathbf{b}^T = \mathbf{1}_d\mathbf{b}^T, \quad \mathbf{J}_d\mathbf{B} = \mathbf{1}_d\mathbf{b}^T, \quad \mathbf{J}_d\mathbf{b}\mathbf{b}^T = b_+\mathbf{1}_d\mathbf{b}^T.$$

17

Latent Lifetimes: Identifiability Crises

Looking at Competing Risks from the point of view of latent lifetimes seems very natural. Statistically respectably, you can specify a parametric model for $\bar{G}(\mathbf{t})$, fit it to the data, test the fit, and test hypotheses, all without the faintest suspicion that there might be anything lacking. A hidden problem with this approach is explored in this chapter.

17.1 The Cox–Tsiatis Impasse

Cox (1959) drew attention to a flaw in the approach to competing risks via latent lifetime models. His discussion concerned various models for failure times with two causes, like those presented by Mendenhall and Hader (1958), including their mixture of exponentials (Section 12.2). He noted that "no data of the present type can be inconsistent with" an independent-risks model. This was extended to the general case of p risks in continuous time by Tsiatis (1975). Specifically, Tsiatis showed that, given any joint survivor function with arbitrary dependence between the component variates, there exists a different joint survivor function in which the variates are independent and which reproduces the sub-densities $f(j, t)$ precisely. Thus, one cannot know, from observations on (C, T) alone, which of the two models is correct – they will both fit the data equally well. This is a bit awkward from the point of view of statistical inference.

17.1.1 Tsiatis's Theorem (Tsiatis, 1975)

Suppose that the set of $\bar{F}(j, t)$ is given for some continuous-time model with dependent risks. Then there exists a unique *proxy model* with independent risks yielding identical $\bar{F}(j, t)$. It is defined by $\bar{G}^*(\mathbf{t}) = \prod_{j=1}^{p} \bar{G}_j^*(t_j)$, where $G_j^*(t) = \exp\{-\int_0^t h(j, s)ds\}$ and the sub-hazard function $h(j, s)$ derives from the given $\bar{F}(j, t)$.

Proof   A constructive proof will be given. We wish to find a function $\bar{G}^*(\mathbf{t}) = \prod_{j=1}^{p} \bar{G}_j^*(t_j)$ such that $f(j, t) = [-\partial \bar{G}^*(\mathbf{t})/\partial t_j]_{t1_p}$ for all (j, t). Hence, we require that

$$f(j, t) = \left[\{-\partial \bar{G}_j^*(t_j)/\partial t_j\}\bar{G}^*(\mathbf{t})/\bar{G}_j^*(t_j)\right]_{t1_p} = \{-d \log \bar{G}_j^*(t)/dt\}\bar{G}^*(t1_p).$$

Summing over j yields

$$f(t) = \{-d \log \bar{G}^*(t1_p)/dt\}\bar{G}^*(t1_p) = -d\bar{G}^*(t1_p)/dt.$$

Integrating, $\bar{F}(t) = \bar{G}^*(t1_p)$ and so

$$-d \log \bar{G}_j^*(t)/dt = f(j, t)/\bar{F}(t) = h(j, t).$$

We have shown so far that, if an independent-risks proxy model exists, then it must have the given form. To complete the proof we have to show that the given form defines $\bar{G}_j^*(t)$ as a valid survivor function, which it obviously does, and also meets the requirement of mimicking the $f(j, t)$. For this last part

$$g^*(j, t) = g_j^*(t) \prod_{k \neq j} \bar{G}_k^*(t) = \{-d \log \bar{G}_j^*(t)/dt\} \prod_{k=1}^{p} \bar{G}_k^*(t)$$

$$= h(j, t) \exp\left\{-\sum_{k=1}^{p} \int_0^t h(k, s)ds\right\} = h(j, t) \exp\left\{-\int_0^t h(s)ds\right\}$$

$$= h(j, t)\bar{F}(t) = f(j, t).$$

The hazard function h_j^* corresponding to $\bar{G}_j^*$ satisfies

$$\bar{G}_j^*(t) = \exp\left\{-\int_0^t h_j^*(s)ds\right\}.$$

So, the proxy model of the theorem is just constructed by using the original sub-hazards $h(j, t)$ for the $h_j^*(t)$. An extended version of this proxy identity can be formulated to accommodate mixed discrete-continuous lifetimes—see Equation (12.2) of Crowder (1997). If the original model has independent risks, then it is its own proxy because in this case $h(j, t) = h_j(t)$, by Gail's theorem (Section 14.4), and so $\bar{G}_j^*(t) = \bar{G}_j(t)$. This hazard condition can hold even without independent risks (Section 14.4), so some dependent-risks systems can provide their own proxy models. This curiosity will be taken up more formally in Section 17.3. Under the classical assumption of independent risks what one is actually estimating is the proxy model, because one observes $h(j, t)$ and assumes that this is $h_j(t)$. Thus, there is an outside chance of being correct, but not one that you would want to bet on.

In the case of proportional hazards

$$\bar{G}_j^*(t) = \exp\left\{-p_j \int_0^t h(s)ds\right\} = \exp\{p_j \log \bar{F}(t)\} = \bar{F}(t)^{p_j},$$

yielding an expression for $\bar{G}^*(t)$ in terms of the single function $\bar{F}(t)$; this was noted previously in Section 14.5.

Tsiatis's theorem establishes that to each dependent-risks model there corresponds a unique independent-risks proxy model with the same sub-survivor functions. More detailed structure will be revealed in Section 17.3: it turns out that each independent-risks model has a whole class of satellite dependent-risks models, and that this class can be partitioned into sets with the same marginals. The following examples illustrate this structure within the limited framework of particular parametric classes of models.

17.1.2 Gumbel's Bivariate Exponential

Referring back to Section 14.1, the sub-hazards are $h(j, t) = \lambda_j + vt$, and so the proxy model has

$$\bar{G}_j^*(t) = \exp\left\{-\int_0^t (\lambda_j + vs)ds\right\} = \exp\{-(\lambda_j t + vt^2/2)\},$$

$$\bar{G}^*(\mathbf{t}) = \exp\{-(\lambda_1 t_1 + vt_1^2/2) - (\lambda_2 t_2 + vt_2^2/2)\}.$$

So, to make predictions about T_j you can use either $\bar{G}_j(t) = \exp(-\lambda_j t)$ or $\bar{G}_j^*(t)$; and unfortunately, the answers will differ. Moreover, you cannot tell which function is the correct one to use purely from (c, t) data, however much of it you have.

There is a 1–1 correspondence between $\bar{G}(\mathbf{t})$ and $\bar{G}^*(\mathbf{t})$ in the example because the parameter set $(\lambda_1, \lambda_2, v)$ is identified by both $\bar{G}(\mathbf{t})$ and $\bar{G}^*(\mathbf{t})$. But it ain't necessarily so, as shown by a second example.

Example (Crowder, 1994)
Consider the form

$$\bar{G}(\mathbf{t}) = \exp\left\{-\lambda_1 t_1 - \lambda_2 t_2 - vt_1 t_2 - \mu_1 t_1^2 - \mu_2 t_2^2\right\}.$$

This is a valid bivariate survivor function if $\lambda_j > 0$ and $\mu_j \geq 0$ for $j = 1, 2$, and $0 \leq v \leq \lambda_1 \lambda_2$. The sub-density and sub-hazard functions are

$$f(j, t) = \{\lambda_j + (v + 2\mu_j)t\}\exp\{-\lambda_+ t - (v + \mu_+)t^2\},$$
$$h(j, t) = \lambda_j + (v + 2\mu_j)t,$$

where $\lambda_+ = \lambda_1 + \lambda_2$ and $\mu_+ = \mu_1 + \mu_2$. The proxy model then has marginals

$$\bar{G}_j^*(t) = \exp\left\{-\int_0^t h(j, s)ds\right\} = \exp\{-\lambda_j t - (v + 2\mu_j)t^2/2\}.$$

Evidently, $\bar{G}^*(\mathbf{t})$ identifies only $(\lambda_1, \lambda_2, v + 2\mu_1, v + 2\mu_2)$, so the set of $\bar{G}^*(\mathbf{t})$ with $v + 2\mu_1 = \tau_1$ and $v + 2\mu_2 = \tau_2$ all share the same independent-risks proxy model

$$\bar{G}^*(\mathbf{t}) = \exp\left\{-\lambda_1 t_1 - \lambda_2 t_2 - \tau_1 t_1^2/2 - \tau_2 t_2^2/2\right\}.$$

Example (Crowder, 1991)

Here is a general class of parametric models in which the non-identifiability behaviour occurs. Consider a joint density of the form

$$g(\mathbf{t}) = \begin{cases} g_\phi(t_2)g_\psi(t_1 - t_2)/2 & \text{on } t_1 > t_2 \\ g_\phi(t_1)g_\psi(t_2 - t_1)/2 & \text{on } t_1 < t_2 \end{cases}$$

where g_ϕ and g_ψ are univariate densities on $(0, \infty)$ involving parameters ϕ and ψ. By the formula $f(j, t) = \int_t^\infty g(\mathbf{r}_j)ds$ (Section 14.1), we have $f(1, t) = g_\phi(t)/2$ and, symmetrically, $f(2, t) = g_\phi(t)/2$. Thus, ψ has gone absent without leave from the sub-distributions. A particular case is obtained when g_ϕ and g_ψ are exponential densities. In that case

$$\bar{G}(\mathbf{t}) = \begin{cases} (1 + \tau)\exp(-\phi t_1) - \tau\exp(-\psi t_1 + \psi t_2 - \phi t_2) & \text{on } t_1 > t_2 \\ (1 + \tau)\exp(-\phi t_2) - \tau\exp(-\psi t_2 + \psi t_1 - \phi t_1) & \text{on } t_1 < t_2 \end{cases}$$

where $\tau = \phi/\{2(\psi - \phi)\}$. It can be verified that $\theta = (\phi, \psi)$ is identified in the joint distribution but that $\bar{F}(1, t) = \bar{F}(2, t) = e^{-\phi t}$, independent of ψ.

17.2 More General Identifiablility Results

The assumption that T_j is continuous in Tsiatis's theorem can be dropped. As before $T = \min\{T_1, \ldots, T_p\}$, and let $\alpha = \sup\{t : \bar{F}(t) > 0\}$ be the upper limit of possible T values.

17.2.1 Miller's Theorem (Miller, 1977)

Suppose that the $F(j, t)$ $(j = 1, \ldots, p)$ have no discontinuities in common and there are no ties among the T_j. Then there exists a set of independent S_j $(j = 1, \ldots, p)$, at least one of which is almost surely finite, such that their sub-survivor functions match the $\bar{F}(j, t)$: $\bar{G}_s(j, t) = \bar{F}(j, t)$ for all (j, t). Also, the S_j distributions are uniquely determined on $[0, \alpha)$.

Proof   See the paper cited.

The assumption that there are no ties can also be dropped. Typical applications occur in Reliability where a piece of equipment is subject to random shocks that can knock out one or more components at the same time. In this extended situation we still have $T = \min(T_1, \ldots, T_p)$, but C is defined more generally as the *failure pattern* or *configuration* (Section 15.2). A simple but significant point is that ties between configurations are eliminated by definition: for example, instead of saying that $\{j, k\}$ and $\{l\}$ have tied, we simply say that $\{j, k, l\}$ has occurred.

Let $\mathbf{R}$ be a failure time vector of length $2^p - 1$ with components R_c, and define C_R and T_R by $T_R = \min_c(R_c) = R_{C_R}$. We quote now a theorem due to Langberg et al. (1978) concerning the existence and form of an independent-risks proxy model for which

$$(C_R, T_R) =_d (C, T),$$

that is, the pair (C_R, T_R) generated from the $\mathbf{R}$-system has the same joint distribution as the original (C, T). In the authors' terminology, systems $\mathbf{R}$ and $\mathbf{T}$ are *equivalent in life length and patterns*, written as $\mathbf{R} =_{LP} \mathbf{T}$, if $P(C_R = c, T_R > t) = \bar{F}(c, t)$ for all configurations c and all $t \geq 0$. The result extends Miller's (1977) by allowing ties and also giving explicit forms for the R_c-distributions. Peterson (1977) covered similar ground but with a different emphasis.

We need some notation for the theorem. Let $D_c = \{\tau_{cl} : l = 1, 2, \ldots\}$ be the set of discontinuities of $\bar{F}(c, t)$, let the discrete sub-hazard contribution at τ_{cl} be

$$h_{cl} = P(C = c, T = \tau_{cl} \mid T > \tau_{cl}-) = \nabla \bar{F}(c, \tau_{cl}) / \bar{F}(\tau_{cl}-),$$

where

$$\nabla \bar{F}(c, \tau_{cl}) = \bar{F}(c, \tau_{cl}-) - \bar{F}(c, \tau_{cl}),$$

and let, for $t \notin D_c$,

$$f^C(c, t) = -d\bar{F}^C(c, t)/dt \text{ and } h^C(c, t) = f^C(c, t)/\bar{F}(t-)$$

be the density and hazard functions of the continuous component of $\bar{F}(c, t)$.

17.2.2 The LPQ Theorem (Langberg, Proschan, and Quinzi, 1978)

There exists a set of independent R_c with $\mathbf{R} =_{LP} \mathbf{T}$ if and only if the D_c are pairwise disjoint on $[0, \alpha)$. In this case the survivor functions of the R_c are uniquely determined on $[0, \alpha)$ as

$$\bar{G}_{Rc}(t) = P(R_c > t) = \exp\left\{-\int_0^t h^C(c, s)ds\right\} \prod_{s=1}^{l_c(t)} (1 - h_{cs}),$$

where $l_c(t) = \max\{l : \tau_{cl} \leq t\}$.

Proof That disjoint D_c are necessary can be seen as follows. Suppose that they are not disjoint, in particular that $\tau \in D_b \cap D_c$. If an independent-risks $\mathbf{R}$-system existed that matched the original sub-survivor functions, we would have $\bar{F}_R(c, t) = \bar{F}(c, t)$ for all (c, t), and so $\bar{F}_R(b, t)$ and $\bar{F}_R(c, t)$ would both be discontinuous at τ. In that case $P(R_b = \tau) > 0$ and $P(R_c = \tau) > 0$, and so $P(E_{bc}) > 0$, where $E_{bc} = \{R_b = R_c = \tau < \text{all other } Rs\}$, since the Rs are independent. Consequently, $\sum_c \bar{F}_R(c, 0) < 1$ because the sum does not include $P(E_{bc})$. Conversely, under matching of the sub-survivor functions,

$$\sum_c \bar{F}_R(c, 0) = \sum_c \bar{F}(c, 0) = 1,$$

and so we have a contradiction.

Now assume that the D_c are disjoint. The form of $\bar{G}_{Rc}(t)$ can be guessed by noting that the proxy model in Tsiatis's theorem is obtained by using the original sub-hazards for the proxy marginal hazards, and then extending this to cover discontinuities via the standard formula for $\bar{F}(t)$ in the case of mixed distributions given in Section 2.5. We just have to prove now that it does the job as advertised, that is, that $\bar{F}_R(c, t) = \bar{F}(c, t)$ for all (c, t). Now,

$$\bar{F}_R(c, t) = \int_t^\infty g_{Rc}^C(s) \left\{ \prod_{b \neq c} \bar{G}_{Rb}(s) \right\} ds + \sum_{s > l(t)} \left\{ \nabla \bar{G}_{Rc}(\tau_{cs}) \prod_{b \neq c} \bar{G}_{Rb}(\tau_{cs}) \right\},$$

where

$$\nabla \bar{G}_{Rc}(\tau_{cs}) = \bar{G}_{Rc}(\tau_{cs}-) - \bar{G}_{Rc}(\tau_{cs}),$$

and $g_{Rc}^C(s) = -d\bar{G}_{Rc}(s)/ds$ is the marginal density defined for $s \notin D_c$. First, from the expression given for $\bar{G}_{Rc}$ in the statement of this theorem,

$$\prod_b \bar{G}_{Rb}(t) = \exp\left\{-\int_0^t h^C(s)ds\right\} \times \prod_b \prod_{s=1}^{l_b(t)} (1 - h_{bs}) = \bar{F}(t),$$

using $h^C(s) = \sum_b h^C(b, s)$ and the fact that the product is over all the discontinuities of $\bar{F}(t) = \sum_b \bar{F}(b, t)$ up to t. The first term on the right-hand side of the expression for $\bar{F}_R(c, t)$ is then equal to

$$\int_t^\infty h_{Rc}^C(s) \left\{ \prod_b \bar{G}_{Rb}(s) \right\} ds = \int_t^\infty h^C(c, s)\bar{F}(s)ds = \int_t^\infty f^C(c, s)ds = \bar{F}^C(c, t),$$

where we have used

$$h_{Rc}^C(s) = -d \log \bar{G}_{Rc}^C(s)/ds = h^C(c, s).$$

For the second term we have, again from the stated form of $\bar{G}_{Rc}$,

$$\nabla \bar{G}_{Rc}(\tau_{cl}) = \exp\left\{-\int_0^{\tau_{cl}} h^C(c, s)ds\right\} \times \left\{ \prod_{s=1}^{l-1}(1 - h_{cs}) - \prod_{s=1}^{l}(1 - h_{cs}) \right\}$$

$$= \exp\left\{-\int_0^{\tau_{cl}} h^C(s)ds\right\} \times \{h_{cl}/(1 - h_{cl})\} \prod_{s=1}^{l}(1 - h_{cs})$$

$$= \bar{G}_{Rc}(\tau_{cl})\{h_{cl}/(1 - h_{cl})\}.$$

Further,

$$h_{cl}/(1 - h_{cl}) = \{\nabla \bar{F}(c, \tau_{cl})/\bar{F}(\tau_{cl}-)\} \div \{1 - \nabla \bar{F}(c, \tau_{cl})/\bar{F}(\tau_{cl}-)$$
$$= \nabla \bar{F}(c, \tau_{cl}) \div \{\bar{F}(\tau_{cl}-) - \nabla \bar{F}(c, \tau_{cl})\}.$$

But,

$$\nabla \bar{F}(\tau_{cl}) = \nabla \sum_b \bar{F}(b, \tau_{cl}) = \nabla \bar{F}(c, \tau_{cl}),$$

since $\nabla \bar{F}(b, \tau_{cl}) = 0$, in consequence of $\tau_{cl} \notin D_b$ when $b \neq c$. So,

$$h_{cl}/(1 - h_{cl}) = \nabla \bar{F}(c, \tau_{cl}) \div \{\bar{F}(\tau_{cl}-) - \nabla \bar{F}(\tau_{cl})\} = \nabla \bar{F}(c, \tau_{cl}) \div \bar{F}(\tau_{cl}).$$

Thus, the second term is equal to

$$\sum_{s>l(t)} \left[\bar{G}_{Rc}(\tau_{cs})\{\nabla \bar{F}(c, \tau_{cs})/\bar{F}(\tau_{cs})\} \prod_{b \neq c} \bar{G}_{Rb}(\tau_{cs}) \right] = \prod_{s>l(t)} \nabla \bar{F}(c, \tau_{cs}),$$

where the preceding expression for $\prod_b \bar{G}_{Rb}(t)$ has been applied. Putting the first and second terms together, $\bar{F}_R(c, t)$ is $\bar{F}(c, t)$, as advertised.

Arjas and Greenwood (1981) extended the LPQ theorem. They allowed the set of configurations to be countable, rather than just finite, and showed that the disjoint D_c condition can be avoided by using a random tie-breaking device. Their presentation is in terms of martingales and compensators, but a rough translation is as follows: First, let $\bar{G}_{Rc}(t)$ be given as in the theorem, and define $\hat{\bar{G}}_{Rc}(t) = \bar{G}_{Rc}(t)$ except when t is a discontinuity point common to more than one D_c. At such a point, say τ, let $d_\tau = \sum_c \nabla \bar{G}_{Rc}(\tau)$ be the total jump and re-allocate the whole of d_τ to just one of the $\hat{\bar{G}}_{Rc}$ at τ. The LPQ theorem now applies to this new system because the modified discontinuity sets are disjoint. Thus, we have a set of independent R_c whose sub-survivor functions match the modified $\bar{F}(c, t)$. Cunningly, Arjas and Greenwood make the modification randomly. They re-allocate d_τ with probabilities $q_{c\tau} = \nabla \bar{G}_{Rc}(\tau)/d_\tau$: $\nabla \hat{\bar{G}}_{Rc}(\tau)$ becomes 0 or d_τ with $P\{\nabla \hat{\bar{G}}_{Rc}(\tau) = d_\tau\} = q_{c\tau}$. The R_c are then conditionally independent, given the particular re-allocation, and, unconditionally,

$$P(R_c > t) = E\{\hat{\bar{G}}_{Rc}(t)\} = E\left\{ \hat{\bar{G}}^C_{Rc}(t) - \sum_{s=1}^{l(t)} \nabla \hat{\bar{G}}_{Rc}(\tau_{cs}) \right\}$$

$$= \bar{G}^C_{Rc}(t) - \sum_{l=1}^{l(t)} \nabla \bar{G}_{Rc}(\tau_{cs}) = \bar{G}_{Rc}(t).$$

The R_c are unconditionally dependent: for instance, if $\tau \in D_b \cap D_c$ then $P(S_c = \tau \mid S_b = t)$ depends on t by being 0 if $t = \tau$ but not otherwise.

Given an **R**-system a failure time vector $\mathbf{S} = (S_1, \ldots, S_p)$ can be constructed such that $\mathbf{R} =_{LP} \mathbf{S}$ as follows. Define $S_j = \min_c\{R_c : c \ni j\}$, so $C_S = c$

if and only if $C_R = c$, that is, configuration c fails first, knocking out all components j with $j \in c$. For example, suppose that an **R**-system comprises R_1, R_2, R_3, $R_{\{1,3\}}$ and $R_{\{1,2,3\}}$. Then we define $S_1 = \min\{R_1, R_{\{1,3\}}, R_{\{1,2,3\}}\}$, $S_2 = \min\{R_2, R_{\{1,2,3\}}\}$, and $S_3 = \min\{R_3, R_{\{1,3\}}, R_{\{1,2,3\}}\}$. Hence,

$$\bar{G}_S(c, t) = P(C_S = c, T_S > t) = P(C_R = c, T_R > t),$$

that is, $\mathbf{R} =_{LP} \mathbf{S}$.

The **S**-system has p components, like the original **T** but unlike the vaguely artificial $2^p - 1$ components of **R**, and the S_j will be dependent in general. Their joint survivor function may be expressed in terms of the **R**-system as follows:

$$\bar{G}_S(\mathbf{s}) = P\left[\cap_{j=1}^p \{S_j > s_j\} \right] = P\left[\cap_{j=1}^p \min_{c \ni j}(R_c) > s_j\} \right]$$
$$= P\left[\cap_c \{R_c > \max_{j \in c}(s_j)\} \right] = \prod_c P\{R_c > \max_{j \in c}(s_j)\}.$$

The marginal survivor function of S_j is then found by setting all the s_k, other than s_j, to zero in $\bar{G}_S(\mathbf{s})$:

$$\bar{G}_{Sj}(s_j) = \prod_{c \ni j} P(R_c > s_j).$$

One can believe that the S_j are dependent in general by staring hard at their definition given above. However, a lurking doubt might remain as to whether they might, in some special circumstance that you can't quite put your finger on, achieve independence. The following lemma can bring express relief to this discomfort.

Lemma
The S_j are either dependent or degenerate.

Proof   The dependence coefficient defined in Section 6.2 is, for the **S**-system,

$$\bar{R}(\mathbf{s}) = P(\mathbf{S} > \mathbf{s}) \div \prod_{j=1}^p P(S_j > s_j) = \prod_c P(R_c > m_c) \div \prod_{j=1}^p \prod_{c \ni j} P(R_c > s_j),$$

where $m_c = \max_{j \in c}(s_j)$. The numerator here is equal to

$$\prod_c \left[\left\{ \prod_{\tau_{cl} \leq m_c} (1 - h_{cl}) \right\} \exp\left\{ -\int_0^{m_c} h^C(c, s)ds \right\} \right],$$

and, using $\prod_{j=1}^{p} \prod_{c \ni j} = \prod_c \prod_{j \in c}$, the denominator is equal to

$$\prod_c \prod_{j \in c} P(R_c > s_j) = \prod_c \prod_{j \in c} \left[\left\{ \prod_{\tau_{cl} \leq s_j} (1 - h_{cl}) \right\} \exp \left\{ -\int_0^{s_j} h^C(c, s) ds \right\} \right],$$

$$= \prod_c \left[\left\{ \prod_l (1 - h_{cl})^{q_{ck}} \right\} \exp \left\{ -\int_0^{m_c} q_c(s) h^C(c, s) ds \right\} \right],$$

where $q_{cl} = \sum_{j \in c} I\{\tau_{cl} \leq s_j\}$ and $q_c(s) = \sum_{j \in c} I\{s \leq s_j\}$. Hence,

$$\bar{R}(s) = \prod_c \left[\left\{ \prod (1 - h_{cl})^{1 - q_{cl}} \right\} \exp \left\{ -\int_0^{m_c} \{1 - q_c(s)\} h^C(cs, s) ds \right\} \right],$$

where the bracketed product is over $\{l : \tau_{cl} \leq m_c\}$. Note that $q_{cl} \geq 1$ on $\{l : \tau_{cl} \leq m_c\}$ and $q_c(s) \geq 1$ on $(0, m_c)$; it follows that $(1 - h_{cl})^{1 - q_{cl}} \geq 1$ and

$$\int_0^{m_c} \{1 - q_c(s)\} h^C(c, s) ds \leq 0.$$

Hence, for $\bar{R}(s)$ to be 1, we must have both $(1 - h_{cl})^{1 - q_{cl}} = 1$ for all (c, l) and

$$\int_0^{m_c} \{1 - q_c(s)\} h^C(c, s) ds = 0$$

for all c, that is, both (1) either $h_{cl} = 0$ or $q_{cl} = 1$, and (2) either $h^C(c, s) = 0$ or $q_c(s) = 1$. In (1), by definition $h_{cl} > 0$ and, since we can choose s_j to make $q_{cl} \neq 1$, the product term must be absent, that is, there can be no discontinuities. In (2), since we can choose s_j to make $q_c(s) \neq 1$, we must have $h^C(c, s) = 0$ for all s. Thus, the only way that $\bar{R}(s)$ can be 1 for all $\mathbf{s}$ is for there to be no discontinuities and $h^C(c, s) = 0$ for all (c, s), in which case $\bar{F}(c, t) = 1$ for all (c, t), a sadly degenerate case.

17.2.3 The Marshall–Olkin Distribution

In Section 7.4 the model was derived from a physical situation and the resulting joint distribution of failure times was investigated. We now show how the LPQ theorem can be used to reverse the argument, that is, derive a physical model from the joint distribution. Thus, we start with

$$\bar{G}(\mathbf{t}) = \exp\{-\lambda_1 t_1 - \lambda_2 t_2 - \lambda_{12} \max(t_1, t_2)\}.$$

Firstly, we verify the fact that the distribution defined by $\bar{G}(\mathbf{t})$ allows ties. This can be established as follows: (1) calculate the continuous and singular components of $\bar{G}(\mathbf{t})$, as demonstrated above; (2) note that any singular component must be concentrated on the subspace $\{\mathbf{t} : t_1 = t_2\}$ of $(0, \infty)^2$, by inspection of $\bar{G}(\mathbf{t})$; (3) calculate that the total probability mass of the singular component is λ_{12}/λ_+, where $\lambda_+ = \lambda_1 + \lambda_2 + \lambda_{12}$. Thus, $P(T_1 = T_2) = \lambda_{12}/\lambda_+$.

"Nextly," we calculate the sub-survivor functions $\bar{F}(c, t)$ for $c = \{1\}$, $\{2\}$ and $\{1, 2\}$. We have

$$\bar{F}(\{1\}, t) = P(t < T_1 < T_2) = \int_{t < t_1 < t_2} g_2(\mathbf{t})d\mathbf{t},$$

where $g_2(\mathbf{t})$ is the density component on the set $\{\mathbf{t} : t_1 < t_2\}$ derived explicitly above. Thus,

$$\bar{F}(\{1\}, t) = \int_t^\infty dt_2 \int_t^{t_2} dt_1 \{\lambda_1(\lambda_2 + \lambda_{12}) \exp(-\lambda_1 t_1 - \lambda_2 t_2 - \lambda_{12} t_2)\}$$
$$= (\lambda_1/\lambda_+)e^{-\lambda_+ t}.$$

Likewise,

$$\bar{F}(\{2\}, t) = (\lambda_2/\lambda_+)e^{-\lambda_+ t} \text{ and } \bar{F}(\{12\}, t) = \bar{G}^S(t\mathbf{1}) = (\lambda_{12}/\lambda_+)e^{-\lambda_+ t},$$

where $\bar{G}^S(\mathbf{t})$ is the singular component. For a check on these calculations note that

$$\bar{F}(t) = \sum_c \bar{F}(c, t) = e^{-\lambda_+ t} = \bar{G}(t\mathbf{1}).$$

Lastly, note that none of the sub-distributions has a discontinuity, so the LPQ theorem applies in simplified form. The sub-survivor functions can be plugged into the formula there for $P(R_C > t)$ to find the probability structure of the **R**-system; the product term in $P(R_C > t)$ does not come into play. Thus,

$$P(R_{\{1\}} > t) = \exp\left\{-\int_0^t (\lambda_1 e^{-\lambda_+ s}/e^{-\lambda_+ s})ds\right\} = e^{-\lambda_1 t};$$

likewise, $P(R_{\{2\}} > t) = e^{-\lambda_2 t}$ and $P(R_{\{1,2\}} > t) = e^{-\lambda_{12} t}$. In this way a characterization of the Marshall–Olkin distribution, namely the one based on the three types of fatal shock with exponential waiting times, has been reconstructed as an **R**-system from the form of $\bar{G}(\mathbf{t})$ alone.

The sub-hazard function for configuration $c = \{j\}$ is the constant $h(j, t) = \lambda_j$ ($j = 1, 2$). This accords with the way in which the distribution is set up. Note, however, that there is an extra sub-hazard here: $h(\{1, 2\}, t) = \lambda_{12}$ for failure configuration $\{1, 2\}$.

17.3 Specified Marginals

An independent-risks proxy model will reproduce the $\bar{F}(c, t)$ for any given $\bar{G}(\mathbf{t})$ but not in general the univariate marginals $\bar{G}_j(t)$, as seen in the example of the preceding section. Hence, if only we could observe the T_j, separately and unmodified, as well as (C, T), we could eventually identify the $\bar{G}_j(t)$

in addition to the $\bar{F}(c, t)$ and thus detect that the independent-risks proxy model was not the true one. For instance, in controlled experimental situations it might be possible to run different components of a system separately but in joint-operation mode. This is more likely to be relevant to physical, engineering applications than to biological ones.

We can compare the marginal survivor functions of the proxy model with those of $\bar{G}(\mathbf{t})$. Thus, $\bar{G}_j^*(t)$ and $\bar{G}_j(t)$ both have form $\exp\{-\int_0^t r(s)ds\}$, but with $r(s) = h(j, s)$ for $\bar{G}_j^*(t)$ and $r(s) = h_j(s)$ for $\bar{G}_j(t)$; $h_j(s)$ here is the marginal hazard function of T_j. Who would have guessed that the coincidence $\bar{G}_j^*(t) = \bar{G}_j(t)$ is equivalent to the Makeham assumption, $h(j, t) = h_j(t)$ (Gail's theorem, Section 14.4.1)?

Evidently, if the Makeham assumption does not hold, there is no independent-risks proxy model that will reproduce both the $\bar{F}(c, t)$ and the $\bar{G}_j(t)$. But is there a dependent-risks proxy model that will do the job? If so, we still have an identifiability problem, though perhaps not one involving independent risks. The following theorem shows that, yes, we still have the problem.

Theorem (Crowder, 1991)

Let a set of univariate marginals $\bar{G}_j(t)$ $(j = 1, \dots, p)$ and a set of sub-survivor functions $\bar{F}(c, t)$ be specified. Suppose that one can find a joint survivor function $\bar{G}(\mathbf{t})$ for which (1) the specified functions are matched, and (2) there exists some set Ω, open in $\mathcal{R}^p$, containing no open subsets of zero probability. Then there exist infinitely many different joint survivor functions satisfying (1) and (2).

Proof   See the paper cited.

The technical proof of this theorem is based on a very simple construction. In Figure 17.1, which illustrates the two-dimensional case, suppose that A, B, C, and D are small neighbourhoods within Ω, each containing probability mass

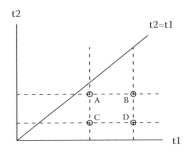

FIGURE 17.1
Construction for theorem.

at least ϵ. Then one can shift probability mass $\epsilon/2$ from A to B, and the same amount from D to C. The point is that none of the $\bar{F}(j, t)$ or $\bar{G}_j(t)$ $(j = 1, 2)$ is changed. (Refer back to Figure 14.1.) For every such shift we get a different $\bar{G}(\mathbf{t})$ with the same sub- and marginal-survivor functions.

Condition (2) of the theorem will be satisfied whenever $\bar{G}(\mathbf{t})$ has a positive density over Ω, plus perhaps some atoms of probability or a singular component of positive probability within a subspace. For the result to be non-empty we have to show that a $\bar{G}(\mathbf{t})$ satisfying (1) and (2) exists. This is done, subject to some coherency conditions on $\bar{G}(\mathbf{t})$, in the paper cited: the argument first uses the LPQ theorem to set up an independent-risks proxy model that matches the sub-survivor functions, and then progressively distorts this joint distribution as described above so that eventually the marginals are also matched.

We can carve up the set of p-dimensional survival distributions into classes whose members have the same sub-survivor functions. One member will be the independent-risks proxy model for the whole class. Unlimited observation of (C, T) will eventually identify the class. It can be further partitioned into sub-classes with the same marginals. Further unlimited observation of the T_j, if possible, will identify the sub-class. One will then know whether or not the data are consistent with an independent-risks model.

In some application areas it has been argued that the failure times must have distributions of certain parametric form, in particular, Weibull: see, for example, Pike (1966) and Peto and Lee (1973) in connection with animal survival times under continuous exposure to a carcinogen, and Galambos (1978, Section 3.12) in connection with the breakdown of systems and the strength of materials. How do we stand now regarding identification? Unfortunately, the theorem shows that even if the $\bar{F}(c, t)$ are given, and parametric forms for the $\bar{G}_j(t)$ are specified, $\bar{G}(\mathbf{t})$ is still not identified. The use of a continuous deformation version of the construction in the theorem will continuously move the functional form of $\bar{G}(\mathbf{t})$ outside any given parametric class without affecting the $\bar{F}(c, t)$ or $\bar{G}_j(t)$.

Lemma

Let a set of univariate marginals $\bar{F}_j(u)$ $(j = 1, \ldots, p)$ and a set of sub-survivor functions $\bar{F}(c, u)$ be specified that satisfy

1. $\sum_{c \ni k} \{\bar{F}(c, u_1) - \bar{F}(c, u_2)\} \leq \bar{F}_k(u_1) - \bar{F}_k(u_2)$ *for each k and $0 < u_1 < u_2$, and*
2. $\sum_c \bar{F}(c, u) \leq \bar{F}_k(u)$ *for each k and $u > 0$.*

Let $\bar{G}(\mathbf{t})$ be a joint survivor function with univariate marginals $\bar{G}_j(t_j)$ $(j = 1, \ldots, p)$ and with sub-survivor functions equal to the specified $\bar{F}(c, u)$. Let $u_1 < u_2 < u_3$, and suppose that, for some k, $\bar{G}_k(u)$ matches the specified $\bar{F}_k(u)$ at u_1 and u_3. Then additional matching of $\bar{G}_k(u)$ to $\bar{F}_k(u)$ at u_2 can be achieved by a modification of $\bar{G}$ that leaves its other univariate marginals and all of its sub-survivor functions unaltered.

Note: Conditions (1) and (2), expressed in terms of failure time components, are

1. $P\{u_1 < T_k \leq u_2, T_k = \min_j(T_j)\} \leq P(u_1 < T_k \leq u_2)$, and
2. $P\{\min_l(T_l) > u\} \leq P(T_k > u)$.

These are just requirements for coherency of the joint distribution.

Proof This is a corrected version of Lemma 1 of Crowder (1991) with a modified proof that allows properly for tied failure times. It applies equally to discrete, continuous, and mixed models. A proof is given in Crowder (2001, Lemma 7.6).

17.4 Discrete Lifetimes

We show in this section that, in the case of discrete failure times, some assessment of dependence between risks can be made, in contrast to the continuous-time case. The possible failure times are $0 = \tau_0 < \tau_1 < \ldots < \tau_m$, and the sub-odds $\bar{h}(c, t)$ (Section 16.2) will make a re-appearance by popular demand. First, a lemma will be found useful.

Lemma (Crowder, 1996)

Let $f^{(1)}(c, t)$ and $f^{(2)}(c, t)$ be the sets of sub-density functions for two discrete-time models that have the same set of configurations. Then the following conditions are equivalent:

1. $f^{(1)}(c, t) = f^{(2)}(c, t)$ for all (c, t);
2. $\bar{h}^{(1)}(c, t) = \bar{h}^{(2)}(c, t)$ for all (c, t);
3. $h^{(1)}(c, t) = h^{(2)}(c, t)$ for all (c, t).

Proof From (1), summing over c, $f^{(1)}(t) = f^{(2)}(t)$. Therefore, $\bar{F}^{(1)}(t) = \bar{F}^{(2)}(t)$ and (2) follows. Conversely, summing (2) over c yields

$$f^{(1)}(t)/\bar{F}^{(1)}(t) = f^{(2)}(t)/\bar{F}^{(2)}(t),$$

which, using $f^{(k)}(t) = \bar{F}^{(k)}(t-) - \bar{F}^{(k)}(t)$, is equivalent to

$$\bar{F}^{(1)}(t-)/\bar{F}^{(1)}(t) = \bar{F}^{(2)}(t-)/\bar{F}^{(2)}(t).$$

It follows, by setting $t = \tau_1, \tau_2, \ldots$ in turn, that $\bar{F}^{(1)}(t) = \bar{F}^{(2)}(t)$ for all t. Hence, from (2), $f^{(1)}(c, t) = f^{(2)}(c, t)$ and so (2)$\Rightarrow$(1). That (1)$\Leftrightarrow$(3) can be proved similarly.

Tsiatis's theorem establishes the existence of an independent-risks proxy model that reproduces the original sub-densities. That is, the existence of

$\bar{G}^*(\mathbf{t})$, of form $\prod_{j=1}^{p} \bar{G}_j^*(t_j)$, such that $g^*(c, t) = f(c, t)$ for all (c, t). In this case, of continuous time without ties, p independent components of $\bar{G}^*$ are used to match p sub-distributions of the original model. When ties are possible, the question arises as to whether a p-variate proxy model can reproduce not only the $f(j, t)$ but also all the $f(c, t)$ for non-simple configurations. Not surprisingly, this is only possible when the original system is of fairly special type, actually, one that obeys (3) of the sub-odds theorem (Section 16.2).

Theorem (Crowder, 1996)
An independent-risks proxy model of basic type exists for $\bar{G}(\mathbf{t})$ if and only if

1. $\bar{h}(c, t) = \prod_{j \in c} \bar{h}(j, t)$ for all (c, t).

In this case, the proxy marginal survivor functions are given uniquely by

2. $\bar{G}_j^*(t) = \prod_{s=1}^{l(t)} \{1 + \bar{h}(j, \tau_s)\}^{-1}$, where $l(t) = \max\{l : \tau_l \leq t\}$.

Proof From the preceding lemma, the proxy assertion, $f(c, t) = g^*(c, t)$ for all (c, t), holds if and only if $\bar{h}(c, t) = \bar{h}^*(c, t)$ for all (c, t). Since the proxy model has independent risks, this in turn implies (1) by (3) of the sub-odds theorem (Section 16.2). Conversely, take an independent-risks model with $\bar{h}$-functions $\bar{h}_j^*(t)$ equal to $\bar{h}(j, t)$ for $j = 1, \ldots, p$. Then, (1) implies that

$$\bar{h}(c, t) = \prod_{j \in c} \bar{h}_j^*(t),$$

which equals $\bar{h}^*(c, t)$ by (2) of the sub-odds theorem, that is, the proxy assertion holds. Finally, the only possible choice for $\bar{h}_j^*(t)$ has been shown to be $\bar{h}(j, t)$. Hence,

$$\bar{h}(j, t) = \bar{h}_j^*(t) = \{\bar{G}_j^*(t-)/\bar{G}_j^*(t)\} - 1,$$

which yields

$$\bar{G}_j^*(t) = \{1 + \bar{h}(j, t)\}^{-1} \bar{G}_j^*(t-),$$

and hence (2) by iteration.

Condition (1) here is just condition (3) of the sub-odds theorem. The present theorem shows that it is necessary and sufficient for the existence of an independent-risks proxy model of basic type in this discrete-time setup. The condition looks a bit like a standard independence criterion, with a product of marginal probabilities on the right-hand side. But it is not—the $\bar{h}(j, t)$ are not probabilities; indeed, both sides of the equation can exceed 1. Crowder (2000) showed that the same condition operates in the general case of mixed discrete-continuous sub-distributions. Various implications were also discussed.

The basic proxy model is constructed by taking its marginal $\bar{h}_j^*(t)$ functions to be the original $\bar{h}(j, t)$. We are whistling the same old Cox–Tsiatis tune, but now more discretely. A couple of examples are given below of dependent-risks models of which one satisfies (1) in the theorem and the other does not.

At first sight there seems to be a conflict between the present theorem and the LPQ theorem. The condition there, that the sub-distributions have no common discontinuities, is here violated. In fact, it would take a rather contrived discrete-time system not to violate the condition. In the discussion following the LPQ theorem, it was shown why the condition was needed there. Looked at again, it also shows why the condition is not needed here. It is because we do not have an $R_{\{1,2\}}$ now, only R_js. So, $\bar{F}(\{1, 2\}, t)$ is only to be matched by a term arising from chance coincidence of R_1 and R_2. This argument also rules out independent-risks proxy models with R_c components where c is not a simple index, unless the rather unlikely condition on the discontinuities is met. Thus, for discrete-time systems not subject to the condition, the present theorem exhausts the possibilities.

We can also generate discrete analogues of the results in Section 17.3. These concern proxy models that match the marginals of $\bar{G}(\mathbf{t})$ as well as its sub-distributions. Firstly, the theorem there is easily dealt with for the discrete case. Examination of its proof, given in Crowder (1991), shows that all we need here is the existence of some k with positive probability at points A, B, C, and D within the set $\{\mathbf{t} : t_k < \min_{l \neq k}(t_l)\}$ such that the lines AB and DC are orthogonal to the t_k axis, and AD and BC are parallel to it. This condition will be met, for instance, by any model that puts positive probability on all points in a lattice set $(\tau_l, \tau_{l+3})^p$ for some integer l. Under the condition, and given the existence of a $\bar{G}(\mathbf{t})$ that matches the given sub-survivor functions and marginals, the theorem says that there are infinitely many such joint survivor functions.

17.4.1 Discrete Freund

Referring back to Section 16.3, the vital condition (1) of the theorem does hold for this model. Thus, an independent-risks proxy model does exist, and this is just composed of the two independent sequences of Bernoulli trials used to set up the model. From (2) of the theorem, the explicit form of the proxy survivor functions is $\bar{G}_j^*(t) = \pi_j^t$.

17.4.2 Discrete Marshall–Olkin

Referring back to Section 16.3, condition (i) of the theorem fails in this case. Thus, there is no independent-risks proxy model with two latent failure times (though there is one with three, as is clear from the way in which the model is set up).

17.4.3 A Test for Independence of Risks (Crowder, 1997)

The theorem given earlier in this section gives the condition $\bar{h}(c, t) = \prod_{j \in c}$ $\bar{h}(j, t)$ as being necessary and sufficient for the existence of a basic independent-risks proxy model. So, if the condition does not obtain, there can be no such proxy model; if it does, an independence proxy exists. In other words, the condition can tell us whether dependence between the risks can be ruled out or not.

To apply this to a random sample, we can compute Kaplan–Meier estimates of the sub-hazards $h(c, t)$ (Section 15.1) and then derive estimates of the sub-odds $\bar{h}(c, t)$ from them. The condition, with these estimates inserted, is then appraised. Obviously, we need observations of multiple-cause failures as well as single-cause ones to do this: without such the left-hand side of the condition can only be evaluated with single-risk configurations, and so equals the right-hand side trivially. A test statistic based on estimates of the ratios

$$\psi_{cl} = \bar{h}_{cl} \Big/ \prod_{j \in c} \bar{h}_{jl}$$

was suggested in Crowder (1997). The condition then boils down to $\log \psi_{cl}$ being zero for all non-simple configurations c. A Wald test was developed for which details can be found in the paper cited. The test was applied to the catheter infection data given in Section 12.1: the result showed strong positive dependence between the two sites.

17.5 Regression Case

In view of the results on identifiability of $\bar{G}(t)$ from observation of (C, T), when T is continuous, it might be thought that one can never obtain hard information about the distributions of individual component failure times or on the dependence structure between them. However, the foregoing results are restricted to the independent, identically distributed case. It turns out that, when there are explanatory variables in the model, identification is possible within a certain framework.

17.5.1 Heckman and Honore's Theorem (Heckman and Honore, 1989)

Suppose that $\bar{G}(t; x)$ has the form $K(\mathbf{v})$, where $\mathbf{v} = (v_1, \ldots, v_p)$ with $v_j = \exp\{-\xi_j(\mathbf{x})H_j(t_j)\}$, and assume that:

1. $\partial K / \partial v_j > 0$ and continuous on $\{\mathbf{v} : 0 < v_j \leq 1; j = 1, \ldots, p\}$;
2. $H_j(t) \to 0$ as $t \to 0$, $H_j(t_0) = 1$ for some t_0, $H'_j(t) = dH_j(t)/dt > 0$ for all t;
3. $\xi_j(\mathbf{x}_0) = 1$ for some $\mathbf{x}_0$, $\{\xi_1(\mathbf{x}), \ldots, \xi_p(\mathbf{x})\}$ covers the range $(0, \infty)^p$ as $\mathbf{x}$ varies.

Then the set of sub-survivor functions $\{\bar{F}(j, t; \mathbf{x}) : j = 1, \ldots, p\}$ determines the joint survivor function $\bar{G}(\mathbf{t}, \mathbf{x})$.

Proof From Tsiatis's lemma (Section 14.1),

$$f(j, t; \mathbf{x}) = [-\partial K / \partial t_j]_{t1_p} = \xi_j(\mathbf{x}) H'_j(t) \exp\{-\xi_j(\mathbf{x}) H_j(t)\} K_j \{\mathbf{v}_t(\mathbf{x})\},$$

where $\mathbf{v}_t(\mathbf{x})$ has jth component $\exp\{-\xi_j(\mathbf{x}) H_j(t)\}$ and $K_j = \partial K / \partial v_j$.

1. Consider the ratio

$$f(j, t; \mathbf{x}) / f(j, t; \mathbf{x}_0) = \{\xi_j(\mathbf{x}) / \xi_j(\mathbf{x}_0)\} \exp[-H_j(t)\{\xi_j(\mathbf{x}) - \xi_j(\mathbf{x}_0)\}]$$
$$\times [K_j \{\mathbf{v}_t(\mathbf{x})\} / K_j \{\mathbf{v}_t(\mathbf{x}_0)\}].$$

   Let $t \to 0$. Then $H_j(t) \to 0$, $\mathbf{v}_t(\mathbf{x}) \to \mathbf{1}_p$ for all $\mathbf{x}$, and the ratio $\to \xi_j(\mathbf{x}) / \xi_j(\mathbf{x}_0) = \xi_j(\mathbf{x})$. Thus, $\xi_j(\mathbf{x})$ is identified as a function of $\mathbf{x}$ from knowledge of $f(j, t; \mathbf{x})$.

2. Set $t = t_0$ and let $\{\xi_1(\mathbf{x}), \ldots, \xi_p(\mathbf{x})\}$ range over $(0, \infty)^p$. Then K is identified as a function of its arguments from

$$\sum \bar{F}(j, t_0; \mathbf{x}) = \bar{G}(t_0 \mathbf{1}_p; \mathbf{x}) = K \{e^{-\xi_1(\mathbf{x})}, \ldots, e^{-\xi_p(\mathbf{x})}\}.$$

3. Fix $\xi_k(\mathbf{x})$ and let the other $\xi_j(\mathbf{x})$ all tend to zero. Then $\bar{G}(t\mathbf{1}_p, \mathbf{x}) \to K(\mathbf{v}_k)$, where $\mathbf{v}_k$ has kth element $e^{-\xi_k(\mathbf{x})H_k(t)}$ with the rest all equal to 1. Thus, $\bar{G}(t\mathbf{1}_p; \mathbf{x})$ is a known function of $\xi_k(\mathbf{x})H_k(t)$ alone, known because K has already been identified. Now $H_k(t)$ is identified because $\xi_k(\mathbf{x})$ has already been identified.

Heckman and Honore introduced their theorem via the univariate proportional hazards model, which has survivor function $\bar{F}(t; \mathbf{x}) = e^{-\xi(\mathbf{x})H(t)}$, $H(t)$ being the integrated hazard function; this is where the argument $v_j = e^{-\xi_j(\mathbf{x})H_j(t_j)}$ in K comes from. In the theorem T_j has marginal survivor function $K_j \{e^{-\xi_j(\mathbf{x})H_j(t_j)}\}$, where K_j is the jth marginal of K; this has proportional hazards form if and only if $K(y) = y^\gamma$ for some γ.

The assumptions $\xi_j(\mathbf{x}_0) = 1$ and $H_j(t_0) = 1$ are just normalizations. If $\xi_j(\mathbf{x}_0)H_j(t_0) = a_j$ in the original K, then it can be replaced by K^* defined by $K^*(\ldots v_j \ldots) = K(\ldots v_j^{a_j} \ldots)$. The assumption in (3), that the $\xi_j(\mathbf{x})$ can vary independently over $(0, \infty)$, is satisfied, for example, by the standard loglinear model $\log \xi_j(\mathbf{x}) = \mathbf{x}^T \beta_j$ provided that $\mathbf{x}$ is unbounded and the vectors β_j can vary independently.

17.5.2 Gumbel's Bivariate Exponential

A regression version for this distribution (Section 14.1) can be obtained by first replacing ν with $\nu \lambda_1 \lambda_2$ (with $0 \le \nu \le 1$ now) and then λ_j with $\exp(\alpha_j + \mathbf{x}^T \beta_j)$ for $j = 1, 2$. Take $\xi_j(\mathbf{x}) = \exp(\mathbf{x}^T \beta_j)$, $\mathbf{x}_0 = \mathbf{0}$, and $H_j(t) = t$ with $t_0 = 1$. Then $v_j = \exp(e^{-\alpha_j} \alpha_j t_j)$ and

$$\bar{G}(\mathbf{t}) = \exp(-\lambda_1 t_1 - \lambda_2 t_2 - \nu \lambda_1 t_1 \lambda_2 t_2) = v_1^{-a_1} v_2^{-a_2}$$
$$\times \exp[-\nu \{\log (v_1^{-a_1})\} \{\log (v_2^{-a_2})\}],$$

where $a_j = e^{-\alpha_j}$ for $j = 1, 2$. This is the $K(\mathbf{v})$ of the theorem.

Heckman and Honoré showed that their theorem goes some way toward covering models of the accelerated-life type, as well as those of the proportional-hazards type at which it is primarily targeted. For such models the univariate survivor functions have form $\bar{F}(t, \mathbf{x}) = e^{-H\{t\xi(\mathbf{x})\}}$ and then the multivariate form $\bar{G}(\mathbf{t}, \mathbf{x}) = K(\mathbf{v})$, with $v_j = \exp[-H_j\{t_j\xi_j(\mathbf{x})\}]$, follows on. By re-expressing v_j as $\exp[-H_j\{-\log e^{-t_j\xi_j(\mathbf{x})}\}]$ we can write $K(\mathbf{v})$ as $K^*(\mathbf{v}^*)$, in which $v_j^* = e^{-t_j\xi_j(\mathbf{x})}$. This is now in the right form for the theorem, and so K^* and the ξ_j will be identified. Further identification of K and the H_j depends on knowing the marginal distributions of K.

The theorem is more deductive than seductive. Its proof follows a correct logical process but reveals a possible weakness for practical application. For instance, one can be persuaded that an ordinary histogram will begin to give a reasonably accurate picture of the underlying density after a moderate number of observations has been recorded. But, looking through the proof of the theorem, one comes away with an impression that $\mathbf{x}$ might have to cover an awful lot of ground before any recognisable picture of the various functional forms emerges.

Abbring and van den Berg (2003) dealt with the case where the dependence between risks arises through a common frailty and extended the result to the case of multiple observation spells per case. Bond and Shaw (2006) presented a different type of identifiability result: they assumed a copula form of dependence and derived bounds on the functions describing covariate effects. Carling and Jacobson (1995) also investigated identification in mixed proportional-hazards models.

17.6 Censoring of Survival Data

In standard Survival Analysis the variate of interest is T_1, the failure time, which is observed unless preceded by censoring, say at time T_2. In general terms this comes naturally under the competing risks umbrella with $p = 2$. The classic problem in this context is whether the censoring inhibits inference about the T_1 distribution. It is thus precisely the problem of identifiability dealt with in this chapter.

We have seen that, in the independent, identically distributed case, the marginal distribution of T_1 is not identifiable in the non-parametric sense; nor is that of T_2 but that is of at most secondary interest. (In the regression case this total lack of identifiability is somewhat ameliorated, as described in the previous section.) In order to achieve identifiability, some knowledge or assumption, external to the data, has to be provided. A very common case is that of fixed-time censoring, where the T_2 values can be regarded as having been fixed in advance of the observation process. More generally, T_2 might be determined independently of T_1 by some random mechanism, so that independent-risks obtains. Then, as shown in Gail's theorem (Section 14.4), the T_1 distribution (and that of T_2) is identified by the observable sub-distributions. In fact, the theorem also shows that identifiability holds under the weaker, Makeham assumption. Thus, it might be the case that T_1 and T_2 are dependent, perhaps through some external factor or conditions affecting both, but that inference is still possible for T_1 without having to consider T_2. In practice, however, postulation of the Makeham condition might be at best optimistic since its proper justification would require deeper knowledge about aspects of the underlying stochastic mechanisms than is actually available.

Williams and Lagakos (1977) discussed the problem of identifying G_1, the marginal distribution function of T_1, in the presence of censoring. They proposed a condition, which they called *constant sum*, under which G_1 is identified by the likelihood function. Kalbfleisch and MacKay (1979) later showed that the condition is equivalent to $h(1, t) = h_1(t)$, that is, the Makeham assumption for T_1 only. The point is that the data identify the sub-hazard functions, $h(1, t)$ and $h(2, t)$, through the likelihood, and then the constant sum condition carries this identification on to $h_1(t)$, and so to G_1 through $G_1(t) = 1 - \exp\{-\int_0^t h_1(s)ds\}$. Actually, of course, any assumed relation giving $h_1(t)$ in terms of the sub-hazards would do the job just as well. But that is the snag—you have to conjure up such a relation from somewhere. Williams and Lagakos showed that the constant sum condition is implied by

$$P(T_1 \in N_1 \mid T_2 \in N_2, T_1 > t_2) = P(T_1 \in N_1 \mid T_1 > t_2)$$

for $t_1 > t_2$, where $N_j = (t_j, t_j + dt_j)$ for $j = 1, 2$. The equality defines survival independent of the conditions producing censoring. They noted that such a condition is an ingredient of nearly all existing statistical methods for assessing censored survival data. Thus, it seems that the constant sum condition, or partial Makeham assumption in our terms, is a pretty useful theoretical property, but maybe not so easy to nail down in practice.

We can extend the above straightforwardly to competing risks proper: to identify a subset of marginal distributions we just need a Makeham-like assumption for these components.

Finally, identification is meant in the non-parametric sense above. If a parametric model is specified for the latent failure times at the outset, none of the discussion above applies! The identifiability problem is now a completely different one, that is, of parameters within a parametric family. This situation is discussed briefly below in Section 17.7. With a parametric model, even though one can isolate a factor comprising the contribution of G_1 to the likelihood, the residual factor might contain some of the parameters of G_1 and so *orthogonality* is not achieved (Section 13.1). In this case the censoring is said to be *informative*. However, as pointed out by Kalbfleisch and MacKay (1979), the G_1 factor is a partial likelihood in the sense of Cox (1975) and therefore potentially capable of yielding worthwhile inferences.

17.7 Parametric Identifiability

Suppose that we have faith in the specified functional form of $\bar{G}(\mathbf{t})$, perhaps through some theoretical argument or just through a pragmatic belief that the specification will suffice for the current purpose. Then, given incomplete (c, t) data, the possibilities are illustrated by the three examples of Section 17.1: (1) complete identifiability of the joint $\bar{G}(\mathbf{t})$ (first example, Gumbel), (2) partial identifiability of the parameters (second example), and (3) complete lack of identifiability (third example, with ϕ set to 1, leaving ψ as the only parameter).

Quite a lot of work has appeared on identifiability within parametric families of distributions. Further published work concerns identifiability within parametric families from T alone, from (C, V), where $V = \max(T_1, \ldots, T_p)$, and from V alone. When the observation is (C, T) it is called the *identified minimum* and T alone is called the *unidentified minimum*, and likewise for (C, V) and V alone. Whereas T is the prime time variate for Competing Risks, V is that for so-called Complementary Risks (Basu and Ghosh, 1980). The distinction can be put in terms of *series* and *parallel systems* (Section 2.6): in the former, the first component failure is fatal to the whole operation, which therefore ceases at time T; in the latter, the operation staggers on until the last component conks out, which occurs at time V. An example, not specifically concerned with identifiability, is Gross et al.'s (1971) maximum likelihood estimation of the Freund (1961) bivariate exponential model from observations of V: in a typical application the component lifetimes are those of two kidneys, and symmetry suggests taking $\lambda_1 = \lambda_2$ and $\mu_1 = \mu_2$.

Proschan and Sullo (1974) calculated the likelihood functions for the bivariate Marshall–Olkin distribution under various observational settings including that of Competing Risks. Identifiability was established as a by-product. They also considered a model that combines the features of the Marshall–Olkin and Freund systems.

Anderson and Ghurye (1977) investigated identifiability within a family $\mathcal{G}$ of univariate probability densities $g(t)$ that are continuous and positive for t exceeding some value t_0, and such that, for any two members, g_1 and g_2 of $\mathcal{G}$, $g_1(t)/g_2(t) \to 0$ or ∞ as $t \to \infty$. Their Theorem 12.1 says that the distribution of $V = \max(T_1, \ldots, T_n)$, where the T_j are independent with densities g_j in $\mathcal{G}$, identifies both n and the g_j. The motivation for this arose from a problem in econometrics. Their proof rests on an examination of the survivor function of V, $\prod \bar{G}_j(t)$, which involves all the individual survivor functions $\bar{G}_j$ ($j = 1, \ldots, n$), and whose log-derivative can be made to reveal the differing rates of approach to zero of the components. Anderson and Ghurye gave examples of the theorem, including the normal distribution, and other examples where the asymptotic condition fails but the conclusion holds, and yet others where the maximum does not identify the components. Their Theorem 3.1 extends their Theorem 2.1 to a certain class of bivariate normal distributions.

Basu and Ghosh (1978) considered the bivariate normal distribution. They showed in their Theorem 1 that the distribution of the identified minimum (C, T) identifies the parameters μ and Σ; the authors plugged a gap in the proof of this result previously given by Nadas (1971). Their Theorem 2 says that the unidentified minimum T identifies the numerical values of the components of μ and Σ but not their ordering. Comparing the two theorems, it seems that the additional C-contribution to the (C, T) distribution tells you which of the two components is more likely to be the minimum and hence differentiates between them. Theorem 3 of the paper gives a similar result to that of Theorem 1 for the trivariate normal under a certain condition on Σ. The authors also mentioned some other distributions where the parameters might or might not be identified by (C, T) or T alone, and opined that these other results are much easier to obtain than for the normal on account of the simple explicit forms for the survivor functions.

Basu and Ghosh (1980) gave a complement of Theorem 2.1 of Anderson and Ghurye (1977): they considered the same family $\mathcal{G}$, this time with t tending to the lower end point of the distribution rather than $+\infty$, and with the minimum T instead of the maximum V. In their second theorem they considered two independent observations, one from each of two distinct gamma distributions, and showed that the distribution of the smaller observation, T, identifies both pairs of gamma parameters up to ordering. Their Theorems 3 and 4 do likewise for the Weibull and Gumbel's (1960) second bivariate exponential, and their Theorem 5 gives a similar result for Gumbel's (1960) first distribution but based on the larger of the pair, V. They also gave a result on partial identification of the Marshall–Olkin distribution from T.

Arnold and Brockett (1983) proved identifiability of the joint distribution of latent failure times from (c, t) data for a certain bivariate Makeham distribution. They also considered a mixture model with proportional hazards: $H_j(t) = \psi_j H_0(t)$. They showed that the joint survivor function $\bar{G}(\mathbf{t})$ is identifiable from (c, t) data in some circumstances. This partial identifiability is reminiscent of Heckman and Honore's (1989) extension to the accelerated life model. Further identifiability results were given for multivariate Pareto, Weibull, and Burr families.

Basu and Klein (1982) gave a survey of some results of the present type.

Part IV

Counting Processes in Survival Analysis

The methodology was introduced primarily for *Event History Analysis* in the mid-1970s and has gained great respect, and wide application, for its power in dealing with the theory in a unified way. The progress of an individual is tracked through time and the instantaneous probabilties of various events that can occur are modelled. The general framework comprises a *counting process*, recording the events, and an *intensity process*, governing the instantaneous probabilities. Ordinary Survival Analysis is a very simple case, with just one possible event, failure, terminating the process. Competing Risks is a moderate extension, allowing one of p terminating events. In this sense, the general counting process approach might seem to be a bit of an overkill for our context, but it facilitates awkward problems such as the inclusion of time-dependent covariates. Also, the technical apparatus smoothes the application of the usual statistical methodology, constructing likelihoods and deriving properties of statistics, particularly for non- and semi-parametric models.

It has to be said that the approach can be rather off-putting for many statisticians. One opens a book on the subject, sees the terms *counting process theory*, *continuous-time martingale*, *filtration*, *predictable process*, *compensator*, *stochastic integration*, *product integral*, and so forth, and feels the need for an aspirin. At this point another of these terms, *stopping time*, may strike a chord. It is true that a certain amount of investment is required to get into the subject, and it will help if one's training has included a course, or three, on Stochastic Processes. That said, however, the application of the theory can be appreciated at arm's length. It is not absolutely necessary to drown in the technicalities to see how the thing works and to see how the ideas can be applied.

The historical development of the subject has been very fully set out in the introductory part of the book by Andersen et al. (1993). Therefore, in this part we will not attempt to duplicate this, only to remark that the origins go back to Aalen (1975), with a competing risks version in Aalen (1976). Also, the pioneering work of Doob (1940 and 1953, Chapter 7) on martingales should not be forgotten. For readable accounts of the subject see the expository papers by Gill (1984), Borgan (1984), and Andersen and Borgan (1985), and for comprehensive treatments, see Andersen et al. (1993) and Fleming and Harrington (1991). Williams's (1991) book is also highly recommended as a friendly and enthusiastic introduction to probability and discrete-time martingales. See also the article "Martingales without Tears" by Lan and Lachin (1995).

To present the basic ideas underlying the approach outlined in this last part, we will first start back at the beginning and develop the machinery in a very informal way, skating over mathematical technicalities. A rigorous treatment would be infested with ifs and buts and arcane conditions: the stickler will regard what follows as a travesty. You can please some of the people some of the time.

18

Some Basic Concepts

18.1 Probability Spaces

The *sample space* Ω is the set of all outcomes under consideration, and subsets of Ω are called *events*. A collection of sets that comprises Ω itself, and the primary events of interest together with unions and complements (and, therefore, intersections too, since the intersection of two sets is the complement of the union of their complements), forms a *field*. A core concept in probability theory is that of a σ-*field* (*sigma field*): this is a field in which unions are allowed to extend to a countable number of events. We will use an overbar to denote the *complement* of a set: $\bar{S} = \Omega - S$. Thus, a σ-field always contains Ω and $\emptyset = \bar{\Omega}$ (the *empty set*), at the very least.

Example: Die Rolling

Suppose that the die is cast (in gaming, not engineering), and that the outcome of interest is the face value uppermost when the die comes to rest. With two rolls of the die there are 36 possible outcomes of the form (R_1, R_2), R_1 being the face value on the first roll and R_2 that on the second; R_1 and R_2 can each take values $1, \ldots, 6$. Consider events of the form $S_r = \{R_1 + R_2 = r\}$. For instance, S_4 contains the three outcomes $\{1, 3\}$, $\{2, 2\}$, and $\{3, 1\}$ of Ω, S_{17} is empty, $\bar{S}_r = \{R_1 + R_2 \neq r\}$, and $S_r \cap S_s = \emptyset$ for $r \neq s$. The non-empty S_r are $S_2, S_3, \ldots, S_{12}$ and these 11 form a *partition* of Ω into disjoint sets. As for the number of distinct sets in the σ-field so generated, I make it $\sum_{j=0}^{11} \binom{11}{j} = 2^{11}$. This is a surprisingly huge number, but that is the nature of σ-fields. The calculation is simpler when one starts with a partition.

Suppose that a σ-field $\mathcal{H}$ is defined on Ω. A *probability measure* P(.) assigns a number between 0 and 1 to each set in $\mathcal{H}$ according to the following rules: $P(\Omega) = 1$, $P(\emptyset) = 0$, and $P(\cup_i A_i) = \sum_i P(A_i)$ for any countable collection of disjoint sets A_i in $\mathcal{H}$. The triple $(\Omega, \mathcal{H}, P)$ then constitutes a *probability space*.

A real-valued *random variable* X defined on a sample space Ω assigns a real number to each outcome ω in Ω, so $X(\omega) = x$ for some real x. Thus, X imposes a structure on Ω corresponding to the different values that it takes. If X is discrete, these sets can be taken to be of the form $\{\omega \in \Omega : X(\omega) = x\}$, or just $\{X = x\}$, meaning the set of outcomes ω in Ω at which X takes the value x. Then X defines a partition of Ω; and when we extend this by bringing in all unions and complements, we obtain $\sigma(X)$, the σ-*field generated by* X. If X is continuous, the primary sets are usually taken to be of the form $\{X \leq x\}$. The σ-field extension here yields all events of the form $\{X \in A\}$, where A is an interval or union of intervals on the real line.

Suppose that Y is a random variable defined on Ω whose value is determined by that of X, that is, Y is a function of X. Then, for each y, the subset $\{Y \leq y\}$ belongs to $\sigma(X)$; this is because $\{Y \leq y\}$ can be expressed as $\{\omega : X(\omega) \in A_y\}$, where A_y is just the set of ω-values in Ω yielding $Y \leq y$, and $A_y \in \sigma(X)$. In this case we say that Y is *measurable* with respect to $\sigma(X)$. Equivalently, $\sigma(Y) \subset \sigma(X)$, meaning that $\sigma(Y)$ is contained within $\sigma(X)$; that is, $\sigma(X)$ defines a finer partition of Ω than does $\sigma(Y)$, each set of $\sigma(Y)$ being a union of sets of $\sigma(X)$. Conversely, if Y is measurable with respect to $\sigma(X)$, Y is constant on ω-sets on which X is constant and so is a function of X.

Example: Die Rolling

Consider the discrete random variable $X = R_1 + R_2$, the sum of the face values after rolling the die twice. The sets in the X-partition of Ω are just the S_r: X takes value r uniquely on S_r, that is, $S_r = \{\omega \in \Omega : X(\omega) = r\}$. The σ-field $\sigma(X)$ can now be generated by repeated application of the set operations, union and complement, to the S_r and their progeny, as illustrated previously. Take $Y = \mid X - 3 \mid$, so Y is a function of X that takes values $0, 1, 2$, and 3; $\{Y \leq 0\} = S_3$, $\{Y \leq 1\} = S_3 \cup S_2 \cup S_4$, $\{Y \leq 2\} = S_3 \cup S_2 \cup S_4 \cup S_1 \cup S_5$, and $\{Y \leq 3\} = \Omega$. So, $\{Y \leq y\} \in \sigma(X)$ for each y, and therefore Y is $\sigma(X)$-measurable.

More generally, if for each y, $\{Y \leq y\} \in \mathcal{H}$, where $\mathcal{H}$ is a σ-field defined on Ω, we say that Y is measurable with respect to $\mathcal{H}$ or just $\mathcal{H}$-*measurable*; equivalently, $\sigma(Y) \subset \mathcal{H}$. If a probability measure $P(.)$ is defined on $\mathcal{H}$, then, provided that Y is $\mathcal{H}$-*measurable*, $P(\{Y \leq y\})$ is defined because $\{Y \leq y\} \in \mathcal{H}$. In this way the probability of any Y-event expressible in terms of basic $\{Y \leq y\}$ sets (by countable unions and complements) is computable.

18.2 Conditional Expectation

Suppose that X and Y are two random variables defined on the same sample space Ω. We assume that a probability measure $P(.)$ is defined on a σ-field of events that includes the one generated by the sets $\{X \leq x, Y \leq y\}$. For $A \in \sigma(X)$ let

$$F(y \mid A) = P(Y \leq y \mid X \in A) = P(Y \leq y \cap X \in A) \div P(X \in A).$$

Then the conditional expectation of Y, given $X \in A$, is defined as

$$E(Y \mid X \in A) = \int y \, dF(y \mid A).$$

If Y is a continuous random variable, $dF(y \mid A)/dy$ is its conditional density function, say $f(y \mid A)$; and then $E(Y \mid X \in A)$ can be written in the more familiar form $\int y \, f(y \mid A) \, dy$. If Y is discrete, $dF(y \mid A)$ is given by differencing as

$$P(Y \le y \mid X \in A) - P(Y \le y- \mid X \in A),$$

and then the conditional expectation is evaluated as a sum over the y-values for which $dF(y \mid A) > 0$.

If $A = \{X = x\}$ the conditional expectation is $E(Y \mid X = x)$. We can write $E(Y \mid X)$ to denote the function of X that takes the value $E(Y \mid X = x)$ at $X = x$. Being a function of X, $E(Y \mid X)$ is $\sigma(X)$-measurable. Regarded as a function of ω, where $\omega \in \Omega$, $E(Y \mid X = x)$ is constant over the set $\{\omega \in \Omega : X(\omega) = x\}$. Likewise, $E(Y \mid X \in A)$ is constant over the set $A \in \sigma(X)$, which is natural because Y has been averaged over A. Again, we can write $E\{Y \mid \sigma(X)\}$ to mean the function that takes value $E(Y \mid X \in A)$ when the particular set A in $\sigma(X)$ is specified. Since $E\{Y \mid \sigma(X)\}$ is constant over the specified set in $\sigma(X)$ it is $\sigma(X)$-measurable. More generally, we can write $E(Y \mid \mathcal{H})$ for conditioning on an arbitrary σ-field $\mathcal{H}$, and then $E(Y \mid \mathcal{H})$ is $\mathcal{H}$-measurable. When $\mathcal{H}$ is the minimal σ-field, comprising just Ω and $\emptyset$, $E(Y \mid \mathcal{H}) = E(Y)$.

If Y is $\sigma(X)$-measurable it is a function of X and then its conditional expectation, given $X = x$, is just $Y(x)$, that is, $E(Y \mid X = x) = Y(x)$ or just $E(Y \mid X) = Y$. Suppose now that Z is a third random variable on Ω, not necessarily $\sigma(X)$-measurable. Then $E(YZ \mid X = x)$ is calculated by averaging YZ over the set $\{X = x\}$. However, since Y is constant over this set, the averaging is effectively only performed over the Z values, that is,

$$E(YZ \mid X = x) = Y(x)E(Z \mid X = x) \text{ or just } E(YZ \mid X) = YE(Z \mid X).$$

More generally, if Y is $\mathcal{H}$-measurable then

$$E(Y \mid \mathcal{H}) = Y \text{ and } E(YZ \mid \mathcal{H}) = YE(Z \mid \mathcal{H}).$$

Suppose that $\mathcal{H}_1$ and $\mathcal{H}_2$ are two σ-fields defined on Ω such that $\mathcal{H}_1 \subset \mathcal{H}_2$, that is, $\mathcal{H}_1$ is a sub-σ-field of $\mathcal{H}_2$. By this is meant that the sets of $\mathcal{H}_1$ are all unions of sets of $\mathcal{H}_2$, that is, $\mathcal{H}_2$ is a refinement of $\mathcal{H}_1$ defining a finer partition of Ω. If Y is $\mathcal{H}_1$-measurable then it is also $\mathcal{H}_2$-measurable, because $\{Y \le y\}$'s belonging to $\mathcal{H}_1$ imples its belonging also to $\mathcal{H}_2$. Again, for any Z defined on Ω, $E(Z \mid \mathcal{H}_1)$ is $\mathcal{H}_1$-measurable, as noted above, and therefore also $\mathcal{H}_2$-measurable by the preceding argument for Y. This fact can be symbolised as

$$E\{E(Z \mid \mathcal{H}_1) \mid \mathcal{H}_2\} = E(Z \mid \mathcal{H}_1).$$

Conversely, if we average first over $\mathcal{H}_2$, to obtain $E(Z \mid \mathcal{H}_2)$, and then average this over $\mathcal{H}_1$, we obtain the same result as we would have had we performed the coarser averaging over $\mathcal{H}_1$ in the first place:

$$E\{E(Z \mid \mathcal{H}_2) \mid \mathcal{H}_1\} = E(Z \mid \mathcal{H}_1).$$

If $\mathcal{H}_1 = \{\Omega, \emptyset\}$, the minimal σ-field, then $E(Z \mid \mathcal{H}_1) = E(Z)$, the averaging being performed over the single subset Ω.

Finally, we need one more definition: if $\mathcal{H}_1$ and $\mathcal{H}_2$ are two σ-fields defined on Ω, we will use $\sigma\{\mathcal{H}_1 \cup \mathcal{H}_2\}$ to mean the smallest σ-field containing them both, that is, the σ-field generated by the sets of $\mathcal{H}_1$ and $\mathcal{H}_2$ together.

18.3 Filtrations

Suppose that we are observing some stochastic process, $\{X(s) : s \geq 0\}$, and suppose that $\mathcal{H}_t$ is a partition of the sample space into sets, each representing a possible *sample path* of X up to time t. Thus, given a set in $\mathcal{H}_t$, we know the values, that is, the sample path, taken by $X(s)$ over the time period $[0, t]$. Assume that, as time goes on, no information is lost, so all that is present in $\mathcal{H}_s$ is included in $\mathcal{H}_t$ for $s \leq t$. In other words, as t increases the partition becomes finer and finer to accommodate the information accruing on the X-values over the extending time period. Now redefine $\mathcal{H}_t$ as the σ-field generated by such a partition, that is, $\mathcal{H}_t = \sigma\{X(s) : 0 \leq s \leq t\}$. Then, for $s < t$, $\mathcal{H}_s \subset \mathcal{H}_t$, and we say that $\{\mathcal{H}_t : t \geq 0\}$, or just $\{\mathcal{H}_t\}$, is an increasing sequence of σ-fields, or *filtration*.

It is possible that $\mathcal{H}_t$ contains more information than the bare histories of the X-process up to time t: it might also record the other events along the way, so that $\mathcal{H}_t \supset \sigma\{X(s) : 0 \leq s \leq t\}$ for each t. Nevertheless, so long as $\mathcal{H}_s \subset \mathcal{H}_t$ for $s < t$, $\{\mathcal{H}_t\}$ is a filtration. Provided that $\mathcal{H}_t$ does contain $\sigma\{X(s) : 0 \leq s \leq t\}$, the process X is said to be *adapted* to $\{\mathcal{H}_t\}$. Conditional on $\mathcal{H}_t$, that is, given a particular set of $\mathcal{H}_t$, there is no randomness left in $\{X(s) : 0 \leq s \leq t\}$ because the sample path up to time t becomes known and therefore fixed. Conditional on $\mathcal{H}_s$, with $s < t$, there is some randomness left in $\{X(s) : 0 \leq s \leq t\}$, namely the sample path over the additional period $(s, t]$.

Assuming that X is adapted to $\{\mathcal{H}_t\}$, the conditional expectation $E\{X(t) \mid \mathcal{H}_s\}$ is defined for $s < t$ by averaging over the part of the sample path still *at random*, that is, the part over the period $(s, t]$. The given set of $\mathcal{H}_s$, on which we are conditioning, is further partitioned by $\mathcal{H}_t$ into subsets that are assigned probabilities by the process specification, and the averaging is performed with respect to these probabilities. Conversely, for $s \geq t$, $E\{X(t) \mid \mathcal{H}_s\} = X(t)$.

To take an example in discrete time, suppose that X is formed as a cumulative sum, $X(t) = \sum_{i=0}^{t} U_i$ for $t = 0, 1, 2, \ldots$, where the U_i are not necessarily independent or identically distributed. Then, for $s < t$, $\mathcal{H}_s = \sigma\{X(r) : 0 \leq r \leq s\}$ fixes $U_0, \ldots, U_s$ and leaves $U_{s+1}, \ldots, U_t$ to be generated by the underlying probability law. In consequence,

$$E\{X(t) \mid \mathcal{H}_s\} = \sum_{i=0}^{s} U_i + \sum_{i=s+1}^{t} E\{U_i \mid \mathcal{H}_s\}.$$

18.4 Martingales in Discrete Time

Consider a discrete-time stochastic process $\{X(t) : t = 0, 1, 2, \dots\}$ and a filtration $\{\mathcal{H}_t : t = 0, 1, 2, \dots\}$ that together satisfy the following conditions:

1. X is adapted to $\{\mathcal{H}_t\}$;
2. $E \mid X(t) \mid < \infty$;
3. $E\{X(t) \mid \mathcal{H}_{t-1}\} = X(t-1)$.

Then X is said to be a *martingale* with respect to $\{\mathcal{H}_t\}$. Since $X(t)$ is $\mathcal{H}_{t-1}$-measurable for $t > 0$, (1), (2), and (3) hold equivalently for $X(t) - X(0)$. Therefore, we can take $X(0) = 0$ without loss of generality in describing properties of martingales. If X is a martingale, the *martingale differences* $d\,X(t) = X(t) - X(t-1)$ satisfy

$$E\{d\,X(t) \mid \mathcal{H}_{t-1}\} = E\{X(t) \mid \mathcal{H}_{t-1}\} - E\{X(t-1) \mid \mathcal{H}_{t-1}\}$$
$$= X(t-1) - X(t-1) = 0.$$

Thus, martingales are zero-drift processes adapted to some filtration. Often, this filtration is just $\sigma\{X(0), \dots, X(t)\}$ itself. For $s < t$, $\mathcal{H}_s \subseteq \mathcal{H}_{t-1}$ so, using a formula given in Section 18.2 (with $\mathcal{H}_1 = \mathcal{H}_s$ and $\mathcal{H}_2 = \mathcal{H}_{t-1}$),

$$E\{d\,X(t) \mid \mathcal{H}_s\} = E\left[\,E\{d\,X(t) \mid \mathcal{H}_{t-1}\} \mid \mathcal{H}_s\,\right] = E\{0 \mid \mathcal{H}_s\} = 0.$$

It follows that

$$E\{X(t) \mid \mathcal{H}_s\} = E\{X(s) + d\,X(s+1) + \cdots + d\,X(t) \mid \mathcal{H}_s\} = E\{X(s) \mid \mathcal{H}_s\} = X(s).$$

In fact, martingales are often more naturally defined as sums, $X(t) = \sum_{s=0}^{t} d\,X(s)$, in which $E\{d\,X(s) \mid \mathcal{H}_{s-1}\} = 0$. The $d\,X(s)$ are zero-mean, with uncorrelated increments because

$$E\{d\,X(s)\} = E\left[E\{d\,X(s) \mid \mathcal{H}_{s-1}\}\right] = E(0) = 0,$$

and for $s < t$,

$$E\{d\,X(s)d\,X(t)\} = E\left[E\{d\,X(s)d\,X(t) \mid \mathcal{H}_{t-1}\}\right] = E\left[d\,X(s)E\{d\,X(t) \mid \mathcal{H}_{t-1}\}\right]$$
$$= E(0) = 0.$$

Thus, for $s \leq t$,

$$\text{cov}\{X(s), X(t)\} = \mathrm{E}\left\{\sum_{j=1}^{s} d\,X(j) \times \sum_{k=1}^{t} d\,X(k)\right\} = \sum_{j=1}^{s} \mathrm{E}\{d\,X(j)^2\}.$$

18.4.1 Likelihood Ratios

Let $U_1, U_2, \ldots$ be a sequence of independent, identically distributed, continuous variates with common density function $f(u)$. Consider the hypotheses $H_0 : f = f_0$ (null) and $H_1 : f = f_1$ (alternative). Then the likelihood ratio for testing H_0 versus H_1, based on observation $(U_1, \ldots, U_n)$, is

$$X(n) = lr(U_1, \ldots, U_n) = \prod_{i=1}^{n} \{f_1(U_i)/f_0(U_i)\}.$$

Taking $\mathcal{H}_n = \sigma(U_1, \ldots, U_n)$, X is adapted to $\{\mathcal{H}_n\}$ and, assuming that $\mathrm{E}_0 \mid f_1(U_i)/f_0(U_i) \mid < \infty$ (where E_0 denotes expectation under H_0), X is a martingale with respect to $\{\mathcal{H}_n\}$ under H_0 because

$$\mathrm{E}_0\{X(n) \mid \mathcal{H}_{n-1}\} = X(n-1)\,\mathrm{E}_0\{f_1(U_n)/f_0(U_n) \mid \mathcal{H}_{n-1}\}$$

$$= X(n-1) \int \{f_1(u)/f_0(u)\} f_0(u) du$$

$$= X(n-1) \int f_1(u) du = X(n-1).$$

The martingale differences here are

$$d\,X(n) = X(n) - X(n-1) = X(n-1)\,[\{f_1(U_n)/f_0(U_n)\} - 1].$$

A process $\{Y(t) : t = 0, 1, 2, \ldots\}$ is said to be *predictable* with respect to $\{\mathcal{H}_t\}$ if $Y(t)$ is $\mathcal{H}_{t-1}$-measurable for each t, that is, the value of $Y(t)$ is determined by events up to time $t - 1$. So, $Y(t)$ can serve as a one-step-ahead prediction of some random variable. For example, if $\mathcal{H}_{t-1}$ is $\sigma\{X(0), \ldots, X(t-1)\}$, $Y(t)$ is just a function of $X(0), \ldots, X(t-1)$. If Z is a stochastic process adapted to $\{\mathcal{H}_t\}$, then we can write

$$d\,Z(t) = Z(t) - Z(t-1) = d\,Y(t) + d\,X(t),$$

where

$$d\,Y(t) = \mathrm{E}\{d\,Z(t) \mid \mathcal{H}_{t-1}\}, \quad d\,X(t) = d\,Z(t) - \mathrm{E}\{d\,Z(t) \mid \mathcal{H}_{t-1}\}.$$

Here,

$$Y(t) = Y(0) + \sum_{s=1}^{t} d\,Y(s)$$

is predictable because $d\,Y(s)$ is $\mathcal{H}_{s-1}$-measurable and therefore $\mathcal{H}_{t-1}$-measurable for $s \leq t$, and the $d\,X(t)$ are martingale differences with respect to $\{\mathcal{H}_t\}$

because $E\{d X(t) \mid \mathcal{H}_{t-1}\} = 0$. Thus follows the famous *Doob decomposition*,

$$Z(t) = Y(t) + X(t),$$

expressing Z as the sum of a predictable process, known as its *compensator*, and a martingale.

18.5 Martingales in Continuous Time

A continuous-time stochastic process $\{X(t) : t \geq 0\}$ is a martingale with respect to the filtration $\{\mathcal{H}_t : t \geq 0\}$ if, for all $t \geq 0$,

1. X is adapted to $\{\mathcal{H}_t\}$;
2. $E \mid X(t) \mid < \infty$;
3. $E\{X(t) \mid \mathcal{H}_s\} = X(s)$ for $s \leq t$.

The theory for martingales in continuous time is much more technically sophisticated than that for discrete time because of the possibility of sample path pathology, for example, discontinuity and the fact that the time points are uncountable, whereas σ-fields, and the probabilities defined on them, deal only with countable collections of sets. In the sequel we will always assume that (2) holds.

Rather informally, we define $d X(t) = X(t) - X(t-)$ and note that

$$E\{d X(t) \mid \mathcal{H}_{t-}\} = E\{X(t) \mid \mathcal{H}_{t-}\} - E\{X(t-) \mid \mathcal{H}_{t-}\} = X(t-) - X(t-) = 0$$

and, for $s < t$,

$$E\{d X(s)d X(t) \mid \mathcal{H}_{t-}\} = E\left[d X(s) \, E\{d X(t) \mid \mathcal{H}_{t-}\}\right] = E\{d X(s) \times 0\} = 0.$$

One of the main results is the *Doob–Meyer decomposition*. This states that a continuous-time stochastic process Z adapted to a filtration $\{\mathcal{H}_t\}$ can be expressed in the form

$$Z(t) = Y(t) + X(t),$$

where $Y(t)$ is predictable, that is, measurable with respect to $\mathcal{H}_{t-}$, and X is a martingale with respect to $\{\mathcal{H}_t\}$.

The *predictable variation process* $\langle X \rangle$ of a martingale X is the compensator of the process $X(t)^2$. The latter has increments

$$d\{X(t)^2\} = X(t)^2 - X(t-)^2 = \{X(t-) + d X(t)\}^2 - X(t-)^2$$

$$= d X(t)^2 + 2X(t-)d X(t).$$

Now,

$$E\{2X(t-)d X(t) \mid \mathcal{H}_{t-}\} = 2X(t-) \, E\{d X(t) \mid \mathcal{H}_{t-}\} = 0,$$

so, by definition of $\langle X \rangle$ as the compensator of $X(t)^2$,

$$d\langle X \rangle(t) = \mathrm{E}\left[d\{X(t)^2\} \mid \mathcal{H}_{t-}\right] = \mathrm{E}\{d\,X(t)^2 \mid \mathcal{H}_{t-}\} = \mathrm{var}\{d\,X(t) \mid \mathcal{H}_{t-}\}.$$

In a similar vein, the *predictable covariation* $\langle X_1, X_2 \rangle$, where X_1 and X_2 are both martingales with respect to $\{\mathcal{H}_t\}$, is the compensator of the process $X_1(t)\,X_2(t)$. This is defined by the increment $\mathrm{E}\{d(X_1 X_2)(t) \mid \mathcal{H}_{t-}\}$. But,

$$\begin{aligned}
d(X_1 X_2)(t) &= X_1(t)\,X_2(t) - X_1(t-)\,X_2(t-) \\
&= \{X_1(t-) + d\,X_1(t)\}\{X_2(t-) + d\,X_2(t)\} - X_1(t-)\,X_2(t-) \\
&= X_1(t-)d\,X_2(t) + X_2(t-)d\,X_1(t) + d\,X_1(t)d\,X_2(t).
\end{aligned}$$

Hence,

$$\begin{aligned}
d\langle X_1, X_2 \rangle(t) &= \mathrm{E}\{d(X_1 X_2)(t) \mid \mathcal{H}_{t-}\} = 0 + 0 + \mathrm{E}\{d\,X_1(t)d\,X_2(t) \mid \mathcal{H}_{t-}\} \\
&= \mathrm{cov}\{d\,X_1(t), d\,X_2(t) \mid \mathcal{H}_{t-}\};
\end{aligned}$$

X_1 and X_2 are *orthogonal* if $\langle X_1, X_2 \rangle(t) = 0$ for all t. Note that $\langle X, X \rangle = \langle X \rangle$ in the notation here.

Let A be predictable and let X be a martingale, both with respect to $\{\mathcal{H}_t\}$. Define the *stochastic integral*

$$Y(t) = \int_0^t A(s)d\,X(s).$$

(It would take too long here to justify stochastic integration; let's just take it on and trust that it can be done.) Then Y is also a martingale with respect to $\{\mathcal{H}_t\}$ because

$$\mathrm{E}\{d\,Y(t) \mid \mathcal{H}_{t-}\} = \mathrm{E}\{A(t)d\,X(t) \mid \mathcal{H}_{t-}\} = A(t)\mathrm{E}\{d\,X(t) \mid \mathcal{H}_{t-}\} = 0.$$

Further, the predictable variation increments of Y are

$$d\langle Y(t) \rangle = \mathrm{var}\{d\,Y(t) \mid \mathcal{H}_{t-}\} = \mathrm{var}\{A(t)d\,X(t) \mid \mathcal{H}_{t-}\} = A(t)^2 d\langle X(t) \rangle,$$

and so the predictable variation process of Y is given as

$$\langle Y(t) \rangle = \int_0^t A(s)^2 d\langle X(s) \rangle.$$

Likewise, the predictable covariation process of the stochastic integrals $\int_0^t A_j(s)d\,X_j(s)$ ($j = 1, 2$) is

$$\int_0^t A_1(s)\,A_2(s)d\langle X_1, X_2 \rangle(s).$$

Also,

$$\text{var}\{Y(t)\} = \text{E}\{Y(t)^2\} = \text{E}\left\{\int_{s=0}^{t}\int_{u=0}^{t} A(s)A(u)d\,X(s)d\,X(u)\right\}$$

$$= \int_{s=0}^{t} A(s)^2\text{E}\{d\,X(s)^2\},$$

and, for $t_1 < t_2$,

$$\text{cov}\{Y(t_1), Y(t_2)\} = \text{E}\left\{\int_{s=0}^{t_1}\int_{u=0}^{t_2} A(s)A(u)d\,X(s)d\,X(u)\right\}$$

$$= \int_{s=0}^{t_1} A(s)^2\text{E}\{d\,X(s)^2\} = \text{var}\{Y(t_1)\}.$$

18.6 Counting Processes

Suppose that $N(t)$ records the number of events observed of some point process over the time interval $[0, t]$: thus, the increment $d\,N(t) = N(t) - N(t-)$ takes the value 1 if a point event occurs at time t and 0 if not. Then N is referred to as a *counting process*. Assume that N is adapted to $\{\mathcal{H}_t\}$. The *intensity function* $\lambda(.)$ of N is essentially a hazard function defined by

$$P\{d\,N(t) = 1 \mid \mathcal{H}_{t-}\} = \lambda(t)dt = d\Lambda(t),$$

where $\Lambda(t)$ is the integrated intensity function, $\Lambda(t) = \int_0^t \lambda(s)ds$. Since $d\,N(t)$ is just a binary variable,

$$\text{E}\{d\,N(t) \mid \mathcal{H}_{t-}\} = d\Lambda(t),$$

and so $\lambda(t)$ and $\Lambda(t)$ are predictable processes, being measurable with respect to $\mathcal{H}_{t-}$.

Let $M(t) = N(t) - \Lambda(t)$. Then M is a martingale with respect to $\{\mathcal{H}_t\}$ because

$$\text{E}\{d\,M(t) \mid \mathcal{H}_{t-}\} = \text{E}\{d\,N(t) - d\Lambda(t) \mid \mathcal{H}_{t-}\} = d\Lambda(t) - d\Lambda(t) = 0;$$

M is then a *martingale counting process*, and Λ is the compensator of the counting process N. The predictable variation process $\langle M\rangle$ has increments

$$d\langle M(t)\rangle = \text{var}\{d\,M(t) \mid \mathcal{H}_{t-}\} = \text{var}\{d\,N(t) \mid \mathcal{H}_{t-}\} = d\Lambda(t)\{1 - d\Lambda(t)\} \approx d\Lambda(t).$$

Thus, $\langle M(t)\rangle = \Lambda(t)$. Further, if A is predictable with respect to $\{\mathcal{H}_t\}$, the stochastic integral $Y(t) = \int_0^t A(s)d\,M(s)$ is a martingale, and its predictable variation process is

$$\langle Y(t)\rangle = \int_0^t A(s)^2 d\langle M(s)\rangle = \int_0^t A(s)^2 d\Lambda(s).$$

In a multivariate counting process, $\{N_1(t), \ldots, N_p(t)\}$, the components $N_j(t)$ are all counting processes defined on the same sample space. Assume that a filtration $\{\mathcal{H}_t\}$ is defined with respect to which the N_j are adapted and that the intensity process for N_j is Λ_j with respect to $\{\mathcal{H}_t\}$. Let $M_j(t) = N_j(t) - \Lambda_j(t)$ be the jth coordinate martingale counting process. Then the predictable covariations are given by the increments

$$d\langle M_j, M_k\rangle(t) = \text{cov}\{dM_j(t), dM_k(t) \mid \mathcal{H}_{t-}\} = \text{cov}\{dN_j(t), dN_k(t) \mid \mathcal{H}_{t-}\}$$

$$= \text{E}\{dN_j(t)dN_k(t) \mid \mathcal{H}_{t-}\} = \text{P}\{dN_j(t) = 1 \cap dN_k(t) = 1 \mid \mathcal{H}_{t-}\},$$

since $dN_j(t)$ and $dN_k(t)$ are binary variates. We will assume throughout that there cannot be more than one jump at a time, so $dN_j(t)$ and $dN_k(t)$ cannot both be non-zero simultaneously for $j \neq k$. In that case the last expression here is 0, and so M_j and M_k are orthogonal for $j \neq k$, $\langle M_j, M_k\rangle = 0$.

18.7 Product Integrals

The last element in our exciting buildup to the counting process approach in Survival Analysis is the concept of the *product integral*. This has much theory and many ramifications attached to it—see the seminal paper by Gill and Johansen (1990). We will view it mainly as a notation that facilitates the construction of likelihood functions and the derivation of properties of estimators and tests.

Consider a failure time variable T with survivor function $S(t) = \text{P}(T > t)$ and hazard function $h(t)$. In the discrete-time case the following expression, obtained by multiplying the successive conditional probabilities together, was given in Section 2.3:

$$S(t) = \prod\{1 - h(\tau_s)\},$$

where the product is over $\{s : \tau_s \leq t\}$, the τ_s being the discrete time points. Suppose now that the τ_s define an increasingly fine grid on the time axis and that we can take a limit as their separations tend to zero (Section 2.5). Then time becomes continuous and we replace the discrete hazard $h(\tau_s)$, which is a conditional probability, with $h(t)dt, h(t)$ being the corresponding continuous-time hazard function, which is a conditional probability density. The right-hand side becomes $\prod\{1 - h(s)ds\}$, where the product is now over $\{s : s \leq t\}$; however, it must be said that such a product over a continuum of s-values is only very informally defined here.

We now reframe this by bringing in N, the *counting process associated with* T, which jumps from 0 to 1 at time T. Thus, $N(t) = I(T \leq t)$, the indicator function, taking value 0 if $T > t$ and 1 if $T \leq t$. If $\mathcal{H}_t$ represents $\sigma\{N(s) : 0 \leq s \leq t\}$, the history generated by the N process, reference to Section 18.6 suggests that we can interpret $h(s)ds$ as $d\Lambda(s)$ by matching λ to h and

$P\{d\,N(t) = 1 \mid \mathcal{H}_{t-}\}$ to $P(T = t \mid \mathcal{H}_{t-})$. Thus, we can write

$$S(t) = \mathcal{P}_{s \leq t}\{1 - d\,\Lambda(s)\},$$

where the *product integral* $\mathcal{P}_{s \leq t}$ is very carefully defined analytically to have sensible mathematical properties: see Gill and Johansen (1990). In the purely continuous case, for instance, this reduces to $S(t) = \exp\{-\Lambda(t)\}$; in the purely discrete case it resumes the original form of a finite product.

19

Survival Analysis

The material from here on is a small selection from that covered, in far greater breadth and depth, in Fleming and Harrington (1991) and Andersen et al. (1992). Here, just enough of the basic ideas are set out to serve our purpose. We just cover the first steps, showing how some simple likelihood functions, estimators, and tests can be constructed and evaluated. The huge amount of methodology that follows on from these humble beginnings will be left for the reader to pursue if sufficiently encouraged by the brief outline here; the books referred to above provide a wealth of material. Our approach, informal and sketchy, might make some purists weep, but they are not part of our target audience.

19.1 A Single Lifetime

19.1.1 The Intensity Process

Consider first a single observed failure time T having a continuous distribution on $(0, \infty)$. Let N be the counting process associated with T, $N(t) = I$ $(T \leq t)$: $N(t)$ counts the number of failures up to time t, in this case only 0 or 1 since we are dealing with a single unit. Now let $d N(t)$ be the increment as defined in Section 18.6. Then $d N(t) = 1$ if N jumps at time t, that is, if $T = t$; otherwise, $d N(t) = 0$. In consequence, taking $\mathcal{H}_t = \sigma\{N(s) : 0 \leq s \leq t\}$,

$$E\{d N(t) \mid \mathcal{H}_{t-}\} = P\{T \in [t, t + dt) \mid \mathcal{H}_{t-}\} = Y(t)h(t)dt,$$

where $h(t)$ is the hazard function of T and $Y(t)$ is the *at risk* indicator: in the absence of censoring

$$Y(t) = I(T \geq t) = 1 - I(T < t) = 1 - N(t-).$$

If $t > T$, $Y(t) = 0$. That $E\{d N(t) \mid \mathcal{H}_{t-}\}$ is correspondingly zero can be seen by noting that the fact that the event $\{T < t\}$ has occurred will be part of the information carried by $\mathcal{H}_{t-}$, thus ruling out the possibility that $T \in [t, t + dt)$.

If $T \geq t$, $Y(t) = 1$ and the relation boils down to the familiar equation

$$P\{T < t + dt \mid T \geq t\} = h(t)dt.$$

In this context $\{Y(t)h(t)\}$ is the *intensity process* of N. The notation for the indicator process $Y(t)$ has become standard in this context, though a little unfortunate for statisticians who tend to think of Y as a response variable in a regression model. Its function is to switch off the hazard when time $t = T$ is passed: after death the intensity becomes zero, a fact of little comfort.

When there is right censoring, say at time U, the observation indicator becomes $Y(t) = I\{\min(T, U) \geq t\}$; it indicates that observation ceases after time T or U, whichever comes first. Note that N jumps from 0 to 1 at time T if $T < U$, but remains forever at 0 if $T > U$. In consequence, $Y(t)$ is no longer equal to $1 - N(t-)$. To make Y predictable, $\mathcal{H}$ must be expanded to accommodate events governed by both T and U. Censoring will be addressed below in Section 19.4 as a competing risk.

The compensator of the counting process N is the integrated intensity process $\Lambda(t)$, given by

$$\Lambda(t) = \int_0^t Y(s)h(s)ds,$$

and $M(t) = N(t) - \Lambda(t)$ is a martingale counting process with respect to $\{\mathcal{H}_t\}$.

19.1.2 Parametric Likelihood Function

Suppose that the hazard function h depends on a parameter (vector) θ, so that

$$\Lambda(t;\theta) = \int_0^t Y(s)h(s;\theta)ds.$$

The likelihood function for θ based on observation of N is then constructed as follows: Either $dN(t) = 1$, indicating that failure is observed at time t, or $dN(t) = 0$ if not. The corresponding conditional probabilities, given the immediately preceding history, are

$$P\{dN(t) = 1 \mid \mathcal{H}_{t-}\} = E\{dN(t) \mid \mathcal{H}_{t-}\} = d\Lambda(t;\theta) = Y(t)h(t;\theta)dt$$

and

$$P\{dN(t) = 0 \mid \mathcal{H}_{t-}\} = 1 - d\Lambda(t;\theta).$$

The conditional likelihood for $dN(t)$ can then be written succinctly, in the usual fashion for a binary variate, as

$$d\Lambda(t;\theta)^{dN(t)}\{1 - d\Lambda(t;\theta)\}^{1-dN(t)} \propto \{Y(t)h(t;\theta)\}^{dN(t)}\{1 - Y(t)h(t;\theta)dt\}^{1-dN(t)},$$

where dt has been dropped from the first factor. The overall likelihood function is now constructed as the product integral of these instantaneous conditional likelihoods over time:

$$L(\theta) = \mathcal{P}_{t\geq 0}[\{d\Lambda(t;\theta)\}^{dN(t)}\{1 - d\Lambda(t;\theta)\}^{1-dN(t)}]$$

$$\propto \mathcal{P}_{t\geq 0}[\{Y(t)h(t;\theta)\}^{dN(t)}\{1 - Y(t)h(t;\theta)dt\}^{1-dN(t)}].$$

Note that $dN(t) = 0$ for all t except possibly at $t = T$; if failure is observed at time T, $dN(T) = 1$.

1. If $T < U$, the only factor not equal to 1 among the $\{Y(t)h(t;\theta)\}^{dN(t)}$ is $h(T;\theta)$. This is because for $t < T$, $Y(t) = 1$ and $dN(t) = 0$, and for $t > T$, $Y(t) = 0$ and $dN(t) = 0$ and 0^0 is taken as 1. The product integral over the other factors yields $\exp\{-\int_0^T Y(t)h(t;\theta)dt\} = S(T;\theta)$.

2. If $T > U$, $\{Y(t)h(t;\theta)\}^{dN(t)} = 1$ for all t because $dN(t) = 0$ for all t. The other factor yields $S(U;\theta)$.

Hence, defining an indicator $D = I(T < U)$ and denoting $\min(T, U)$ by W,

$$L(\theta) \propto h(W;\theta)^D S(W;\theta);$$

so $L(\theta)$ reduces to the familiar form in this most basic case of a single failure time.

19.2 Independent Lifetimes

Consider now the case of n independent, not necessarily identically distributed, failure times, $T_1, \ldots, T_n$; some of the T_i may be censoring times. Define counting processes $N_i(t) = I(T_i \leq t, \text{observed})$ and observation indicator processes $Y_i(t)$ for $i = 1, \ldots, n$. Also, consider the multivariate counting process $\{N_1(t), \ldots N_n(t)\}$ and the associated filtration given by $\mathcal{H}_t = \sigma\{\cup_{i=1}^n \mathcal{H}_{it}\}$, where $\mathcal{H}_{it}$ is the history generated by events involving unit i. Then N_i has compensator

$$\Lambda_i(t;\theta) = \int_0^t Y_i(s)h_i(s;\theta)ds$$

with respect to $\{\mathcal{H}_t\}$. The conditional probability that N_i jumps from 0 to 1 at time t, given the immediately preceding history, is

$$P\{dN_i(t) = 1 \mid \mathcal{H}_{t-}\} = E\{dN_i(t) \mid \mathcal{H}_{t-}\} = d\Lambda_i(t;\theta) = Y_i(t)h_i(t;\theta)dt.$$

We assume that $dN_+(t)$ is 0 or 1, that is, that there are no ties among the T_i. Then the conditional probability that none of the N_i jumps at time t is

$$P\{dN_+(t) = 0 \mid \mathcal{H}_{t-}\} = 1 - E\{dN_+(t) \mid \mathcal{H}_{t-}\} = 1 - d\Lambda_+(t;\theta),$$

where $dN_+(t) = \sum_{i=1}^n dN_i(t)$, and so forth; in general, replacement of an index by $+$ will be used to indicate summation over that index. Thus, the conditional likelihood contribution for time t is

$$\prod_{i=1}^{n}\{d\Lambda_i(t;\theta)\}^{dN_i(t)} \times \{1 - d\Lambda_+(t;\theta)\}^{1-dN_+(t)},$$

a multinomial expression analogous to the binomial one for a single lifetime. The observation indicator $Y_i(t)$ is equal to 1 until T_i is either observed or right-censored, and then equal to 0 after that. The likelihood function for observation of $N_1, \ldots, N_n$ is then obtained as the product integral

$$L(\theta) = \mathcal{P}_{t\geq 0}\left[\prod_{i=1}^{n}\{d\Lambda_i(t;\theta)\}^{dN_i(t)} \times \{1 - d\Lambda_+(t;\theta)\}^{1-dN_+(t)}\right]$$

$$\propto \mathcal{P}_{t\geq 0}\left[\prod_{i=1}^{n}\{Y_i(t)h_i(t;\theta)\}^{dN_i(t)} \times \left\{1 - \sum_{i=1}^{n}Y_i(t)h_i(t;\theta)dt\right\}^{1-dN_+(t)}\right].$$

Referring back to the case of a single lifetime, $\mathcal{P}_{t\geq 0}\{Y_i(t)h_i(t;\theta)\}^{dN_i(t)}$ reduces to $h_i(W_i;\theta)^{D_i}$, where $W_i = \min(T_i, U_i)$ and $D_i = I(T_i < U_i)$, U_i being the right-censoring time for unit i. Likewise, the other factor yields

$$\exp\left\{-\int_{t\geq 0}\sum_{i=1}^{n}Y_i(t)h_i(t;\theta)dt\right\} = \exp\left\{-\sum_{i=1}^{n}\int_{0}^{W_i}h_i(t;\theta)dt\right\} = \prod_{i=1}^{n}S_i(W_i,\theta).$$

Hence,

$$L(\theta) \propto \prod_{i=1}^{n}\{h_i(T_i;\theta)^{D_i}S_i(W_i;\theta)\}.$$

In the independent, identically distributed case, in which $h_i(t;\theta) = h(t;\theta)$ for all i, $N_+(t)$ is a counting process with compensator $\Lambda_+(t;\theta) = \int_0^t q(s) h(s;\theta)ds$, where $q(t) = \sum_{i=1}^{n}Y_i(t)$ is the number of T_is greater than or equal to t; $q(t)$ is the number of individuals at risk at time $t-$, that is, the size of the risk queue (cf., q_l in Section 4.1).

19.3   Competing Risks

To deal with p competing risks we set up a p-variate counting process for each individual: in $\{N_{i1}(t), \ldots, N_{ip}(t)\}$, N_{ij} refers to individual i ($i = 1, \ldots, n$) and risk j ($j = 1, \ldots, p$). Thus, at failure time $t = T_i$, when (C_i, T_i) is observed, the C_ith component, N_{iC_i}, jumps from 0 to 1 and the other N_{ij} remain at 0. We assume in this section that there is no censoring. The framework can be set in terms of filtrations $\mathcal{H}_{it}$ and an overall filtration $\mathcal{H}_t$ containing them.

The corresponding compensators are

$$\Lambda_i(j, t; \theta) = \int_0^t Y_i(s) h_i(j, s; \theta) ds,$$

where the $h_i(j, s; \theta)$ are the sub-hazards for individual i. The conditional probability that N_{ij} jumps from 0 to 1 at time t, given the immediately preceding history, is

$$P\{dN_{ij}(t) = 1 \mid \mathcal{H}_{t-}\} = E\{dN_{ij}(t) \mid \mathcal{H}_{t-}\} = d\Lambda_i(j, t; \theta) = Y_i(t) h_i(j, t; \theta) dt;$$

the conditional probability that none of the N_{ij} jumps at time t is

$$P\{dN_{++}(t) = 0 \mid \mathcal{H}_{t-}\} = 1 - E\{dN_{++}(t) \mid \mathcal{H}_{t-}\} = 1 - d\Lambda_+(+, t; \theta).$$

Thus, the conditional likelihood contribution for time t is the multinomial expression

$$\prod_{i=1}^{n} \prod_{j=1}^{p} \{d\Lambda_i(j, t; \theta)\}^{dN_{ij}(t)} \times \{1 - d\Lambda_+(+, t; \theta)\}^{1 - dN_{++}(t)}.$$

The overall likelihood function is then

$$L(\theta) = \mathcal{P}_{t \geq 0} \left[\prod_{i=1}^{n} \prod_{j=1}^{p} \{d\Lambda_i(j, t; \theta)\}^{dN_{ij}(t)} \times \{1 - d\Lambda_+(+, t; \theta)\}^{1 - dN_{++}(t)} \right]$$

$$\propto \mathcal{P}_{t \geq 0} \left[\prod_{i=1}^{n} \prod_{j=1}^{p} \{Y_i(t) h_i(j, t; \theta)\}^{dN_{ij}(t)} \right.$$

$$\left. \times \left\{ 1 - \sum_{i=1}^{n} \sum_{j=1}^{p} Y_i(t) h_i(j, t; \theta) dt \right\}^{1 - dN_{++}(t)} \right].$$

In the case of pure competing risks with no ties, only one of $(N_{i1}, \ldots, N_{ip})$ can jump at any one time; and when this happens for the first time, observation ceases on individual i. If the outcome for the ith individual is (C_i, T_i) the only non-zero value among the $dN_{ij}(t)$ (for $j = 1, \ldots, p$) is that for $j = C_i$ and $t = T_i$, when $dN_{iC_i}(T_i) = 1$ and $Y_i(T_i) = 1$. Thus, the term $\mathcal{P}_{t \geq 0} \prod_{j=1}^{p} \{Y_i(t) h_i(j, t; \theta)\}^{dN_{ij}(t)}$ in $L(\theta)$ reduces to $h_i(C_i, T_i; \theta)$. In the other term $dN_{++}(t)$ is also equal to 0 for all $t \geq 0$ except at each T_i, when it takes value 1; we assume that $dN_{++}(t)$ cannot be greater than 1, that is, that there are no ties among the T_i. Also, $Y_i(t)$ is equal to 1 on $[0, T_i]$ and to 0 on (T_i, ∞). Thus,

$$L(\theta) \propto \prod_{i=1}^{n} \{h_i(C_i, T_i; \theta)\} \times \exp\left\{ -\sum_{i=1}^{n} \sum_{j=1}^{p} \int_{t \geq 0} Y_i(t) h_i(j, t; \theta) dt \right\}$$

$$= \prod_{i=1}^{n} \{h_i(C_i, T_i; \theta)\} \times \exp\left\{ -\sum_{i=1}^{n} \int_{0}^{T_i} h_i(t; \theta) dt \right\}$$

$$= \prod_{i=1}^{n} \{h_i(C_i, T_i; \theta) S_i(T_i; \theta)\} = \prod_{i=1}^{n} f_i(C_i, T_i; \theta).$$

19.4 Right-Censoring

Some notes were made in Section 17.6, essentially setting the censoring mechanism in context as an additional competing risk. The known identifiability results for competing risks can then be applied to determine whether censoring obstructs inference about the target failure time distribution. The simplest cases yielding non-obstruction are where the censoring time U is fixed in advance, such as in a study of prespecified time period, or where it is generated by some stochastic mechanism that runs independently of the one generating the failure time T. Even in this second situation, however, censoring might be *informative* in the sense that the U-distribution could depend on a parameter that also appears in the T-distribution.

The explicit effect of censoring on the competing-risks likelihood is as follows. Consider first a prespecified censoring time, with $C_i = 0$ in our notation. Then, in the penultimate expression,

$$\prod_{i=1}^{n} \{h_i(C_i, T_i; \theta) S_i(T_i; \theta)\},$$

any factor with $C_i = 0$ is just replaced by 1, and the likelihood contribution from that individual is, familiarly, $S_i(T_i; \theta)$ instead of $f_i(T_i; \theta)$. (Actually, this is a slight simplification: in the product-integral derivation it is $h(0, t; \theta) dt$ that is equal to 1, entailing an infinite value of $h(0, t; \theta)$.) In stochastic censoring this could likewise be included as an additional competing risk, extending the factor in the original expression for $L(\theta)$ to

$$\prod_{i=1}^{n} \prod_{j=0}^{p} \{d \Lambda_i(j, t; \theta)\}^{d N_{ij}(t)},$$

so $\prod_{j=1}^{p}$ has become $\prod_{j=0}^{p}$. If the ith individual failure time is censored, the resulting contribution is $f_i(0, T_i; \theta)$. The omission of such factors from $L(\theta)$ is effectively the same as setting them equal to 1, and this yields $S_i(T_i; \theta)$ in place of $f_i(T_i; \theta)$ in the final expression. If the censoring hazard $h_i(0, t; \theta)$ does not depend on θ nothing is lost in $L(\theta)$ by omitting it; if it does, its omission makes $L(\theta)$ a partial likelihood in a broad sense.

In general, censoring is a difficult area, see, for example, Andersen et al. (1993, III.2.2), and it would drag us too far off course to deal with it in any great detail.

20

Non- and Semi-Parametric Methods

20.1 Survival Times

Consider a random sample of times, ordered as $W_1 < W_2 < \cdots < W_n$. Some of these will be observed failure times T and others may be censoring times U; as before, $W_i = \min(T_i, U_i)$ and, by convention, $W_0 = 0$ and $W_{n+1} = \infty$. Let N_i and Y_i be the associated individual counting and observation processes, respectively. The predictable process $q(t) = \sum_{i=1}^{n} Y_i(t)$ indicates the number of individuals *at risk* just before time t: in the absence of censoring $q(t) = n - i$ on $(T_i, T_{i+1}]$ for $i = 0, \ldots, n$. Denote the individual intensity processes by Λ_i, with $d\Lambda_i(t) = Y_i(t)h(t)dt$.

Let $M_+(t) = N_+(t) - \Lambda_+(t)$ be the corresponding martingale counting process. Then, from $dN_+(t) = d\Lambda_+(t) + dM_+(t)$, the natural (conditional moment) estimator of $d\Lambda_+(t)$ is $dN_+(t)$, since $E\{dM_+(t) \mid \mathcal{H}_{t-}\} = 0$ confers upon $dM_+(t)$ the status of a zero-mean error term. But $d\Lambda_+(t) = q(t)h(t)dt$, so the corresponding estimator for the integrated hazard function $H(t) = \int_0^t h(s)ds$ is $\int_0^t q(s)^{-1}dN_+(s)$. To avoid the problem caused here when $q(s) = 0$ the *Nelson–Aalen estimator* is defined as

$$\tilde{H}(t) = \int_0^t q(s)^- dN_+(s);$$

here $q(s)^- = q(s)^{-1}$ when $q(s) > 0$ and $q(s)^- = 0$ when $q(s) = 0$, using a notation analogous to that for a generalised inverse matrix. The integral for $\tilde{H}(t)$ actually reduces to a sum:

$$\tilde{H}(t) = \sum_t q(T_i)^-,$$

where $\sum_t$ denotes summation over $\{i : T_i \le t, \ T_i < U_i\}$. Thus, $\tilde{H}$ is a step function: as t increases, $\tilde{H}$ steps up by $q(t)^-$ as each observed failure time is reached; $\tilde{H}(t)$ is zero from $t = 0$ up until the first observed failure time.

331

We have

$$\tilde{H}(t) = \int_0^t q(s)^- dN_+(s) = \int_0^t q(s)^- \{d\Lambda_+(s) + dM_+(s)\}$$

$$= \int_0^t q(s)^- q(s)h(s)ds + \int_0^t q(s)^- dM_+(s).$$

Thus,

$$\tilde{H}(t) - H(t) = \int_0^t \{q(s)^- q(s) - 1\}h(s)ds + \int_0^t q(s)^- dM_+(s).$$

But $q(s)^- q(s) = I\{q(s) > 0\}$, so

$$E\{q(s)^- q(s)\} = P\{q(s) > 0\} = 1 - P\{q(s) = 0\},$$

and then

$$E\{\tilde{H}(t)\} = H(t) - \int_0^t P\{q(s) = 0\}h(s)ds,$$

the expectation of the second term, a stochastic integral with respect to the martingale M_+, being zero. Thus, $\tilde{H}(t)$ is approximately unbiased for $H(t)$ if $P\{q(s) = 0\}$ is small on $[0, t)$, and this will be true when $P(T_n \leq t)$ is small. In fact, $P(T_n \leq t) \to 0$ as $n \to \infty$ for fixed t when the underlying failure-time distribution has positive probability beyond t.

For the predictable variation process of $\tilde{H}(t)$ we have

$$d\langle\tilde{H}\rangle(t) = E\{d\tilde{H}(t)^2 \mid \mathcal{H}_{t-}\} = E\left[\{q(t)^- dN_+(t)\}^2 \mid \mathcal{H}_{t-}\right]$$

$$= \{q(t)^-\}^2 E\{dN_+(t)^2 \mid \mathcal{H}_{t-}\} = \{q(t)^-\}^2 d\Lambda_+(t) = q(t)^- dH(t),$$

and then $\langle\tilde{H}\rangle(t) = \int_0^t q(s)^- dH(s)$.

An estimator of the survivor function may be constructed directly as

$$\tilde{S}(t) = \exp\{-\tilde{H}(t)\} = \exp\left\{-\sum_t q(T_i)^-\right\}.$$

For large $q(T_i)$ values,

$$\tilde{S}(t) = \prod_t \exp\{-q(T_i)^-\} \approx \prod_t \{1 - q(T_i)^-\},$$

where $\prod_t$ is the product over $\{i : T_i \leq t, \ T_i < U_i\}$. This is the Kaplan–Meier estimator. An alternative derivation goes via the product integral route:

$$\tilde{S}(t) = \exp\left\{-\int_0^t d\tilde{H}(s)\right\} = \mathcal{P}_{s \leq t}\{1 - d\tilde{H}(s)\} = \mathcal{P}_{s \leq t}\{1 - q(s)^- dN_+(s)\}$$

$$= \prod_t \{1 - q(T_i)^-\}.$$

20.2 Competing Risks

The extension of the above to the case of p competing risks is effected by defining a p-variate counting process $(N_{+1}, \ldots, N_{+p})$, as in Section 19.3. We define $H(j, t) = \int_0^t h(j, s)ds$, the integrated sub-hazard function, and the estimator of Nelson–Aalen type,

$$\tilde{H}(j, t) = \int_0^t q(s)^- d N_{+j}(s).$$

The use of $q(s)^-$ here reflects an assumption that all individuals in the risk set are equally liable to failure from cause j at time s. Now, N_{+j} jumps at the times of observed j-type failures and then

$$\tilde{H}(j, t) = \sum_j q(T_i)^-,$$

where the summation is over $\{i : T_i \leq t, \ T_i < U_i, \ C_i = j\}$.
Estimates for the other risk functions follow from

$$\tilde{H}(t) = \sum_{j=1}^p \tilde{H}(j, t), \quad \tilde{S}(t) = \exp\{-\tilde{H}(t)\}, \quad \tilde{h}(j, t) = q(t)^{-1} d N_{+j}(t),$$

$$\tilde{f}(j, t) = \tilde{h}(j, t)\tilde{S}(t), \quad \tilde{S}(j, t) = \int_t^\infty \tilde{f}(j, s)ds = \sum_j' q(T_i)^{-1} \tilde{S}(T_i),$$

where $\sum_j'$ represents summation over $\{i : T_i > t, \ T_i < U_i, \ C_i = j\}$.
We can now define $M_{+j}(t) = N_{+j}(t) - \Lambda_+(j, t)$ and then, recalling that $d\Lambda_+(j, t) = q(t)h(j, t)$,

$$\tilde{H}(j, t) - H(j, t) = \int_0^t \{q(s)^- \{d M_{+j}(s) + d\Lambda_+(j, s)\} - \int_0^t h(j, s)ds$$

$$= \int_0^t \{q(s)^- q(s) - 1\}h(j, s)ds + \int_0^t q(s)^- d M_{+j}(s)$$

$$= -\int_0^t I\{q(s) = 0\}h(j, s)ds + \int_0^t q(s)^- d M_{+j}(s).$$

Hence, as for $\tilde{H}$ in Section 20.1,

$$E\{\tilde{H}(j, t)\} = H(j, t) - \int_0^t P\{q(s) = 0\}h(j, s)ds,$$

showing that $\tilde{H}(j, s)$ is approximately unbiased for $H(j, t)$.
The special case of censored survival data with a single cause of failure can be dealt with in a competing-risks framework as follows: Take $p = 2$, with $j = 0$ for censoring and $j = 1$ for observed failure. Then $\tilde{H}(1, t) = \sum_1 q(T_i)^{-1}$,

where $\sum_1$ is summation over $\{i : T_i \leq t, \ C_i = 1\}$; and it is here assumed that all T_i are observed, for example by nominally taking the U_i to be infinite. For the corresponding survivor function we have, under the Makeham condition (Section 14.5),

$$\tilde{S}(1, t) = \exp\{-\tilde{H}(1, t)\} = \mathcal{P}_{s\leq t}\{1 - d\tilde{H}(1, s)\} = \mathcal{P}_{s\leq t}\{1 - q(s)^- dN_{+1}(s)\}$$
$$= \prod_1 \{1 - q(T_i)^-\} = \prod_1 \{1 - q(T_i)^{-1}\},$$

the Kaplan–Meier estimator.

20.3 Large-Sample Results

20.3.1 Consistency

One of the more transparent sufficient conditions for consistency of $\tilde{H}$ for H is that the size of the risk set at all times considered grows without limit as $n \to \infty$: $\inf_{0\leq s\leq t} q(s) \to \infty$. (The inf entails uniform divergence to infinity, that is, simultaneously at all points within the range.) With a monotone decreasing risk set over times it suffices that its size increases in this way at the end of the period. The result (Andersen et al., 1993, Theorem IV.1.1) is that $\sup_{0\leq s\leq t} | \tilde{H}(s) - H(s) | \to 0$ as $n \to \infty$.

20.3.2 Asymptotic Normality

Suppose that $n^{-1}q(s) \to_p y(s)$ uniformly on $[0, t]$, and define $\sigma^2(t) = \int_0^t y(s)^{-1}h(s)ds$. Then, under some conditions (Andersen et al., 1993, Theorem IV.1.2), $n^{1/2}(\tilde{H}-H) \to_d Z$, a Gaussian martingale with $Z(0) = 0$ and $\mathrm{cov}\{Z(s_1), Z(s_2)\}$ $\sigma^2\{\min(s_1, s_2)\}$. The variance function can be consistently estimated by

$$\tilde{\sigma}^2(t) = n \int_0^t \{q(s)^-\}^2 dN_+(s) = n \sum_t \{q(T_i)^-\}^2;$$

to see how this matches up to $\sigma^2(t)$ note that $\mathrm{E}\{dN_+(s) \mid \mathcal{H}_{s-}\} = q(s)h(s)ds$. Corresponding results for the Kaplan–Meier estimator are given in Andersen et al. (1993, Section IV.3.2).

20.3.3 Confidence Intervals

Approximate pointwise confidence interval for $H(t)$ can be derived from the asymptotic result

$$n^{1/2}\{\tilde{H}(s) - H(s)\}/\tilde{\sigma}(s) \to_d N(0, 1).$$

For example, this yields approximate 95% confidence limits $\tilde{H}(s) \pm 1.96\tilde{\sigma}(s)/ n^{1/2}$ for $H(s)$. The normal approximation may be improved by applying a transformation such as working in terms of $\log \tilde{H}(s)$; the corresponding variance may then be calculated via the delta method.

The above result gives an interval for a single time point. For simultaneous confidence bands, applying to H over a whole interval, see Andersen et al. (1993, Section IV.1.3.2).

20.4 Hypothesis Testing

20.4.1 Single-Sample Case

Suppose that the null hypothesis specifies the integrated hazard function as H_0. Take as test statistic $n^{1/2}\{\tilde{H}(t) - H_0(t)\}/\tilde{\sigma}(t)$ and apply its asymptotic standard normal distribution to obtain appropriate critical values.

Instead of focussing on the single time t, one could look at the maximum discrepancy,

$$\sup_{0 \le s \le t} n^{1/2} \mid \tilde{H}(s) - H_0(s) \mid = n^{1/2} \max_j \mid \tilde{H}(T_j) - H_0(T_j) \mid,$$

where the T_j are the observed failure times. A null distribution for this statistic can be obtained by simulation, generating samples from the null survivor function $S_0(s) = \exp\{-H_0(s)\}$ via the probability integral transform.

A general class of test statistics can be constructed as follows: Let $H_0^*(t) = \int_0^t I\{q(s) > 0\} d H_0(s)$ and define the process $Z(t) = \int_0^t K(s) d\{\tilde{H}(s) - H_0^*(s)\}$, where K is a specified predictable non-negative weight function. Then Z is a martingale with predictable variation process

$$\langle Z(t) \rangle = \int_0^t K^2(s) d \langle \tilde{H} - H_0^* \rangle(s) = \int_0^t K^2(s) q(s)^- d H_0(s).$$

The corresponding standardised statistic is $Z_0(t)/\langle Z_0 \rangle(t)^{1/2}$; this has approximate distribution N(0,1) under H_0.

A particular example is obtained by taking $K(s) = q(s)$. Then,

$$Z(t) = \int_0^t q(s) \left[q(s)^- d N_+(s) - I\{q(s) > 0\} d H_0^*(s) \right]$$

$$= \int_0^t I\{q(s) > 0\} d N_+(s) - \int_0^t q(s) d H_0(s) = N_+(t) - E_+(t),$$

where $E_+(t) = E_0\{N_+(t)\}$, and

$$\langle Z \rangle(t) = \int_0^t q(s)^2 q(s)^- d H_0(s) = \int_0^t q(s) d H_0(s) = E_+(t).$$

The corresponding standard normal deviate is $\{N_+(t) - E_+(t)\}/E_+(t)^{1/2}$.

20.4.2 Several Samples

Let $T_i^{(g)}$ be the failure or censoring time and let $D_i^{(g)}$ be the censoring indicator (1 if failure observed, 0 if right-censored) for individual i in group g ($i = 1, \ldots, n_g$). We assume that the individual $(T_i^{(g)}, D_i^{(g)})$ are all independent and that the censoring is independent. Let $N_i^{(g)}(t)$ be the individual counting processes, let $Y_i^{(g)}(t)$ be the observation indicator processes, and let $h^{(g)}(t)$ be the hazard function for individuals in group g. Define

$$N_+^{(g)}(t) = \sum_{i=1}^{n} N_i^{(g)}(t), \quad d\Lambda_+^{(g)}(t) = \sum_{i=1}^{n} Y_i^{(g)}(t)h^{(g)}(t)dt = q^{(g)}(t)h^{(g)}(t)dt,$$

$$M_i^{(g)}(t) = N_i^{(g)}(t) - \Lambda_i^{(g)}(t);$$

the $M_i^{(g)}$ are martingales with respect to the filtration generated by the $N_i^{(g)}$.

The Nelson–Aalen estimator for the integrated hazard function in the gth group is

$$\tilde{H}^{(g)}(t) = \int_0^t q^{(g)}(s)^- dN_+^{(g)}(s) = \sum_t q^{(g)}(T_i^{(g)})^-,$$

where $\sum_t$ sums over $\{i : T_i^{(g)} \leq t, \text{ observed}\}$. Under the null hypothesis, that the $h^{(g)}$ are all equal to a common hazard function h, a pooled estimator for the common integrated hazard function is

$$\tilde{H}(t) = \int_0^t q^{(+)}(s)^- dN_+^{(+)}(s).$$

To compare $\tilde{H}^{(g)}$ with $\tilde{H}$, construct a modified version of $\tilde{H}$ that selects values for which $q^{(g)}(s) > 0$: let

$$\tilde{H}_*^{(g)}(t) = \int_0^t I\{q^{(g)}(s) > 0\} q^{(+)}(s)^- dN_+^{(+)}(s)$$

and examine

$$\tilde{H}^{(g)}(t) - \tilde{H}_*^{(g)}(t) = \int_0^t I\{q^{(g)}(s) > 0\}\{q^{(g)}(s)^- dN_+^{(g)}(s) - q^{(+)}(s)^- dN_+^{(+)}(s)\}.$$

Now,

$$d\tilde{H}^{(g)}(s) - d\tilde{H}_*^{(g)}(s) = I\{q^{(g)}(s) > 0\}\left[q^{(g)}(s)^-\{dM_+^{(g)}(s) + d\Lambda_+^{(g)}(s)\}\right.$$

$$\left. -q^{(+)}(s)^-\{dM_+^{(+)}(s) + d\Lambda_+^{(+)}(s)\}\right].$$

Under the null hypothesis, $q^{(g)}(s)^- d\Lambda_+^{(g)}(s) = h(s)ds$ and $q^{(+)}(s)^- d\Lambda_+^{(+)}(s) = h(s)ds$. In that case,

$$d\tilde{H}^{(g)}(s) - d\tilde{H}_*^{(g)}(s) = I\{q^{(g)}(s) > 0\}\{q^{(g)}(s)^- dM_+^{(g)}(s) - q^{(+)}(s)^- dM_+^{(+)}(s)\},$$

and so $\tilde{H}^{(g)}(t) - \tilde{H}_*^{(g)}(t)$ is a martingale.

As for the single-sample case, we can construct test statistics as

$$Z^{(g)}(t) = \int_0^t K_g(s)\{d\,\tilde{H}^{(g)}(s) - d\,H_*^{(g)}(s)\}.$$

Different choices for the predictable weighting functions K_g, together with subsequent combination of the $Z^{(g)}$, produce different overall test statistics. A particular class of tests is generated by taking $K_g(s) = q^{(g)}(s)K(s)$, where K is a predictable function of $(N_+^{(+)}, q^{(+)})$. Then,

$$
\begin{aligned}
Z^{(g)}(t) &= \int_0^t q^{(g)}(s)K(s)\{d\,\tilde{H}^{(g)}(s) - d\,H_*^{(g)}(s)\} \\
&= \int_0^t K(s)\{d\,N_+^{(g)}(s) - q^{(g)}(s)q^{(+)}(s)^- d\,N_+^{(+)}\} \\
&= \int_0^t K(s)d\,N_+^{(g)}(s) - \int_0^t K(s)q^{(g)}(s)q^{(+)}(s)^- d\,N_+^{(+)}.
\end{aligned}
$$

The first term on the right-hand side is the (weighted) observed number of failures in group g during $[0, t]$, and the second is an estimate of its null expected value. Also note that $\sum_g Z^{(g)}(t) = 0$.

To calculate the null covariances among the $Z^{(g)}$ we start with

$$
\begin{aligned}
Z^{(g)}(t) &= \int_0^t q^{(g)}(s)K(s)\{d\,\tilde{H}^{(g)}(s) - d\,\tilde{H}_*^{(g)}(s)\} \\
&= \int_0^t q^{(g)}(s)K(s)\{q^{(g)}(s)^- d\,M_+^{(g)}(s) - q^{(+)}(s)^- d\,M_+^{(+)}(s)\} \\
&= \sum_l \int_0^t K(s)\{\delta_{lg} - q^{(g)}(s)q^{(+)}(s)^-\}d\,M_+^{(l)}(s).
\end{aligned}
$$

Now, assuming that $N_+^{(l)}$ and $N_+^{(m)}$ have no jumps in common for $l \neq m$,

$$
\begin{aligned}
d\langle M_+^{(l)}, M_+^{(m)}\rangle(t) &= \mathrm{E}\big[\{d\,N_+^{(l)}(t) - d\,\Lambda_+^{(l)}(t)\} \times \{d\,N_+^{(m)}(t) - d\,\Lambda_+^{(m)}(t)\} \mid \mathcal{H}_{t-}\big] \\
&= \delta_{lm}\mathrm{var}\{d\,N_+^{(l)}(t) \mid \mathcal{H}_{t-}\} = \delta_{lm}d\,\Lambda_+^{(l)}(t) = \delta_{lm}q^{(l)}(t)h(t)dt.
\end{aligned}
$$

Hence, the predictable covariation processes are

$$
\begin{aligned}
\langle Z^{(g)}, Z^{(h)}\rangle(t) &= \sum_l \int_0^t K^2(s)\{\delta_{lg} - q^{(g)}(s)q^{(+)}(s)^-\} \\
&\quad \times \{\delta_{lh} - q^{(h)}(s)q^{(+)}(s)^-\}q^{(l)}(s)h(s)ds \\
&= \int_0^t K^2(s)\{\delta_{gh}q^{(g)}(s) - q^{(g)}(s)q^{(h)}(s)q^{(+)}(s)^-\}h(s)ds.
\end{aligned}
$$

The covariances among the $Z^{(g)}(t)$ are then

$$C_{gh}(t) = \mathrm{cov}\{Z^{(g)}(t), Z^{(h)}(t)\} = \mathrm{E}\langle Z^{(g)}, Z^{(h)}\rangle(t),$$

which may be estimated by replacing $h(s)ds$ by $q(s)^- dN_+(s)$ in the above integral.

Setting $t = \infty$, and combining the $Z^{(g)}(\infty)$ into a chi-square statistic of familiar form $Z^T C^- Z$, produces overall test statistics. (The fact that the covariance matrix has rank one fewer than the number of groups, since $Z^{(+)}(t) = 0$, has to be allowed for: C^- is a generalised inverse.) The particular choice $K(s) = I\{q^{(+)}(s) > 0\}$ yields the famous *log-rank test* (Peto and Peto, 1972).

20.5 Regression Models

20.5.1 Intensity Models and Time-Dependent Covariates

The Cox proportional hazards (PH) model for competing risks (Section 15.3) specifies the sub-hazard function $h(j, t; \mathbf{x})$ to be of form $\psi_j(\mathbf{x}; \beta_j)h_0(j, t)$ in which $\psi_j(\mathbf{x}; \beta_j)$ is a positive function of the covariate $\mathbf{x}$ and parameter β_j, often taken to be $\exp(\mathbf{x}^T \beta_j)$, and $h_0(j, t)$ is a baseline sub-hazard. The β_j $(j = 1, \ldots, p)$ are sub-vectors, not necessarily disjoint, of an overall regression vector β. In the present context the model would be written as

$$d\Lambda(j, t; \beta_j) = Y(t)\psi_j(\mathbf{x}; \beta_j)\lambda_0(j, t)dt.$$

This is a type of *multiplicative intensity model*. Had we begun with the form $\psi_j(\mathbf{x}; \beta_j) + h_0(j, t)$ for the sub-hazard, an *additive intensity model* would have resulted. As usual in regression the inferences will normally be made conditionally on the $\mathbf{x}$ values in the sample. Otherwise, their distribution has to be accommodated in the likelihood function, which might or might not be a simple matter.

The covariate vector is assumed to be constant over time in the above, but this is not necessary and is often violated in practice. Time-dependent covariates were discussed in Section 4.3. The form is $\mathbf{x}(t)$, with sample path over time governed by some stochastic mechanism that may be connected with that generating the failure time itself. Broadly speaking, if $\mathbf{x}(t)$ is determined by the past history, that is, is measurable with respect to $\mathcal{H}_{t-}$, it can be included in the likelihood without too much trouble: its predictability, like that of $Y(t)$, allows one to insert its value into the likelihood contribution at time t. Note that the filtration $\{\mathcal{H}_t\}$ might have to be augmented to accommodate the $\mathbf{x}$ process.

Oakes (1981) and Arjas (1985) have made important points concerning the definition of *time* in Survival Analysis, one of which can be illustrated as follows. Suppose that certain time-varying conditions affect all individuals in a study, environmental factors being a case in point. Further, suppose that different individuals enter at different dates (staggered entry) and that a PH model with unspecified baseline hazard $h_0(t)$ is adopted that is intended to account for some of the unrecorded time-varying conditions. Then, the usual time shift, making each individual start at $t = 0$, disrupts the synchrony in $h_0(t)$ between them. In effect, times in $h_0(t)$ and the $x_i(t)$ are different.

20.5.2 Proportional Hazards Model

The Cox model for competing risks is defined by

$$d \Lambda_i(j, t) = Y_i(t) \, h_i(j, t) \, dt = Y_i(t) \, \psi_j(x_i(t); \beta_j) \, h_0(j, t; \phi) \, dt,$$

where the covariate vector x may vary with t and the baseline hazard may depend on a parameter vector ϕ. To construct a likelihood function $L(\theta)$, where $\theta = (\beta, \phi)$, we set up a multivariate counting process as in Section 19.3. The log-likelihood can be written down as

$$\log L(\theta) = \sum_{i=1}^{n} \log h_i(c_i, t_i) - \sum_{i=1}^{n} \sum_{j=1}^{p} \int_{t \geq 0} Y_i(t) h_i(j, t) dt$$

$$= \sum_{i=1}^{n} \sum_{j=1}^{p} \left[\int_{t \geq 0} Y_i(s) \log\{h_i(j, s; \phi)\} d N_{ij}(s) - \int_{t \geq 0} Y_i(s) h_i(j, s; \phi) ds \right].$$

The theoretical manipulations now follow from seeing that the score function, $d \log L(\theta)/d\theta$, has components that are linear combinations of stochastic integrals with respect to martingales $M_{ij}(t) = N_{ij}(t) - \Lambda_i(j, t)$ and are therefore themselves martingales.

To tackle the semi-parametric version, in which the $h_0(j, t; \phi)$ are treated as unspeakable nuisance functions, we adopt Cox's partial likelihood function, $P(\beta)$, as the basis for inference. It is then found that the partial score vector $d P(\beta)/d\beta$ is a martingale and the rest, as they say, is geography.

20.5.3 Martingale Residuals

Various types of residuals can be defined for assessing model fit, both overall and for individual cases. In addition to those discussed in Section 3.4, residuals associated with the counting process approach are described and exploited by Fleming and Harrington (1991, Section 4.5).

In a random sample of failure times, the process $M_i(t) = N_i(t) - \Lambda_i(t; \theta)$ is a martingale specific to individual i. At the end of the study, at time τ say, $N_i(\tau)$ will be 1 if failure has been observed and 0 if not. The differences, $N_i(\tau) - \Lambda_i(\tau; \hat{\theta})$ are known as *martingale residuals*: they can be plotted and tested in various ways.

The possible values assumed by $N_i(t)$ in standard Survival Analysis are 0 and 1, so the range of values for $M_i(t)$ is $(-\infty, 1)$ and such martingale residuals can tend to have negatively skewed sampling distributions. Some calculations, along the lines that produce *deviance residuals* in the context of *generalised linear models*, produce a transformed version of $M_i(t)$ that has a more symmetric distribution. The transformation (Fleming and Harrington, 1991, Equation 5.14) has the effect of shrinking large negative values toward

0 and expanding values near 1 away from 0. Other types described in that book include *martingale transform residuals,* of which a particular species is *score residuals,* and the authors show how to use such residuals to assess influence of individual cases and model adequacy. Additional material on residuals is to be found in Chapter 7 of Andersen et al. (1993). *Schoenfeld residuals* are particularly aimed at detecting non-proportionality in the PH model. In Schoenfeld (1982) and Grambsch and Therneau (1994) they are used for plotting and the construction of test statistics.

Appendix A

Terms, Notations, and Abbreviations

Various terms, notations, and abbreviations are used in this book. As many as I can think of are assembled here for reference.

AL: accelerated life

β: beta, regression coefficient (vector), usually in linear combination with covariate $\mathbf{x}$ as $\mathbf{x}^T \beta$

DFR: decreasing failure rate

$E(T)$: expected value (mean) of T

EM: expectation-maximisation (algorithm)

$f(t)$, $\bar{F}(t)$, $h(t)$: density, survivor, and hazard functions

$f(j, t)$, $\bar{F}(j, t)$, $h(j, t)$, $\bar{h}(j, t)$: sub-density, sub-survivor, sub-hazard, and sub-odds functions

IFR: increasing failure rate

iid: independently, identically distributed

KM: Kaplan–Meier (estimator)

LIDA: *Lifetime Data Analysis* (journal)

lr, llr: likelihood ratio, log-likelihood ratio

MB: multivariate Burr (distribution)

mle: maximum likelihood estimate(s)/estimator/estimation

MRL: mean residual life

MW: multivariate Weibull (distribution)

$N(\mu, \sigma^2)$: normal distribution with mean μ and variance σ^2

NPML: non-parametric maximum likelihood

θ: theta, general parameter (vector)

PH: proportional hazards

PO: proportional odds

$U(a, b)$: uniform distribution on the interval (a, b)

var(T): variance of T

Appendix B

Basic Likelihood Methods

Likelihood plays a central role in Statistics. It is normally the tool of first choice for developing methods for real applications. Only occasionally does it fail to yield effective techniques, the exceptional cases tending to be either mathematically pathological in some way or where it is difficult to obtain the likelihood function in a useable form from the underlying model. The following brief notes focus on Frequentist methods, reflecting the slant of the book.

B.1 Likelihood, Score and Information Functions

Denote by $p(D; \theta)$ the probability, or density, function of the data D expressed in terms of parameter θ. Regarded as a function of θ, $p(D; \theta)$ is known as the *likelihood function*. In particular, for *random samples*, when $D = (y_1, ..., y_n)$ in which the y_is are *iid*, $p(D; \theta) = \prod_{i=1}^{n} p(y_i; \theta)$. Let $l(\theta) = \log p(D; \theta)$ denote the *log-likelihood function*. Then $l'(\theta) = d \log p(D; \theta)/d\theta$ is the *score function*, and $-l''(\theta) = -d^2 \log p(D; \theta)/d\theta^2$ is the *information function*. For the multi-parameter case, $\theta = (\theta_1, \theta_2, ...)$, $l'(\theta)$ is the *score vector* with jth component $\partial l(\theta)/\partial \theta_j$, $l''(\theta)$ is the *Hessian matrix* with (j, k)th element $\partial^2 l(\theta)/\partial \theta_j \partial \theta_k$, and then $-l''(\theta)$ is the *information matrix*.

B.2 Maximum Likelihood Estimation

The maximum likelihood estimator (*mle*) $\hat{\theta}$ is obtained by maximizing $p(D; \theta)$, or equivalently $l(\theta)$, over θ. If $l(\theta)$ is differentiable with respect to θ, $\hat{\theta}$ is a solution of the *likelihood equation* $l'(\theta) = 0$. At a maximum, the matrix $l''(\hat{\theta})$ will be negative definite, or just $l''(\hat{\theta}) < 0$ for the single-parameter case.

A reasonable conjecture is that, as the information increases, the *mle* should get closer to the true, underlying parameter value θ_0. The *asymptotic theory* states that, as $n \to \infty$, $\hat{\theta} \to \theta_0$ in a probability sense: this property of an estimator is called *consistency*. It can be proved (under some technical regularity conditions) that

1. for sufficiently large n, $\hat{\theta}$ exists and is unique, and it is consistent;
2. as $n \to \infty$ the distribution of $l''(\hat{\theta})^{-1/2}(\hat{\theta} - \theta_0)$ tends toward the standard normal.

So, roughly speaking, in large samples we can take $\hat{\theta}$ as being approximately normally distributed with mean θ_0 and covariance matrix $-l''(\hat{\theta})^{-1}$. The square roots of the diagonal elements of $-l''(\hat{\theta})^{-1}$ thus provide estimates of standard errors for the individual parameter components.

The proof of (2) runs along the following lines: By Taylor expansion of the score function about θ_0,

$$l'(\hat{\theta}) \approx l'(\theta_0) + l''(\theta_0)(\hat{\theta} - \theta_0).$$

But $l'(\hat{\theta}) = 0$ (the likelihood equation) so

$$\hat{\theta} \approx \theta_0 - l''(\theta_0)^{-1}l'(\theta_0).$$

In regular cases, $l'(\theta_0)$ has mean 0 and covariance matrix I_0, where $I_0 = E\{-l''(\theta_0)\}$ is the *Fisher information matrix*. (Remember that $l'(\theta_0)$ is a function of the data D, so its mean and covariance are with respect to repeated samples of D.) Also, its distribution tends to normal as $n \to \infty$. For example, when D is a random sample, $l'(\theta_0)$ is the sum of n *iid* quantities, and so the basic *central limit theorem* kicks in. Further, $-l''(\theta_0)$ stabilises to its expectation I_0 as $n \to \infty$, by the *law of large numbers*. The resulting asymptotic distribution for $\hat{\theta}$ is therefore normal with mean θ_0 and covariance matrix

$$\mathrm{var}\{-l''(\theta_0)^{-1}l'(\theta_0)\} \approx I_0^{-1}\mathrm{var}\{-l'(\theta_0)\}I_0^{-1} = I_0^{-1},$$

and I_0 is estimated by $-l''(\hat{\theta})$. Incidentally, even when I_0 can be written down explicitly, there is evidence that it is better to use $-l''(\hat{\theta})$ (Efron and Hinkley, 1978).

B.3 The Delta Method

The maximum likelihood estimator *mle* of a function of θ, say $\tau(\theta)$, is $\tau(\hat{\theta})$. Assume that the distribution of $\hat{\theta}$ approaches normality in the manner indicated above. Then that of $\tau(\hat{\theta})$ approaches normal with mean $\tau(\theta_0)$ and (for scalar θ) variance $\tau'(\theta_0)^2 I_0^{-1}$; this is just the large-sample variance of $\hat{\theta}$ times the squared derivative of $\tau(\theta)$ at θ_0. The derivation of these results comes from another Taylor expansion:

$$\tau(\hat{\theta}) \approx \tau(\theta_0) + (\hat{\theta} - \theta_0)\tau'(\theta_0).$$

In the case of a vector parameter θ, the variance formula is $\tau'(\theta_0)^T I_0^{-1}\tau'(\theta_0)$, where $\tau'(\theta)$ has jth component $\partial\tau(\theta)/\partial\theta_j$. The sample estimate of $\mathrm{var}\{\tau(\hat{\theta})\}$ is $\tau'(\hat{\theta})^T\{-l''(\hat{\theta})^{-1}\}\tau'(\hat{\theta})$.

The delta method is based on asymptotics: there is a sequence of random variables indexed by n, here $\hat{\theta}$, converging in distribution, not necessarily to the normal. It is not correct simply to use the result as an approximation for a single random variable, for example, for calculating the mean of the square root of a Poisson variate. The term *delta* comes from a faintly old-fashioned notation in which the increments in the Taylor expansion used to prove the result are written as $\delta\theta$.

The delta method is sometimes used to transform standard errors of *mle*. For example, suppose that a probability parameter p is to be expressed as $(1 + e^{-\theta})^{-1}$, with θ unrestricted in maximising the likelihood. If s_θ is returned as the standard error of $\hat{\theta}$ then

$$se(\hat{p}) = s_\theta \mid \frac{d}{d\theta}(1 + e^{-\theta})^{-1} \mid_{\theta=\hat{\theta}} .$$

However, the asymptotic normal approximation for $\hat{\theta}$, which makes s_θ a meaningful quantity, might not hold very well for $\hat{p}$: one might find that $\hat{p} \pm 1.96 se(\hat{p})$ embarrassingly spans values 0 or 1. For safety, it is best to transform the range $\hat{\theta} \pm 1.96 se(\hat{\theta})$ back to the p scale *en bloc*.

B.4 Likelihood Ratio Tests

Suppose that we have a *null hypothesis* $H_0 : \theta = \theta_0$ and an *alternative hypothesis* $H_1 : \theta \neq \theta_0$, where θ is a scalar parameter and θ_0 is a specified value. If the *likelihood ratio*, $lr(D) = p(D; \hat{\theta})/p(D; \theta_0)$, greatly exceeds 1, then it suggests that θ_0 is an unlikely value for θ in the light of data D. Formally, under H_0 (i.e., assuming that θ_0 is the true value), the large-sample distribution of $2\log\{lr(D)\}$ is χ_1^2. So, if H_0 is true, we should obtain a typical χ_1^2 value for $2\log\{lr(D)\}$. Otherwise, a large value will be taken to indicate significant evidence against θ_0. How significant is conveyed by the p-value, which is the probability of obtaining a value at least as extreme as the one from the data D: the smaller the p-value, the stronger the evidence against H_0.

The situation can be generalised. Suppose now that θ is a vector parameter and that H_0 and H_1 restrict its value in some ways. For example, H_0 might say that a particular component of θ is zero, as when setting a regression coefficient to zero in order to omit the associated covariate from the model. The likelihood ratio comparison of the two hypotheses can now be formed as $lr(D) = p(D; \hat{\theta}_1)/p(D; \hat{\theta}_0)$, where $\hat{\theta}_0$ and $\hat{\theta}_1$ are the *mle* under H_0 and H_1, respectively: each hypothesis is given its best shot, as it were.

Under certain circumstances a large-sample chi-square distribution is obtained for $2\log\{lr(D)\}$. These are, roughly speaking, that H_0 is essentially H_1 plus some further constraints: in effect, H_1 defines an *enhanced model* over that of H_0. The common case is where H_0 fixes some of the parameters previously allowed to be free by H_1. Then the aforesaid chi-square distribution has degrees of freedom equal to the number of additional constraints imposed by H_0.

B.5 Large-Sample Equivalents of the Likelihood Ratio Test

Consider the situation where θ is scalar (one-dimensional) and $H_0 : \theta = \theta_0$ is a *simple hypothesis*, meaning that θ_0 is a specified value. By Taylor expansion about $\hat{\theta}$,

$$l(\theta_0) \approx l(\hat{\theta}) + l'(\hat{\theta})(\theta_0 - \hat{\theta}) + \frac{1}{2} l''(\hat{\theta})(\theta_0 - \hat{\theta})^2.$$

Now, $l'(\hat{\theta}) = 0$ (the likelihood equation) and $l(\hat{\theta}) - l(\theta_0) = \log\{lr(D)\}$. The third term on the right-hand side is $-\frac{1}{2}(\hat{\theta} - \theta_0)^2/\text{var}(\hat{\theta})$, say $-\frac{1}{2}W$. Then we have $W \approx 2\log\{lr(D)\}$: W, known as the *Wald statistic*, is thus asymptotically equivalent to the log-likelihood ratio statistic. The asymptotic distribution of $W^{1/2} = (\hat{\theta} - \theta_0)/se(\hat{\theta})$ is standard normal; it follows that the asymptotic distribution of $2\log\{lr(D)\}$ is χ_1^2.

The result can be generalised to null hypotheses of form $H_0 : \tau(\theta) = 0$: the suggested test statistic is $W = \tau(\hat{\theta})^2/v_\tau$, where $v_\tau = \{\tau'(\hat{\theta})\}^2\{-l''(\hat{\theta})^{-1}\}$ is an estimate of the asymptotic variance of $\tau(\hat{\theta})$ by the delta method. The test based on W is known as the *Wald test*. In a further generalization, to the case of a vector parameter, the *Wald statistic* for testing $H_0 : \tau(\theta) = 0$ can be written as $W = \tau(\hat{\theta})^T V_\tau^{-1}\tau(\hat{\theta})$, where $V_\tau = \tau'(\hat{\theta})^T\{-l''(\hat{\theta})^{-1}\}\tau'(\hat{\theta})$.

Recall from above

$$\hat{\theta} \approx \theta_0 - l''(\theta_0)^{-1}l'(\theta_0).$$

Thus,

$$(\hat{\theta} - \theta_0)^2\{-l''(\theta_0)\} \approx -l''(\theta_0)^{-1}l'(\theta_0)^2,$$

or

$$W \approx l'(\theta_0)^2/\text{var}\{l'(\theta_0)\}.$$

The right-hand side is the *score test statistic*, which is asymptotically equivalent to W and therefore to the lr statistic.

For $lr(D)$ we need to calculate *mle* under both H_0 and H_1; for the Wald statistic we need the *mle* only under H_1; for the score statistic we need the *mle* only under H_0, or just θ_0 in the case of a simple null hypothesis. Thus, less calculation is involved for the Wald and score tests than for the lr test. However, $lr(D)$ is invariant to parameter transformation, for example, it would make no difference if the likelihood were parametrised in terms of $\log \theta$, say, rather than θ itself. This is not true of the Wald and score tests. On the other hand, in finite samples it could be that one parametrisation gives higher power than another for a particular alternative.

Appendix C

Some Theory for Partial Likelihood

Cox (1975) provided some more formal justification for the likelihood-like properties of $P(\beta)$, Oakes (1981) considered further aspects, and Wong (1986) continued with the task at a more rigorous level. However, the most elegant treatment is via martingale counting processes: Andersen and Gill (1982) provided a general theory; Gill (1984) and Andersen and Borgan (1985) contain very readable accounts of this rather technical area. An introduction to this topic is given in Part IV, but here we will just follow Cox and give an informal justification for inference based on the partial likelihood.

In general terms a partial likelihood is set up as follows: Suppose that $\theta = (\beta, \phi)$, where β is the parameter of interest and ϕ is a nuisance parameter; in the proportional hazards model ϕ is h_0, the baseline hazard function, of infinite dimension. Suppose further that the accumulating data sequence can be formulated as $D_l = (A_1, B_1, \ldots, A_l, B_l)$ for $l = 1, 2, \ldots$. Then the likelihood function can be written as

$$L(\theta) = p_\theta(D_m) = \prod_{l=1}^{m} p_\theta(A_l, B_l \mid D_{l-1}) = \prod_{l=1}^{m} p_\theta(A_l \mid B_l, D_{l-1})\, p_\theta(B_l \mid D_{l-1})$$

$$= \prod_{l=1}^{m} p_\theta(A_l \mid B_l, D_{l-1}) \times \prod_{l=1}^{m} p_\theta(B_l \mid D_{l-1}) = P(\theta)Q(\theta),$$

say, where p_θ is a generic notation for a probability mass function or density. Now, this formulation will be useful if $P(\theta)$ is a function only of β, the parameter of interest, and if $Q(\theta)$ does not depend on β. In that case we can ignore the *constant* factor $Q(\theta)$ in the likelihood factorisation $L(\theta) = P(\beta)Q(\phi)$ and just use $P(\beta)$ for inference about β. Normally, however, $Q(\theta)$ will depend to some extent on β and therefore contain some residual information about β. Cox's justification for ignoring $Q(\theta)$, in the case of the proportional hazards model, is that this residual information is unavailable because β and ϕ are inextricably entangled in $Q(\theta)$. This type of argument has a long history (e.g., Kalbfleisch and Sprott, 1970).

To see how this factorisation is applied to the proportional-hazards setup, consider data of the type considered in Section 4.2. Thus, the distinct observed failure times are $t_1 < \cdots < t_m$, and the censoring times during $[t_l, t_{l+1})$ are t_{ls} $(s = 1, \ldots, s_l)$; also, assuming no ties, the individual who failed at time t_l has label i_l, and those censored during $[t_l, t_{l+1})$ have labels i_{ls} $(s = 1, \ldots, s_l)$. Let $C_l = \{t_{ls}, i_{ls} : s = 1, \ldots, s_l\}$ comprise the full record of censored cases during $[t_l, t_{l+1})$, that is, the censoring times and the individuals involved. Now take $A_l = \{i_l\}$ and $B_l = \{C_{l-1}, t_l\}$ in the above factorisation of $L(\theta)$. Then, $p_\theta(A_l \mid B_l, D_{l-1})$ is just the conditional probability of Section 4.2, and so $P(\beta)$ here is the corresponding partial likelihood.

The asymptotic properties of maximum partial likelihood estimators are generally similar to those of ordinary maximum likelihood estimators, as pointed out by Cox (1975). The *partial score function* is

$$\mathbf{U}(\beta) = \partial \log P(\beta)/\partial \beta = \sum_{l=1}^{m} \partial \log p_\beta(A_l \mid B_l, D_{l-1})/\partial \beta = \sum_{l=1}^{m} \mathbf{U}_l(\beta).$$

Under the usual regularity conditions we can differentiate the identity $\int p_\theta(A_l \mid B_l, D_{l-1}) dA_l = 1$ with respect to β under the integral sign (or summation sign in the case of discrete A_l, as it is in the case of the proportional hazards partial likelihood) to obtain

$$\mathrm{E}_\beta\{\partial \log p_\beta(A_l \mid B_l, D_{l-1})/\partial \beta \mid B_l, D_{l-1}\} = \mathrm{E}_\beta\{\mathbf{U}_l(\beta) \mid B_l, D_{l-1}\} = \mathbf{0},$$

where E_β indicates that the expectation is taken with respect to the same β value as in $p_\beta(.)$. In formal terms this shows that the $\mathbf{U}_l(\beta)$ form a sequence of martingale differences adapted to the filtration $\{B_l, D_{l-1}\}$; see Part IV. Hence, $\mathrm{E}_\beta\{\mathbf{U}_l(\beta)\} = \mathbf{0}$ and so $\mathrm{E}_\beta\{\mathbf{U}(\beta)\} = \mathbf{0}$ for each β: $\mathbf{U}(\beta) = \mathbf{0}$ is then an *unbiased estimating equation* for β.

Denote by β_0 the true value of β, and by E_0 expectation taken under β_0, and let $\mu(\beta) = \mathrm{E}_0\{\mathbf{U}(\beta)\}$. Suppose that, under β_0, (1) $\mathbf{U}(\beta)$ obeys a law of large numbers for each β, to the effect that it stabilises around $\mu(\beta)$ as $m \to \infty$, (2) $\mu(\beta)$ stays clear of $\mathbf{0}$ whenever $\beta \neq \beta_0$, and (3) either $\mathbf{U}(\beta)$ or $\mu(\beta)$ obeys some suitable continuity condition. Then, from the general theory of estimating equations (Crowder, 1986), the maximum partial likelihood estimator $\tilde{\beta}$ will be *consistent* for β_0. In particular, for the basic PH model with $\psi_i = \exp(\mathbf{x}_i^T \beta)$, $\log P(\beta)$ is a differentiable convex function, so conditions (2) and (3) will be within grasp.

The asymptotic distribution of $\tilde{\beta}$ can be investigated via the mean value theorem: we have, expanding about the true value β_0,

$$\mathbf{0} = \mathbf{U}(\tilde{\beta}) = \mathbf{U}(\beta_0) - \mathbf{V}(\beta_*)(\tilde{\beta} - \beta_0),$$

where $\mathbf{V}(\beta) = -\partial^2 \log P(\beta)/\partial \beta^2$ and β_* lies between β_0 and β. Rearranging, approximating $\mathbf{V}(\beta_*)$ by $\mathbf{V}(\beta_0)$, and assuming that $\mathbf{V}(\beta_0)$ stabilises toward its expected value as $n \to \infty$, we have

$$\tilde{\beta} - \beta_0 \approx \mathbf{M}^{-1}\mathbf{U}(\beta_0),$$

where $\mathbf{M} = \mathrm{E}_0\{\mathbf{V}(\beta_0)\}$ must be non-singular; $\mathbf{M}$ could be described as the expected partial information matrix. The asymptotic distribution of $\tilde{\beta} - \beta_0$ is thus that of $\mathbf{M}^{-1}\mathbf{U}(\beta_0)$.

Since the realised value of $\mathbf{U}_l$ is determined by $\{B_k, D_{k-1}\}$, when $l < k$ we have

$$\mathrm{E}_\beta\{\mathbf{U}_l(\beta)\mathbf{U}_k(\beta)^T \mid B_k, D_{k-1}\} = \mathbf{U}_l(\beta) \times \mathrm{E}_\beta\{\mathbf{U}_k(\beta)^T \mid B_k, D_{k-1}\}$$
$$= \mathbf{U}_l(\beta) \times \mathbf{0}^T = \mathbf{0},$$

so $\mathrm{E}_\beta\{\mathbf{U}_l(\beta)\mathbf{U}_k(\beta)^T\} = \mathbf{0}$, that is, the $\mathbf{U}_l(\beta)$ are uncorrelated; this is a standard property of any martingale difference sequence. Again, by differentiating the identity $\int p_\beta(A_l \mid B_l, D_{l-1})d\,A_l = 1$ a second time with respect to β under the integral sign, we obtain

$$\mathbf{0} = \mathrm{E}_\beta\{\partial^2 \log p_\beta(A_l \mid B_l, D_{l-1})/\partial\beta^2 + \mathbf{U}_l(\beta)\mathbf{U}_l(\beta)^T \mid B_l, D_{l-1}\},$$

that is,

$$\mathrm{E}_\beta\{\mathbf{V}_l(\beta) \mid B_l, D_{l-1}\} = \mathrm{E}_\beta\{\mathbf{U}_l(\beta)\mathbf{U}_l(\beta)^T \mid B_l, D_{l-1}\} = \mathrm{var}\{\mathbf{U}_l(\beta) \mid B_l, D_{l-1}\}.$$

Hence,

$$\mathbf{M} = \mathrm{E}_0\{\mathbf{V}(\beta_0)\} = \sum \mathrm{E}_0\{\mathbf{V}_l(\beta_0)\} = \sum \mathrm{E}_0\{\mathbf{U}_l(\beta_0)\mathbf{U}_l(\beta_0)^T\}$$
$$= \mathrm{E}_0\{\mathbf{U}(\beta_0)\mathbf{U}(\beta_0)^T\} = \mathrm{var}_0\{\mathbf{U}(\beta_0)\},$$

where the fact that the $\mathbf{U}_l(\beta_0)$ are uncorrelated under β_0 has been used. To summarize, $\mathbf{U}(\beta_0)$ is the sum of m terms $\mathbf{U}_l(\beta_0)$, uncorrelated under β_0, and has covariance matrix $\mathbf{M}$. Under a mild condition to prevent dominance of any one term in the sum, for example, that the variances of the $\mathbf{U}_l(\beta_0)$ are bounded uniformly over l, $\mathbf{M}^{-1/2}\mathbf{U}(\beta_0)$ will tend as $m \to \infty$ to a normal distribution with mean $\mathbf{0}$ and unit covariance matrix. It then follows from the above that $\tilde{\beta} \sim_d N(\beta_0, \mathbf{M}^{-1})$ and, in practice, $\mathbf{M}$ can be estimated as $\mathbf{V}(\tilde{\beta})$.

Appendix D

Numerical Optimisation of Functions

Suppose that we have a data set D and a model involving a parameter vector θ. Denote the likelihood function by $L(\theta)$. Maximum likelihood estimates are found by maximizing $L(\theta)$ over θ. In regular cases this can be done by differentiating $\log L$ with respect to θ to obtain the *score function*, a vector of length $\dim(\theta)$, and then the *likelihood equations* result from setting the score function equal to the zero vector. Sometimes, these equations can be solved explicitly for at least some θ-components, thus expressing a subset of the maximum likelihood estimates in terms of the rest and thus reducing the dimension of the maximisation problem.

For complicated likelihood functions it is often more convenient to apply a function optimisation routine, such as `optim` in R, directly, rather than differentiating analytically to obtain the likelihood equations and then solving them numerically. The direct scheme is obviously less trouble to implement and less error prone. It provides fewer of those golden opportunities for algebraic and program-coding slips that we all so eagerly grasp. It is also possibly more accurate numerically, provided that the sub-routine to compute $L(\theta)$ on call has been carefully coded to minimise cancellation errors and the like. This is because gradient-based algorithms compute derivatives by finite differencing, the operation being controlled to give values that might well be more accurate than one's own coding.

In using general optimisation algorithms, it is always worthwhile to take some care with the parametrisation. For instance, scale the parameters so that the effect on the log-likelihood of changing any of their values by, say, 0.1 is about the same. Even more importantly, give them some elbow room, for example, for a positive parameter α evaluate it in the code as $\exp(\alpha')$, and if α is constrained to lie between 0 and 1, re-parametrise in terms of $\alpha' = \log\{\alpha/(1-\alpha)\}$: now α' is free to roam the wide-open spaces. An alternative, brute-force method, is to return a huge value to `optim` when it proposes an invalid parameter set; one hopes that this will make it go and look elsewhere in the parameter space. Ross (1970) gave useful advice on this general topic.

To give a brief outline of the direct-optimisation approach, suppose that a general, twice-differentiable function $G(\theta)$ is to be minimised over θ; for maximum likelihood take $G(\theta) = -\log L(\theta)$. We begin with an initial estimate

θ_1 of the minimum point and then update it to $\theta_2 = \theta_1 + s_1 \mathbf{d}_1$, where $\mathbf{d}_1$ is the step direction vector and s_1 is the step length. The procedure is iterated, taking θ_2 as the new initial estimate and so on. A sequence $\{\theta_1, \theta_2, \theta_3, \ldots\}$ is thus generated that, we hope, will converge rapidly to the minimum point sought. Termination of the search will be suggested when the sequences $\{\theta_j\}$ and $\{G(\theta_j)\}$ have stopped moving significantly, and $\mathbf{g}_j$, denoting the gradient vector $\mathbf{G}' = dG/d\theta$ evaluated at θ_j, is near zero.

Since we wish to move downhill, thinking of $G(\theta)$ as a surface above the θ-plane, $\mathbf{d}_j$ should be closely related to $-\mathbf{g}_j$. In general, $\mathbf{d}_j$ is taken as $-\mathbf{A}_j \mathbf{g}_j$, where $\mathbf{A}_j$ is a matrix used to modify the step direction in some advantageous way. By Taylor expansion, and using $\theta_{j+1} = \theta_j - s_j \mathbf{A}_j \mathbf{g}_j$,

$$G(\theta_{j+1}) \approx G(\theta_j) + G'(\theta_j)^T (\theta_{j+1} - \theta_j) = G(\theta_j) - s_j \mathbf{g}_j^T \mathbf{A}_j \mathbf{g}_j,$$

so that $G(\theta_{j+1}) < G(\theta_j)$ if $\mathbf{A}_j$ is positive definite and $\mathbf{g}_j \neq \mathbf{0}$. The scaling factor s_j is used to adjust the step length to locate a minimum, or just to achieve a reduction in $G(\theta)$, along the step direction; it is determined by a linear search in this direction. If $\mathbf{A}_j$ is positive definite, then a reduction is ensured for small enough s_j, but a larger value of s_j might give a larger reduction and hence quicker minimisation. There are various standard choices for $\mathbf{A}_j$.

1. $\mathbf{A}_j = \mathbf{I}$ (unit matrix) yields the method of *steepest descents*. Unfortunately, over successive steps the search path can come to resemble hem-stitching, that is, a zig-zag, crossing and re-crossing the valley floor, approaching the minimum point painfully slowly. The problem here is one of re-tracing one's steps, that is, the search direction not breaking fresh ground, not moving into sub-spaces unspanned by previous search directions.

2. $\mathbf{A}_j = (\mathbf{G}_j'')^{-1}$, where $\mathbf{G}_j'' = \mathbf{G}''(\theta_j)$, yields *Newton's*, or the *Newton–Raphson*, method. This arises from the observation that, if θ_{j+1} is actually the minimum point, then

$$0 = \mathbf{G}'(\theta_{j+1}) \approx \mathbf{G}'(\theta_j) + \mathbf{G}_j''(\theta_{j+1} - \theta_j),$$

by Taylor expansion. Hence,

$$\theta_{j+1} \approx \theta_j - (\mathbf{G}_j'')^{-1} \mathbf{G}'(\theta_j) = \theta_j - \mathbf{A}_j \mathbf{g}_j.$$

If $G(\theta)$ is a positive definite quadratic function of θ, $\mathbf{G}''(\theta)$ is a constant matrix, the Taylor expansion is exact, and the minimum will be reached in one step from any starting point. However, for a non-quadratic function, if $\mathbf{G}_j''$ is not positive definite at θ_j this search direction might be totally useless. Even if $\mathbf{G}_j''$ is positive definite, its computation and inversion at each step is costly.

3. In the celebrated *Levenburg–Marquardt compromise* between Newton–Raphson and steepest descents, $\mathbf{A}_j$ is taken as $(\mathbf{G}_j'' + \lambda_j \mathbf{I})^{-1}$, where λ_j is a tuning constant. For sufficiently large λ_j, $\mathbf{G}_j'' + \lambda_j \mathbf{I}$ will be positive definite. Statisticians will hear an echo of this in *Ridge Regression*.

4. Another variant is *Fisher's scoring method* in which $\mathbf{G}(\theta)$ depends on sample values of random variables and $\mathbf{A}_j$ is taken as $\{E(\mathbf{G}_j'')\}^{-1}$. In certain special situations, such as when $G(\theta)$ is the log-likelihood function for an exponential family model, this can yield a worthwhile simplification (McCullagh and Nelder, 1989, Section 2.5).

5. *Conjugate direction* methods set out to cure the deficiency mentioned in (1) by making the search direction $\mathbf{d}_j$ orthogonal to the previous ones. For instance, we might modify the basic steepest-descents process by taking $\mathbf{d}_1 = -s_1\mathbf{g}_1$, then $\mathbf{d}_2 = -s_2\mathbf{g}_2 + t_1\mathbf{d}_1$, with t_1 chosen to make $\mathbf{d}_1^T\mathbf{d}_2 = 0$, that is, $t_1 = s_1\mathbf{d}_1^T\mathbf{g}_2/\mathbf{d}_1^T\mathbf{d}_1$, and so on. There are many cunning variants, each with its own scheme for generating the $\mathbf{d}_j$, and each with its own strengths and weaknesses.

6. In taking $-\mathbf{A}_j\mathbf{g}_j$ for the step direction, rather than just $-\mathbf{g}_j$, one is effectively altering the metric of the θ space. For a positive definite quadratic function, the contours of constant G are elliptical, and taking $\mathbf{d}_j \propto -(\mathbf{G}_j'')^{-1}\mathbf{g}_j$ effectively makes them circular, so that the steepest descent direction points straight at the minimum. Davidon (1959) coined the term *variable metric* for methods in which the metric-correcting transformation $\mathbf{A}_j$ is updated at each step. As the neighbourhood of the minimum is approached, $\mathbf{A}_j$ should tend to $(\mathbf{G}_j'')^{-1}$ since G will be approximately quadratic there, which is when Newton's method comes into its own. Fletcher and Powell (1963) developed Davidon's idea and, with a particular updating scheme for $\mathbf{A}_j$, the method found fame as the *DFP algorithm*. Much work has followed, and there is now a vast literature on sophisticated *quasi-Newton methods*. The book by Gill, Murray, and Wright (1981) gives a pretty comprehensive survey up to 1980.

In the so-called *orthogonal* case (Cox and Reid, 1987) the log-likelihood breaks down into a sum of contributions involving separate subsets of parameters:

$$G(\theta) = -\log L(\theta) = G^{(1)}(\theta^{(1)}) + \cdots + G^{(r)}(\theta^{(r)}),$$

say. Then $\mathbf{G}_j''$, or any of its variants here, will be block-diagonal. As a result, the r component steps in $\mathbf{d}_j$ will be computed independently. Thus, the minimisation will proceed almost as if the contributions were being dealt with separately; almost, because the determination of s_j, the step length adjustment, will introduce some degree of compromise. However, if storage is not a problem, minimisation of $G(\theta)$ in one go will be hardly any slower than giving each contribution special treatment.

There is an unspoken assumption in all of this that the value located by the optimisation algorithm is the maximum likelihood estimate. In practice, however, it is not uncommon for the likelihood function to have local maxima to which an algorithm can be easily drawn. Such problems can be addressed by more sophisticated methods such as *simulated annealing* (Brooks and Morgan,

1995), which are more likely to locate global maxima. However, the existence of multiple maxima can indicate that the asymptotic theory, on which the usefulness of maximum likelihood estimation relies, is not effective for the situation; for instance, the sample size might be too small or the likelihood might be non-regular. This is a real problem, and there are no simple, general solutions. (In theory, it is not a problem for the Bayesian approach.)

To summarise, powerful methods for function optimisation have been developed over many years by the professionals, expert numerical analysts. Algorithms for conjugate directions and variable metric methods are implemented in many computer packages. The availability of the function optim in R, for instance, has enabled statisticians to catch up with numerical methods dating from the early 1960s. An excellent general book on numerical methods, described in simple terms and giving actual computer routines to implement them, is that of Press et al. (1990).

Only gradient-based methods have been described here. There is a completely different class of search methods based on moving a simplex around the parameter space. Among these the *Nelder–Mead* version is celebrated. Whereas a quasi-Newton method such as *BFGS* is the hare, Nelder–Mead is the tortoise: the former is usually quicker but the latter is ploddingly robust.

References

Books

Andersen, P.K., Borgan, O., Gill, R.D., and Keiding, N. (1993) *Statistical Models Based on Counting Processes*. Springer-Verlag, New York.

Bedford, T., and Cooke, R.M. (2001) *Probabilistic Risk Analysis: Foundations and Methods*. Cambridge University Press, Cambridge.

Chiang, C.L. (1968) *Introduction to Stochastic Processes in Biostatics*. Wiley, New York.

Collett, D. (2003) *Modelling Survival Data in Medical Research, Second Edition*. Chapman & Hall/CRC Press, Boca Raton, FL.

Cook, R.J., and Lawless, J.F. (2007) *The Statistical Analysis of Recurrent Events*. Springer, New York.

Cox, D.R. (1962) *Renewal Theory*. Methuen, London.

Cox, D.R., and Oakes, D. (1984) *Analysis of Survival Data*. Chapman & Hall, London.

Crowder, M.J., Kimber, A.C., Smith, R.L., and Sweeting, T.J. (1991) *Statistical Analysis of Reliability Data*. Chapman & Hall, London.

David, H.A., and Moeschberger, M.L. (1978) *The Theory of Competing Risks*. Griffin, London.

Elandt-Johnson, R.C., and Johnson, N.L. (1980) *Survival Models and Data Analysis*. Wiley, New York.

Fleming, T.R., and Harrington, D.P. (1991) *Counting Processes and Survival Models*. Wiley, New York.

Groenboom, P., and Wellner, J.A. (1992) *Nonparametric Maximum Likelihood Estimators for Interval Censoring and Deconvolution*. Birkhauser, Boston.

Hougaard, P. (2000) *Analysis of Multivariate Survival Data*. Springer-Verlag, New York.

Kalbfleisch, J.D., and Prentice, R.L. (1980) *The Statistical Analysis of Failure Time Data*. Wiley, New York.

Kalbfleisch, J.D., and Prentice, R.L. (2002) *The Statistical Analysis of Failure Time Data, Second Edition*. Wiley, New York.

Kemeny, J.G., and Snell, J.L. (1960) *Finite Markov Chains*. Van Nostrand, Princeton, NJ.

King, J.R. (1971) *Probability Charts for Decision Making*. Industrial Press, New York.

Lawless, J.F. (1982) *Statistical Models and Methods for Lifetime Data*. Wiley, New York.

Lawless, J.F. (2003) *Statistical Models and Methods for Lifetime Data*, Second Edition. Wiley, New York.

Martz, H.F. and Waller, R.A. (1982) *Bayesian Reliability Analysis*. Wiley, New York.

McCullagh, P., and Nelder, J.A. (1989) *Generalized Linear Models*, Second Edition. Chapman & Hall/CRC Press, Boca Raton, FL.

Nelson, W.B. (1982) *Applied Life Data Analysis*. Wiley, New York.

Nelson, W.B. (1990) *Accelerated Testing: Statistical Models, Test Plans and Data Analysis*. Wiley, New York.

Sobczyk, K., and Spencer, Jr., B.F. (1992) *Random Fatigue: From Data to Theory*. Academic Press, New York.

Therneau, T.M., and Grambsch, P.M. (2000) *Modelling Survival Data: Extending the Cox Model*. Springer-Verlag, New York.

Williams, D. (1991) *Probability with Martingales*. Cambridge University Press, Cambridge.

Wolstenholme, L. (1998) *Reliability Modelling: a Statistical Approach*. Chapman & Hall/CRC Press, Boca Raton, FL.

Univariate Survival and Reliability

Aalen, O.O. (1975) *Statistical Inference for a Family of Counting Processes*. Ph.D. Thesis, University of California, Berkeley.

Aalen, O.O. (1995) Phase type distributions in survival analysis. *Scand. J. Statist.* 22, 447–463.

Albert, J.R.G., and Baxter, L.A. (1995) Applications of the EM algorithm to the analysis of life length data. *Appl. Statist.* 44, 323–348.

Allen, W.R. (1963) A note on the conditional probability of failure when hazards are proportional. *Operations Res.* 11, 658–659.

Armitage, P. (1959) The comparison of survival curves. *J. Roy. Statist. Soc.* A 122, 279–300.

Barlow, W.E., and Prentice, R.L. (1988) Residuals for relative risk regression. *Biometrika* 75, 65–74.

Boardman, T.J., and Kendell, P.J. (1970) Estimation in compound exponential failure models. *Technometrics* 12, 891–900.

Breslow, N., and Crowley, J. (1974) A large sample study of the life table and product limit estimates under random censorship. *Ann. Statist.* 2, 437–453.

Buckley, J., and James, I. (1979) Linear regression with censored data. *Biometrika* 66, 429–436.

Caroni, C., Crowder, M.J., and Kimber, A.C. (2010) Proportional hazards models with discrete frailty. *LIDA* 16, 374–384.

Cox, D.R. (1972) Regression models and life tables (with discussion). *J. Roy. Statist. Soc.* B 34, 187–220.

Cox, D.R. (1975) Partial likelihood. *Biometrika* 62, 269–276.

Crowder, M.J. (1990) On some nonregular tests for a modified Weibull distribution. *Biometrika* 77, 499–506.

Crowder, M.J. (1996) Some tests based on extreme values for a parametric survival model. *J. Roy. Statist. Soc.* 58, 417–424.

Crowder, M.J. (2000) Tests for a family of survival models based on extremes. In *Recent Advances in Reliability Theory*, Eds. N. Limnios and M. Nikulin, Birkhauser, Boston, 307–321.

Davison, A.C., and Snell, E.J. (1991) Residuals and diagnostics. In *Statistical Theory and Modelling: In Honour of Sir David Cox*, Eds. D.V. Hinkley, N. Reid, and E.J. Snell, Chapman & Hall, London, 83–106.

Desmond, A.F., and Chapman, G.R. (1993) Modelling task completion data with inverse Gaussian mixtures. *Applied Statistics* 42, 603–613.

Gompertz, B. (1825) On the nature of the function expressive of the law of human mortality. *Phil. Trans. Roy. Soc.* (London) 115, 513–583.

Grambsch, P.M., and Therneau, T.M. (1994) Proportional hazards tests and diagnostics based on weighted residuals. *Biometrika* 81, 515–526.

Hill, D.L., Saunders, R., and Purushottam, W.L. (1980) Maximum likelihood estimation for mixtures. *Can. J. Statist.* 8, 87–93.

Hougaard, P. (1984) Life table methods for heterogeneous populations: distributions describing the heterogeneity. *Biometrika* 71, 75–83.

Hougaard, P. (1986a) Survival models for heterogeneous populations derived from stable distributions. *Biometrika* 73, 387–396.

Huang, J. (1996) Efficient estimation for the proportional hazards model with interval censoring. *Ann. Statist.* 24, 540–568.

Huang, J., and Wellner, J.A. (1995) Asymptotic normality of the NPMLE of linear functionals for interval censored data: Case I. *Statist. Neerlandica* 49, 153–163.

Jewell, N.P. (1982) Mixtures of exponential distributions. *Ann. Statist.* 10, 479–484.

Kalbfleisch, J.D., and Prentice, R.L. (1973) Marginal likelihoods based on Cox's regression and life model. *Biometrika* 60, 267–279.

Kaplan, E.L., and Meier, P. (1958) Nonparametric estimation from incomplete observations. *J. Amer. Statist. Assoc.* 53, 457–481.

Kimber, A.C., and Hansford, A.R. (1993) A statistical analysis of batting in cricket. *J. Roy. Statist. Soc.* A 156, 443–455.

Laird, N. (1978) Nonparametric maximum likelihood estimation of a mixing distribution. *J. Amer. Statist. Assoc.* 73, 805–811.

Lawless, J.F. (1983) Statistical methods in reliability (with comments). *Technometrics* 25, 305–335.

Lawless, J.F., Crowder, M.J., and K.-A Lee (2009) Analysis of reliability and warranty claims in products with age and usage scales. *Technometrics* 51, 14–24.

Lesperance, M.L., and Kalbfleisch, J.D. (1992) An algorithm for computing the nonparametric MLE of a mixing distribution. *J. Amer. Statist. Assoc.* 87, 120–126.

Lindsay, B.G. (1983a) The geometry of mixture likelihoods: A general theory. *Ann. Statist.* 11, 86–94.

Lindsay, B.G. (1983b) The geometry of mixture likelihoods, Part II: The exponential family. *Ann. Statist.* 11, 783–792.

Maguluri, G., and Zhang, C.-H. (1994) Estimation in the mean residual lifetime regression model. *J. Roy. Statist. Soc.* 56, 477–489.

Makeham, W.M. (1860) On the law of mortality and the construction of annuity tables. *J. Inst. Actuaries* 8, 301–310.

McGilchrist, C.A., and Aisbett, C.W. (1991) Regression with frailty in survival analysis. *Biometrics* 47, 461–466.

Meier, P. (1975) Estimation of a distribution function from incomplete observations. In *Perspectives in Probability and Statistics*, Ed. J. Gani, Applied Probability Trust, Sheffield, 67–87.

Miller, R.G. (1976) Least squares with censored data. *Biometrika* 63, 449–464.

Nadas, A. (1970a) On proportional hazard functions. *Technometrics* 12, 413–416.

North, P., and Cormack, R.M. (1981) On Seber's method for estimating age-specific bird survival rates from ringing recoveries. *Biometrics* 37, 103–112.

Oakes, D. (1981) Survival times: Aspects of partial likelihood. *Int. Statist. Rev.* 49, 235–264.

Oakes, D., and Desu, T. (1990) A note on residual life. *Biometrika* 77, 409–410.

Peacock, S.T., Keller, A.Z., and Locke, N.J. (1985) Reliability predictions for diesel generators. *Reliability '85*, 4C/5/1–4C/5/12.

Peto, R. (1972) Contribution to discussion of paper by D.R. Cox. *J. Roy. Statist. Soc.* B 34, 205–207.

Peto, R. (1973) Experimental survival curves for interval-censored data. *Applied Statistics* 22, 86–91.

Peto, R., and Lee, P. (1973) Weibull distributions for continuous carcinogenesis experiments. *Biometrics* 29, 457–470.

Peto, R., and Peto, J. (1972) Asymptotically efficient rank invariant test procedures (with discussion). *J. Roy. Statist. Soc.* 135, 185–206.

Pike, M.C. (1966) A method of analysis of a certain class of experiments in carcinogenesis. *Biometrics* 22, 142–161.

Prentice, R.L., and El Shaarawi, A. (1973) A model for mortality rates and a test of fit for the Gompertz force of mortality. *Applied Statistics* 22, 301–314.

Prentice, R.L., and Breslow, N.E. (1978) Retrospective studies and failure time models. *Biometrika* 65, 153–158.

Schoenfeld, D. (1982) Partial residuals for the proportional hazards regression model. *Biometrika* 69, 239–241.

Simar, L. (1976) Maximum likelihood estimation of a compound Poisson process. *Ann. Statist.* 4, 1200–1209.

Singpurwalla, N.D., and Youngren, M.A. (1990) Models for dependent lifelengths induced by common environments. In *Topics in Statistical Dependence*, Eds. H.W. Block, A.R. Sampson, and T.H. Savits, Institute of Mathematical Statistics, Hayward, CA, 435–441.

Singpurwalla, N.D. (1995) Survival in dynamic environments. *Statistical Science* 10, 86–103.

Smith, R.L. (1991) Weibull regression models for reliability data. *Rel. Eng. System Safety* 34, 35–57.

Turnbull, B.W. (1974) Nonparametric estimation of a survivorship function with doubly censored data. *J. Amer. Statist. Assoc.* 69, 169–173.

Turnbull, B.W. (1976) The empirical distribution function with arbitrarily grouped, censored and truncated data. *J. Roy. Statist. Soc.* B 38, 290–295.

Vaupel, J.W., Manton, K.G., and Stallard, E. (1979) The impact of heterogeneity in individual frailty on the dynamics of mortality. *Demography* 16, 439–454.

Weinberg, C.R., and Gladen, B.C. (1986) The beta-geometric distribution applied to comparative fecundability studies. *Biometrics* 42, 547–560.

Whitmore, G.A., Crowder, M.J., and Lawless, J.F. (1998) Failure inference from a marker process based on a bivariate Wiener model. *Lifetime Data Analysis* 4, 229–251.

Wong, W.H. (1986) Theory of partial likelihood. *Ann. Statist.* 14, 88–123.

Multivariate Survival

Aalen, O.O., and Johansen, S. (1978) An empirical transition matrix for non-homogeneous Markov chains based on censored observations. *Scand. J. Statist.* 5, 141–150.

Akritas, M.G., and Van Heilegon, I. (2003) Estimation of bivariate and marginal distributions with censored data. *J. Roy. Statist. Soc.* 65, 457–471.

Ali, M.M., Mikhail, N.N., and Haq, M.S. (1978) A class of bivariate distributions including the bivariate logistic. *J. Multiv. Anal.* 8, 405–412.

Balka, J., Desmond, A.F., and McNicholas, P.D. (2009) Review and implementation of cure models based on first hitting times for Wiener processes. *Lifetime Data Analysis* 15, 147–176.

Baxter, L.A. (1994) Estimation from quasi-life tables. *Biometrika* 81, 567–577.

Bhattacharyya, G.K., and Fries, A. (1982) Fatigue-failure models: Birnbaum–Saunders vs. Gaussian. *IEEE Trans. Rel.* R-31, 439–441.

Birnbaum, Z.W., and Saunders, S.C. (1969) A new family of life distributions. *J. Appl. Prob.* 6, 319–327.

Cai, J. (1999) Hypothesis testing of hazard ratio parameters in marginal models for multivariate failure time data. *Lifetime Data Analysis* 5, 39–53.

Cai, J., and Kim, J. (2003) Nonparametric quantile estimation with correlated failure time data. *Lifetime Data Analysis* 9, 357–371.

Cai, J., and Prentice, R.L. (1997) Regression estimation using multivariate failure time data and a common baseline hazard function model. *Lifetime Data Analysis* 3, 197–213.

Chen, M.-C. and Bandeen-Roche, K. (2005) A diagnostic for association in bivariate survival models. *Lifetime Data Analysis* 11, 245–264.

Chiao, C.-H., and Hamada, M. (1995) Experiments with degradation data for improving reliability and for achieving robust reliability. *IIQP Research Report* RR-95-10, University of Waterloo, Canada.

Clayton, D. (1978) A model for association in bivariate life tables and its application in epidemiological studies of familial tendency in chronic disease incidence. *Biometrika* 65, 141–151.

Clayton, D., and Cuzick, J. (1985) Multivariate generalisations of the proportional hazards model (with discussion). *J. Roy. Statist. Assoc.* A 148, 82–117.

Cook, R.D., and Johnson, M.E. (1981) A family of distributions for modelling non-elliptically symmetric multivariate data. *J. Roy. Statist. Soc.* B 43, 210–218.

Cowan, R. (1987) A bivariate exponential distribution arising in random geometry. *Ann. Inst. Statist. Math.* 39, 103–111.

Cowling, B.J., Hutton, J.L., and Shaw, J.E.H. (2006) Joint modelling of event counts and survival times. *Applied Statistics* 55, 31–39.

Cox, D.R. (1962) *Renewal Theory.* Methuen, London.

Cox, D.R.(1999) Some remarks on failure times, surrogate markers, degradation, wear, and the quality of life. *Lifetime Data Analysis* 5, 307–314.

Crowder, M.J. (1985) A distributional model for repeated failure time measurements. *J. Roy. Statist. Soc.* B 47, 447–452.

Crowder, M.J. (1989) A multivariate distribution with Weibull connections. *J. Roy. Statist. Soc.* B 51, 93–107.

Crowder, M.J. (1991) A statistical approach to a deterioration process in reinforced concrete. *Applied Statistics* 40, 95–103.

Crowder, M.J. (1996) Keep timing the tablets: Statistical analysis of pill dissolution rates. *Applied Statistics* 45, 323–334.

Crowder, M.J. (1998) A multivariate model for repeated failure time measurements. *Scand. J. Statist.* 25, 53–67.

Crowder, M.J., and Kimber, A.C. (1997) A score test for the multivariate Burr and other Weibull mixture distributions. *Scand. J. Statist.* 24, 419–432.

Crowder, M.J., and Lawless, J.F. (2007) On a scheme for predictive maintenance. *European J. Oper. Res.* 176, 1713–1722.

Crowder, M.J., and Hand, D.J. (2005) On loss distributions from installment-repaid loans. *Lifetime Data Analysis* 11, 545–564.

Crowder, M.J., and Stephens, D.A. (2003) On the analysis of quasi-life tables. *Lifetime Data Analysis* 9, 345–355.

Crowder, M.J., Hand, D.J., and Krzanowski, W. (2007) On optimal intervention for customer lifetime value. *European J. Oper. Res.* 183, 1550–1559.

Cuzick, J. (1981) Boundary crossing probabilities for stationary Gaussian processes and Brownian motion. *Trans. Amer. Math. Soc.* 263, 469–492.

Doksum, K.A., and Hoyland, A. (1992) Models for variable-stress accelerated life testing experiments based on Wiener processes and the inverse Gaussian distribution. *Technometrics* 34, 74–82.

Drezner, Z., and Wesolowsky, G.O. (1989) On the computation of the bivariate normal integral. *J. Statist. Comput. Simul.* 35, 101–107.

Durbin, J. (1985) The first-passage density of a continuous Gaussian process to a general boundary. *J. Appl. Prob.* 22, 99–122.

Fan, J., Hsu, L., and Prentice, R.L. (2000) Dependence estimation over a finite bivariate failure time region. *Lifetime Data Analysis* 6, 343–355.

Farewell, D., and Henderson, R. (2010) Longitudinal perspectives on event history analysis. *Lifetime Data Analysis* 16, 102–117.

Freund, J.E. (1961) A bivariate extension of the exponential distribution. *J. Amer. Statist. Assoc.* 56, 971–977.

Genest, C., and MacKay, J. (1986) The joy of copulas: bivariate distributions with uniform marginals. *American Statistician* 40, 280–283.

Genest, C., and Rivest, L-P. (1993) Statistical inference for bivariate Archimedean copulas. *J. Amer. Statist. Assoc.* 88, 1034–1043.

Gesch, C.B., Hammond, S.M., Hampson, S.E., Eves, A., and Crowder, M.J. (2002) Influence of supplementary vitamins, minerals and essential fatty acids on the antisocial behaviour of young adult prisoners. *Brit. J. Psychiatry* 181, 22–28.

Ghurye, S.G. (1987) Some multivariate lifetime distributions. *Adv. Appl. Prob.* 19, 138–155.

Giampieri, G., Davis, M., and Crowder, M.J. (2005) Analysis of default data using hidden Markov chains. *Quantitative Finance* 5, 27–34.

Grambsch, P.M., and Therneau, T.M. (1994) Proportional hazards tests and diagnostics based on weighted residuals. *Biometrika* 81, 515–526.

Gray, R.J. (2003) Weighted estimating equations for linear regression analysis of clustered failure time data. *Lifetime Data Analysis* 9, 123–138.

Gumbel, E.J. (1960) Bivariate exponential distributions. *J. Amer. Statist. Assoc.* 55, 698–707.

Hanley, J.A., and Parnes, M.N. (1983) Nonparametric estimation of a multivariate distribution in the presence of censoring. *Biometrics* 39, 129–139.

Hougaard, P. (1986b) A class of multivariate failure time distributions. *Biometrika* 73, 671–678.

Hougaard, P. (1987) Modelling multivariate survival. *Scand. J. Statist* 14, 291–304.

Hougaard, P. (1999) Multi-state models: A review. *Lifetime Data Analysis* 5, 239–264.

Johnson, N.L., and Kotz, S. (1972) *Distributions in Statistics: Continuous Multivariate Distributions.* Wiley, New York.

Jones, M.P., and Yoo, B. (2005) Linear signed-rank tests for paired survival data subject to a common censoring time. *Lifetime Data Analysis* 11, 351–365.

Jung, S-H., and Jeong, J-H. (2003) Rank tests for clustered survival data. *Lifetime Data Analysis* 9, 21–33.

Karim, Md. R., Yamamoto, W., and Suzuki, K. (2001) Statistical analysis of marginal count failure data. *Lifetime Data Analysis* 7, 173–186.

Karimi, A.R., Ramachandran, K., Buenfeld, N., and Crowder, M.J. (2005) Probabilistic analysis of reinforcement corrosion with spatial variability. In *Safety and Reliability of Engineering Systems and Structures, Proceedings of the 9th Conference on Structural Safety and Reliability*, Eds. G. Augusti, G.I. Schueller, and M. Ciampoli, Rome, Italy.

Kotz, S., and Singpurwalla, N.D. (1999) On a bivariate distribution with exponential marginals. *Scand. J. Statist.* 26, 451–464.

Lawless, J.F., and Crowder, M.J. (2004) Covariates and random effects in a gamma process model with application to degradation and failure. *Lifetime Data Analysis* 10, 213–227.

Lawrance, A.J., and Lewis, P.A.W (1983) Simple dependent pairs of exponential and uniform random variables. *Operations Research* 31, 1179–1197.

Lee, M-L.T., and Whitmore, G.A. (2006) Threshold regression for survival analysis: Modeling event times by a stochastic process reaching a boundary. *Statist. Science* 21, 501–513.

Lehmann, E.L. (1966) Some concepts of dependence. *Ann. Math. Statist.* 37, 1137–1153.

Li, Q.H., and Lagakos, S.W. (2004) Comparisons of test statistics arising from marginal analyses of multivariate survival data. *Lifetime Data Analysis* 10, 389–405.

Liang, K-Y., Self, S.G., and Chang, Y-C. (1993) Modelling marginal hazards in multivariate failure time data. *J. Roy. Statist. Soc.* B 55, 441–453.

Liang, K-Y., Self, S.G., Bandeen-Roche, K.J., and Zeger, S.L. (1995) Some recent developments for regression analysis of multivariate failure time data. *Lifetime Data Analysis* 1, 403–415.

Lu, C.J., and Meeker, W.Q. (1993) Using degradation measurements to estimate a time-to-failure distribution. *Technometrics* 35, 161–174.

Lu, W. (2007) Tests of independence for censored bivariate failure time data. *Lifetime Data Analysis* 13, 75–90.

Lu, S-E., and Wang, M-C. (2005) Marginal analysis for clustered failure time data. *Lifetime Data Analysis* 11, 61–79.

Marshall, A.W., and Olkin, I. (1967a) A generalized bivariate exponential distribution. *J. Appl. Prob.* 4, 291–302.

Marshall, A.W., and Olkin, I. (1967b) A multivariate exponential distribution. *J. Amer. Statist. Assoc.* 62, 30–44.

Marshall, A.W., and Olkin, I. (1988) Families of multivariate distributions. *J. Amer. Statist. Assoc.* 83, 834–841.

McGilchrist, C.A., and Aisbett, C.W. (1991) Regression with frailty in survival analysis. *Biometrics* 47, 461–466.

Molenberghs, G., Verbeke, G., and Demetrio, C.G.B. (2007) An extended random-effects approach to modeling repeated, overdispersed count data. *Lifetime Data Analysis* 13, 513–531.

Oakes, D. (1982) A model for association in bivariate survival data. *J. Roy. Statist. Soc.* B 44, 414–422.

Oakes, D. (1986) Semiparametric inference in a model for association in bivariate survival data. *Biometrika* 73, 353–361.

Oakes, D. (1989) Bivariate survival models induced by frailties. *J. Amer. Statist. Assoc.* 84, 487–493.

Park, C., and Padgett, W.J. (2005) Accelerated degradation models for failure based on geoemetric Brownian motion and gamma processes. *Lifetime Data Analysis* 11, 511–527.

Peng, Y., Taylor, J.M.G., and Yu, B. (2007) A marginal regression model for multivariate failure time data with a surviving fraction. *Lifetime Data Analysis* 13, 351–369.

Petersen, J.H. (1998) An additive frailty model for correlated life times. *Biometrics* 54, 646–661.

Pons, O. (1986) A test of independence between two censored survival times. *Scand. J. Statist.* 13, 173–185.

Quale, C.M., and Van der Laan, M.J. (2000) Inference with bivariate truncated data. *Lifetime Data Analysis* 6 391–408.

Raftery, A.E. (1984) A continuous multivariate exponential distribution. *Comm. Statist. Theory Meth.* 13, 947–965.

Raftery, A.E. (1985) Some properties of a new continuous bivariate exponential distribution. *Statistics and Decisions, Supplement Issue 2*, 53–58.

Robinson, M.E., and Crowder, M.J. (2000) Bayesian methods for a growth-curve degradation model with repeated measures. *Lifetime Data Analysis* 6, 357–374.

Scholtens, D., and Betensky, R.A. (2006) A computationally simple bivariate survival estimator for efficacy and safety. *Lifetime Data Analysis* 12, 365–387.

Segal, M.R., Neuhaus, J.M., and James, I.R. (1997) Dependence estimation for marginal models of multivariate survival data. *Lifetime Data Analysis* 3, 251–268.

Shaked, M. (1982) A general theory of some positive dependence notions. *J. Multiv. Anal.* 12, 199–218.

Shih, J.H. (1998) Modeling multivariate discrete failure time data. *Biometrics* 54, 1115–1128.

Singpurwalla, N.D., and Youngren, M.A. (1993). Multivariate distributions induced by dynamic environments. *Scand. J. Statist.* 20, 251–261.

Sobczyk, K. (1986) Modelling random fatigue crack growth. *Eng. Fract. Mech.* 24, 609–623.

Spencer, B.F., and Tang, J. (1988) A Markov process model for fatigue crack growth. *J. Eng. Mech.* 114, 2134–2157.

Suzuki, K., Yamamoto, W., Karim, Md.R., and Wang, L. (2000) Data analysis based on warranty database. In *Recent Advances in Reliability Theory*, Eds. N. Limnios and M. Nikulin, Birkhauser, Boston, 213–227.

Takahasi, K. (1965) Note on the multivariate Burr's distribution. *Ann. Inst. Statist. Math.* 17, 257–260.

Tang, J., and Spencer, Jr., B.F. (1989) Reliability solution for the stochastic fatigue crack growth problem. *Eng. Fract. Mech.* 34, 419–433.

Tortorella, M. (1996) Life estimation from pooled discrete renewal counts. In *Lifetime Data: Models in Reliability and Survival Analysis*. Ed. N.P. Jewell, Kluwer Academic Publishers, 331–338.

Tseng, S.T., Hamada, M.S., and Chiao, C.H. (1994) Using degradation data from a fractional factorial experiment to improve fluorescent lamp reliability. *IIQP Research Report* RR-94-05, University of Waterloo, Canada.

Tsiatis, A.A., and Davidian M. (2004) Joint modelling of longitudinal and time-to-event data: an overview. *Statist. Sci.* 14, 793–818.

Virkler, D.A., Hillberry, B.M., and Goel, P.K. (1979) The statistical nature of fatigue crack propagation. *J. Eng. Mat. and Technology*, ASME 101, 148–153.

Walker, S.G., and Stephens, D. (1999) A multivariate family of distributions on $(0, \infty)^p$. *Biometrika* 86, 703–709.

Whitmore, G.A. (1995) Estimating degradation by a Wiener diffusion process subject to measurement error. *Lifetime Data Analysis* 1, 307–319.

Yilmaz, Y.E., and Lawless, J.F. (2011) Likelihood ratio procedures and tests of fit in parametric and semiparametric copula models with censored data. *Lifetime Data Analysis* 17, 386–408.

Yin, G., and Cai, J. (2004) Additive hazards model with multivariate failure time data. *Biometrika* 91 801–818.

Zeng, D., and Cai, J. (2010) Additive transformation models for clustered failure time data. *Lifetime Data Analysis* 16, 333–352.

Zhang, S., Zhang, Y., Chaloner, K., and Stapleton, J.T. (2010) A copula model for bivariate hybrid censored survival data with application to the MACS study. *Lifetime Data Analysis* 16, 231–249.

Zucchini, W., and MacDonald, I.L. (2009) *Hidden Markov Models for Time Series: An Introduction Using R.* Chapman & Hall/CRC Press, Boca Raton, FL.

Competing Risks

Aalen, O.O. (1976) Nonparametric inference in connection with multiple decrement models. *Scand. J. Statist.* 3, 15–27.

Abbring, J.H., and van den Berg, G.J. (2003) The identifiability of the mixed proportional hazards competing risks model. *J. Roy. Statist. Soc.* B 65, 701–710.

Altshuler, B. (1970) Theory for the measurement of competing risks in animal experiments. *Math. Biosci.* 6, 1–11.

Andersen, J., Goetghebeur, E., and Ryan, L. (1996) Missing cause of death information in the analysis of survival data. *Statist. Med.* 15, 1191–2201.

Anderson, T.W., and Ghurye, S.G. (1977) Identification of parameters by the distribution of a maximum random variable. *J. Roy. Statist. Soc.* B 39, 337–342.

Arjas, E., and Greenwood, P. (1981) Competing risks and independent minima: a marked point process approach. *Adv. Appl. Prob.* 13, 669–680.

Arjas, E., and Haara, P. (1984) A marked point process approach to censored failure data with complicated covariates. *Scand. J. Statist.* 11, 193–209.

Arnold, B., and Brockett, P. (1983) Identifiability for dependent multiple decrement/competing risk models. *Scand. Actuar. J.* 10, 117–127.

Bagai, I., Deshpande, J.V., and Kocher, S.C. (1989a) A distribution-free test for the equality of failure rates due to two competing risks. *Comm. Statist. Theor. Meth.* 18, 107–120.

Bagai, I., Deshpande, J.V., and Kocher, S.C. (1989b) Distribution-free tests for stochastic ordering in the competing risks model. *Biometrika* 76, 775–781.

Bancroft, G.A., and Dunsmore, I.R. (1976) Predictive distributions in life tests under competing causes of failure. *Biometrika* 63, 195–217.

Barlow, R.E., and Proschan, F. (1975) Importance of system components and fault tree events. *Stoch. Proc. App.* 3, 153–173.

Basu, A.P. (1981) Identifiability problems in the theory of competing and complementary risks: A survey. In *Statistical Distributions in Scientific Work*, Eds. C. Taillie, G.P. Patil, and B. Baldesari, Reidel Publ. Co., Dortrecht, Holland, 335–348.

Basu, A.P., and Ghosh J.K. (1978) Identifiability of the multinormal and other distributions under competing risks model. *J. Multiv. Anal.* 8, 413–429.

Basu, A.P., and Ghosh J.K. (1980) Identifiability of distributions under competing risks and complementary risks model. *Comm. Statist.—Theory and Meth.* A9, 1515–1525.

Basu, A.P., and Klein, J.P. (1982) Some recent results in competing risks theory. In *Survival Analysis*, Vol 2, Eds. J. Crowley and R.A. Johnson, IMS Monograph Series, Hayward, CA.

Beck, G. (1979) Stochastic survival models with competing risks and covariates. *Biometrics* 35, 427–438.

Berkson, J., and Elveback, L. (1960) Competing exponential risks, with particular reference to the study of smoking and lung cancer. *J. Amer. Statist. Assoc.* 55, 415–428.

Berman, S.M. (1963) Note on extreme values, competing risks and semi-Markov processes. *Ann. Math. Statist.* 34, 1104–1106.

Bernoulli, D. (1760, 1765) Essai d'une nouvelle analyse de la mortalite causee par la petite Verole, et des avantages de l'inoculation pour la prevenir. *Mem. de l'Academie Royale de Science* 1760, 1–45.

Beyersmann, J., Dettenkofer, M., Bertz, H., and Schumacher, M. (2007) A competing risks analysis of bloodstream infection after stem-cell transplantation using sub-distribution hazards and cause-specific hazards. *Statist. Med.* 26, 5360–5369.

Block, H.W., and Basu, A.P. (1974) A continuous bivariate exponential distribution. *J. Amer. Statist. Assoc.* 69, 1031–1037.

Boag, J.W. (1949) Maximum likelihood estimates of the proportion of patients cured by cancer therapy. *J. R. Statist. Soc.* B11, 15–44.

Bond, S.J., and Shaw, J.E.H. (2006) Bounds on the covariate-time transformation for competing-risks survival analysis. *Lifetime Data Analysis* 12, 285–303.

Carling, K., and Jacobson, T. (1995) Modeling unemployment duration in a dependent competing risks framework: Identification and estimation. *Lifetime Data Analysis* 1, 111–122.

Carriere, K.C., and Kochar, S.C. (2000) Comparing sub-survival functions in a competing risks model. *Lifetime Data Analysis* 6, 85–97.

Chiang, C.L. (1961a) A stochastic study of the life table and its applications: III. The follow-up study with the consideration of competing risks. *Biometrics* 17, 57–78.

Chiang, C.L. (1961b) On the probability of death from specific causes in the presence of competing risks. *Proc. 4th Berk. Symp.* 4, 169–180.

Chiang, C.L. (1964) A stochastic model of competing risks of illness and competing risks of death. In *Stochastic Models in Medicine and Biology*, Ed. J. Gurland, University Wisconsin Press, Madison, 323–354.

Chiang, C.L. (1970) Competing risks and conditional probabilities. *Biometrics* 26, 767–776.

Cornfield, J. (1957) The estimation of the probability of developing a disease in the presence of competing risks. *Amer. J. Public Health* 47, 601–607.

Couper, D., and Pepe, M.S. (1997) Modelling prevalence of a condition: chronic graft-versus-host disease after bone marrow transplantation. *Statist. Med.* 16, 1551–1571.

Cox, D.R. (1959) The analysis of exponentially distributed lifetimes with two types of failure. *J. Roy. Statist. Soc.* B 21, 411–421.

Craiu, R.V., and Reiser, B. (2006) Inference for the dependent competing risks model with masked causes of failure. *Lifetime Data Analysis* 12, 21–33.

Crowder, M.J. (1991) On the identifiability crisis in competing risks analysis. *Scand. J. Statist.* 18, 223–233.

Crowder, M.J. (1994) Identifiability crises in competing risks. *Int. Statist. Rev.* 62, 379–391.

Crowder, M.J. (1995) Analysis of fracture toughness data. *Lifetime Data Analysis* 1, 59–71.

Crowder, M.J. (1996) On assessing independence of competing risks when failure times are discrete. *Lifetime Data Analysis* 2, 195–209.

Crowder, M.J. (1997) A test for independence of competing risks with discrete failure times. *Lifetime Data Analysis* 3, 215–223.

Crowder, M.J. (2000) Characterisations of competing risks in terms of independent-risks proxy models. *Scand. J. Statist.* 27, 57–64.

Cutler, S., and Ederer, F. (1958) Maximum utilization of the life table method in analyzing survival. *J. Cron. Dis.* 8, 699–712.

D'Alembert, J. (1761) Sur l'application du Calcul des Probabilites a l'inoculation de la petite Verole. *Opuscules* II, 26–95.

David, H.A. (1957) Estimation of means of normal populations from observed minima. *Biometrika* 44, 282–286.

David, H.A. (1970) On Chiang's proportionality assumption in the theory of competing risks. *Biometrics* 26, 336–339.

David, H.A. (1974) Parametric approaches to the theory of competing risks. In *Reliability and Biometry: Statistical Analysis of Lifelength*, Eds. F. Proschan and R.J. Serfling, Society for Industrial and Applied Mathematics, Philadelphia, PA, 275–290.

Davis, T.P., and Lawrance, A.J. (1989) The likelihood for competing risk survival analysis. *Scand. J. Statist.* 16, 23–28.

Desu, M.M., and Narula, S.C. (1977) Reliability estimation under competing causes of failure. In *The Theory and Applications of Reliability*, Vol. 2, Eds. C.P. Tsokos and I.N. Shimi, Academic Press, New York, 471–481.

Dewanji, A. (1992) A note on a test for competing risks with missing failure type. *Biometrika* 79, 855–857.

Dewanji, A., and Sengupta, D. (2003) Estimation of competing risks with general missing pattern in failure types. *Biometrics* 59, 1063–1070.

Dinse, G.E. (1982) Nonparametric estimation for partially complete time and type of failure. *Biometrics* 38, 417–431.

Dinse, G.E. (1986) Nonparametric prevalence and mortality estimators for animal experiments with incomplete cause of death. *J. Amer. Statist. Assoc.* 81, 328–336.

Dinse, G.E. (1988) Estimating tumour incidence rates in animal carcinogeneity experiments. *Biometrics* 44, 405–415.

DiRienzo, A.G. (2003) Nonparametric comparison of two survival-time distributions in the presence of dependent censoring. *Biometrics* 59, 497–504.

Dixon, S.N., Darlington, G.A., and Desmond, A.F. (2011) A competing risks model for correlated data based on the subdistribution hazard. *Lifetime Data Analysis* 17, 473–495.

Doganaksoy, N. (1991) Interval estimation from censored and masked failure data. *IEEE Trans. Reliab.* 40, 280–285.

Ebrahimi, N. (1996) The effects of misclassification of the actual cause of death in competing risks analysis. *Statist. Med.* 15, 1557–1566.

Elandt-Johnson, R. (1976) Conditional failure time distributions under competing risk theory with dependent failure times and proportional hazard rates. *Scand. Actuar. J.* 59, 37–51.

Elveback, L. (1958) Estimation of survivorship in chronic disease: the actuarial method. *J. Amer. Statist. Assoc.* 53, 420–440.

Farragi, D., and Korn, E.L. (1996) Competing risks with frailty models when treatment affects only one failure type. *Biometrika* 83, 467–471.

Faulkner, J.E., and McHugh, R.B. (1972) Bias in observable cancer age and life-time of mice subject to spontaneous mammary carcinogenesis. *Biometrics* 28, 489–498.

Finch, P.D. (1977) On the crude analysis of survivorship data. *Austral. J. Statist.* 19, 1–21.

Fine, J.P., and Gray, R.J. (1999) A proportional hazards model for the subdistribution of a competing risk. *J. Amer. Statist. Assoc.* 94, 496–509.

Fisher, L., and Kanarek, P. (1974) Presenting censored survival data when censoring and survival times may not be independent. In *Reliability and Biometry: Statistical Analysis of Lifelength*, Eds. F. Proschan and R.J. Serfling, Society for Industrial and Applied Mathematics, Philadelphia, PA, 303–326.

Fix, E., and Neyman, J. (1951) A simple stochastic model of recovery, relapse, death and loss of patients. *Human Biology* 23, 205–241.

Flinn, C.J., and Heckman, J.J. (1982) New methods for analyzing structural models of labor force dynamics. *J. Econometrics* 18, 115–168.

Flinn, C.J., and Heckman, J.J. (1983a) Are unemployment and out of the labor force behaviourally distinct labor force states? *J. Labor Econ.* 1, 28–42.

Flinn, C.J., and Heckman, J.J. (1983b) The likelihood function for the multistate-multi-episode model in "models for the analysis of labor force dynamics." In *Advances in Econometrics*, Vol. 3, Eds. R. Bassman and G. Rhodes, JAI Press, Greenwich, CT, 115–168.

Gail, M. (1975) A review and critique of some models used in competing risks analysis. *Biometrics* 31, 209–222.

Gail, M., Lubin, J.H., and Rubenstein, L.V. (1981) Likelihood calculations for matched case-control studies and survival studies with tied death times. *Biometrika* 68, 703–707.

Gasemyr, J., and Natvig, B. (1994) Bayesian estimation of component lifetimes based on autopsy data. *Statistical Research Report No. 8*, Institute of Mathematics, University of Oslo.

Gaynor, J.J., Feuer, E.J., Tan, C.C., Wu, D.H., Little, C.R., Straus, D.J., Clarkson, B.D., and Brennan, M.F. (1993) On the use of cause-specific failure and conditional failure probabilities: Examples from clinical oncology data. *J. Amer. Statist. Assoc.* 88, 400–409.

Goetghebeur, E., and Ryan, L. (1990) A modified log rank test for competing risks with missing failure type. *Biometrika* 77, 207–211.

Goetghebeur, E., and Ryan, L. (1995) Analysis of competing risks survival data when some failure types are missing. *Biometrika* 82, 821–833.

Grambauer, N., Schumacher, M., and Beyersmann, J. (2010) Proportional subdistribution hazards modeling offers a summary analysis, even if misspecified. *Statist. Med.* 29, 875–884.

Greenwood, M. (1926) The natural duration of cancer. In *Reports on Public Health and Medical Subjects*, No. 33, 1–26, HMSO, London.

Greville, T.N.E. (1948) Mortality tables analyzed by cause of death. *Record, Amer. Inst. Actuaries* 37, 283–294.

Greville, T.N.E. (1954) On the formula for the L-function in a special mortality table eliminating a given cause of death. *Trans. Soc. Act.* 6, 1–5.

Gross, A.J. (1970) Minimization of misclassification of component failure in a two-component system. *IEEE Trans. Reliab.* 19, 120–122.

Gross, A.J. (1973) A competing risk model: a one organ subsystem plus a two organ subsystem. *IEEE Trans. Reliability* 22, 24–27.

Gross, A.J., Clark, V.A., and Liu, V. (1971) Estimation of survival parameters when one of two organs must function for survival. *Biometrics* 27, 369–377.

Guess, F.M., Usher, J.S., and Hodgson, T.J. (1991) Estimating system and component reliabilities under partial information on cause of failure. *J. Statist. Plan. Inf.* 29, 75–85.

Guttman, I., Lin, D.K.J., Reiser, B., and Usher, J.S. (1995) Dependent masking and system life data: Bayesian inference for two-component systems. *Lifetime Data Anal.* 1, 87–100.

Heckman, J.J., and Honore, B.E. (1989) The identifiability of the competing risks model. *Biometrika* 76, 325–330.

Herman, R.J., and Patell, R.K.N. (1971) Maximum likelihood estimation for multi-risk model. *Technometrics* 13, 385–396.

Hoel, D.G. (1972) A representation of mortality data by competing risks. *Biometrics* 28, 475–488.

Holt, J.D. (1978) Competing risk analyses with special reference to matched pair experiments. *Biometrika* 65, 159–165.

Horrocks, J., and Thompson, M.E. (2012) Modelling event times with multiple outcomes using the Wiener process with drift. *Lifetime Data Analysis* 10, 29–49.

Huang, X., Li, G., Elashoff, R.M., and Pan, J. (2011) A general joint model for longitudinal measurements and competing risks survival data with heterogeneous random effects. *Lifetime Data Analysis* 17, 80–100.

Jeong, J-H., and Fine, J.P. (2006) Direct parametric inference for the cumulative incidence function. *Applied Statist.* 55, 1–14.

Kalbfleisch, J.D., and Mackay, R.J. (1979) On constant-sum models for censored survival data. *Biometrika* 66, 87–90.

Karia, S.R., and Deshpande, J.V. (1997) *Bounds for Hazard Gradients and Rates in the Competing Risks Set Up.* Technical Report 285, Department of Statistics, University of Michigan, Ann Arbor.

Karn, M.N. (1931) An inquiry into various death rates and the comparative influence of certain diseases on the duration of life. *Ann. Eugenics* 4, 279–326.

Karn M.N. (1933) A further study of methods of constructing life tables when certain causes of death are eliminated. *Biometrika* 25, 91.

Kay, R. (1986) Treatment effects in competing risks analysis of prostate cancer data. *Biometrics* 42, 203–211.

Keyfitz, N., Preston, S.H., and Schoen, R. (1972) Inferring probabilities from rates: extension to multiple decrement. *Skand. Aktuar.* 55, 1–13.

Kimball, A.W. (1958) Disease incidence estimation in populations subject to multiple causes of death. *Bull. Int. Statist. Inst.* 36, 193–204.

Kimball, A.W. (1969) Models for the estimation of competing risks from grouped data. *Biometrics* 25, 329–337.

Kimball, A.W. (1971) Model I vs. Model II in competing risk theory. *Biometrics* 27, 462–465.

Klein, J.P., and Andersen, P.K. (2005) Regression modeling of competing risks data based on pseudovalues of the cumulative incidence function. *Biometrics* 61, 223–229.

Klein, J.P., and Basu, A.P. (1981) Weibull accelerated life tests when there are competing causes of failure. *Comm. Statist. Theor. Meth.* A 10, 2073–2100.

Klein, J.P., and Moeschberger, M.L. (1988) Bounds on net survival probabilities for dependent competing risks. *Biometrics* 44, 529–538.

Kochar, S.C., and Proschan, F. (1991) Independence of time and cause of failure in the multiple dependent competing risks model. *Statistica Sinica* 1, 295–299.

Kochar, S.C., Lam, K.F., and Yip, P.S.F. (2002) Generalized supremum tests for the equality of cause specific hazard rates. *Lifetime Data Analysis* 8, 277–288.

Kodell, R.L., and Chen, J.J. (1987) Handling cause of death in equivocal cases using the EM algorithm. With rejoinder. *Comm. Statist.* A 16, 2565–2585.

Krall, J.M., and Hickman, J.C. (1970) Adjusting multiple decrement tables. *Trans. Soc. Act.* 22, 163–179.

Kulathinal, S.B., and Gasbarra, D. (2002) Testing equality of cause-specific hazard rates corresponding to m competing risks among K groups. *Lifetime Data Analysis* 8, 147–161.

Lagakos, S.W. (1978) A covariate model for partially censored data subject to competing causes of failure. *Appl. Statist.* 27, 235–241.

Lagakos, S.W., and Louis, T.A. (1988) Use of tumour lethality to interpret tumorigenicity experiments lacking cause of death data. *Appl. Statist.* 37, 169–179.

Lagakos, S.W., and Williams, J.S. (1978) Models for censored survival analysis: a cone class of variable-sum models. *Biometrika* 65, 181–189.

Langberg, N.A., and Proschan, F. (1979) A reliability growth model involving dependent components. *Ann. Prob.* 7, 1082–1087.

Langberg, N., Proschan, F., and Quinzi, A.J. (1977) Converting dependent models into independent ones, with applications in reliability. In *The Theory and Applications of Reliability*, Vol. 1, Eds. C.P. Tsokos and I.N. Shimi, Academic Press, New York, 259–276.

Langberg, N., Proschan, F., and Quinzi, A.J. (1978) Converting dependent models into independent ones, preserving essential features. *Ann. Prob.* 6, 174–181.

Langberg, N., Proschan, F., and Quinzi, A.J. (1981) Estimating dependent life lengths, with applications to the theory of competing risks. *Ann. Statist.* 9, 157–167.

Langberg, N., and Shaked, M. (1982) On the identifiability of multivariate life distribution functions. *Ann. Prob.* 10, 773–779.

Larson, M.G., and Dinse, G.E. (1985) A mixture model for the regression analysis of competing risks data. *Appl. Statist.* 34, 201–211.

Lee, L., and Thompson, Jr., W.A. (1974) Results on failure time and pattern for the series system. In *Reliability and Biometry: Statistical Analysis of Lifelength*, Eds. F. Proschan and R.J. Serfling, Society for Industrial and Applied Mathematics, Philadelphia, PA, 291–302.

Lin, D.K.J., and Guess, F.M. (1994) System life data analysis with dependent partial knowledge on the exact cause of system failure. *Microelec. and Reliab.* 34, 535–544.

Lindkvist, H., and Belyaev, Y. (1998) A class of non-parametric tests in the competing risks model for comparing two samples. *Scand. J. Statist.* 25, 143–150.

Lu, K., and Tsiatis, A.A. (2001) Multiple imputation methods for estimating regression coefficients in the competing risks model with missing cause of failure. *Biometrics* 57, 1191–1197.

Lu, K., and Tsiatis, A.A. (2005) Comparison between two partial likelihood approaches for the competing risks model with missing cause of failure. *Lifetime Data Analysis* 11, 29–40.

Lu, W., and Peng, L. (2008) Semiparametric analysis of mixture regression models with competing risks data. *Lifetime Data Analysis* 14, 231–252.

Makeham, W.M. (1874) On an application of the theory of the composition of decremental forces. *J. Inst. Actuaries* 18, 317–322.

Matthews, D.E. (1984) Efficiency considerations in the analysis of a competing risk problem. *Can. J. Statist.* 12, 207–210.

Mendenhall, W., and Hader, R.J. (1958) Estimation of parameters of mixed exponentially distributed failure time distributions from censored life test data. *Biometrika* 45, 504–520.

Miller, D.R. (1977) A note on independence of multivariate lifetimes in competing risk models. *Ann. Statist.* 5, 576–579.

Miyakawa, M. (1984) Analysis of incomplete data in competing risks model. *IEEE Trans. Reliability* 33, 293–296.

Moeschberger, M.L. (1974) Life tests under dependent competing causes of failure. *Technometrics* 16, 39–47.

Moeschberger, M.L., and David, H.A. (1971) Life tests under competitive causes of failure and the theory of competing risks. *Biometrics* 27, 909–933.

Moeschberger, M.L., and Klein, J.P. (1995) Statistical methods for dependent competing risks. *Lifetime Data Analysis* 1, 195–204.

Nadas, A. (1970b) On estimating the distribution of a random vector when only the smallest co-ordinate is observable. *Technometrics* 12, 923–924.

Nadas, A. (1971) The distribution of the identified minimum of a normal pair determines the distribution of the pair. *Technometrics* 13, 201–202.

Nair, V.J. (1993) Bounds for reliability estimation under dependent censoring. *Int. Statist. Rev.* 61, 169–182.

Nelson, W.B. (1970) Hazard plotting methods for analysis of life data with different failure modes. *J. Qual. Tech.* 2, 126–149.

Reiser, B., Guttman, I., Lin, D.K.J., Guess, F.M., and Usher, J.S. (1995) Bayesian inference for masked system lifetime data. *Appl. Statist.* 44, 79–90.

Pepe, M.S. (1991) Inference for events with dependent risks in multiple endpoint studies. *J. Amer. Statist. Assoc.* 86, 770–778.

Pepe, M.S., and Mori, M. (1993) Kaplan–Meier, marginal or conditional probability curves in summarizing competing risks failure time data? *Statist. Med.* 12, 737–751.

Peterson, A.V. (1976) Bounds for a joint distribution function with fixed sub-distribution functions: Application to competing risks. *Proc. Nat. Acad. Sci. USA* 73, 11–13.

Peterson, A.V. (1977) Expressing the Kaplan–Meier estimator as a function of empirical sub-survival functions. *J. Amer. Statist. Assoc.* 72, 854–858.

Pike, M.C., and Roe, F.J.C. (1963) An actuarial method of analysis in an experiment in two-stage carcinogenesis. *Brit. J. Cancer* 17, 605–610.

Pike, M.C. (1970) A note on Kimball's paper "Models for the estimation of competing risks from grouped data." *Biometrics* 26, 579–581.

Prentice, R.L., Kalbfleisch, J.D., Peterson, A.V., Flournoy, N., Farewell, V.T., and Breslow, N.E. (1978) The analysis of failure times in the presence of competing risks. *Biometrics* 34, 541–554.

Proschan, F., and Sullo, P. (1974) Estimating parameters of a bivariate exponential distribution in several sampling situations. In *Reliability and Biometry: Statistical Analysis of Lifelength*, Society for Industrial and Applied Mathematics, Philadelphia, PA, 423–440.

Rachev, S.T., and Yakovlev, A.Y. (1985) Some problems of the competing risks theory. *Fifth Int. Summer School on Prob. Theory and Math. Statist., Varna 1985.* Publ: Bulgarian Acad. Sci.

Racine-Poon, A.H., and Hoel, D.G. (1984) Nonparametric estimation of the survival function when the cause of death is uncertain. *Biometrics* 40, 1151–1158.

Sampford, M.R. (1952) The estimation of response-time distributions: II. Multi-stimulus distributions. *Biometrics* 8, 307–369.

Scheike, T.H., and Zhang, M-J. (2008) Flexible competing risks regression modeling and goodness-of-fit. *Lifetime Data Analysis* 14, 464–483.

Seal, H.L. (1954) The estimation of mortality and other decremental probabilities. *Skand. Aktuarietidskr.* 37, 137–162.

Seal, H.L. (1977) Studies in the history of probability and statistics. XXXV Multiple decrements or competing risks. *Biometrika* 64, 429–439.

Sen, P.K. (1974) Nonparametric tests for interchangeability under competing risks. *Inst. of Statist.* Mimeo Series No. 905, University of North Carolina.

Serio, C.D. (1997) The protective impact of a covariate on competing failures with an example from a bone marrow transplantation study. *Lifetime Data Analysis 3,* 99–122.

Shen, Y., and Cheng, S.C. (1999) Confidence bands for cumulative incidence curves under the additive risk model. *Biometrics 55,* 1093–1100.

Slud, E.V., and Rubenstein, L.V. (1983). Dependent competing risks and summary survival curves. *Biometrika 70,* 643–649.

Solari, A., Salmaso, L., El Barmi, H., and Pesarin, F. (2008) Conditional tests in a competing risks model. *Lifetime Data Analysis 14,* 154–166.

Sun, L., Liu, J., Sun, J., and Zhang, M. (2006) Modeling the subdistribution of a competing risk. *Statist. Sinica 16,* 1367–1385.

Tsiatis, A. (1975) A nonidentifiability aspect of the problem of competing risks. *Proc. Nat. Acad. Sci. USA 72,* 20–22.

Tsiatis, A.A., Davidian, M., and McNeney, B. (2002) Multiple imputation methods for testing different treatment differences in survival distributions with missing cause of failure. *Biometrika 89,* 238–244.

Usher, J.S., and Hodgson, T.J. (1988) Maximum likelihood analysis of component reliability using masked system life-test data. *IEEE Trans. Reliab. 37,* 550–555.

Usher, J.S., and Guess, F.M. (1989) An iterative approach for estimating component reliability from masked system life data. *Qual. Reliab. Eng. Inst. 5,* 257–261.

Wang, O. (1977) A competing risk model based on the life table procedure in epidemiologic studies. *Int. J. Epidemiology 6,* 153–159.

Williams, J.S. and Lagakos, S.W. (1976) Independent and dependent censoring mechanisms. *Proc. 9th Int. Biometric Conf.,* Vol. 1, 408–427.

Williams, J.S., and Lagakos, S.W. (1977) Models for censored survival analysis: Constant-sum and variable-sum models. *Biometrika 64,* 215–224.

Yashin, A.I., Manton, K.G., and Stallard, E. (1986) Dependent competing risks: a stochastic process model. *J. Math. Biol. 24,* 119–164.

Zheng, M., and Klein, J.P. (1995) Estimates of marginal survival for dependent competing risks based on an assumed copula. *Biometrika 82,* 127–138.

Counting Processes

Andersen, P.K., and Gill, R.D. (1982) Cox's regression model for counting processes: a large sample study. *Ann. Statist. 10,* 1100–1120.

Andersen, P.K., and Borgan, O. (1985) Counting process models for life history data: A review. *Scand. J. Statist. 12,* 97–158.

Arjas, E. (1985) Contribution to discussion of Andersen and Borgan. *Scand. J. Statist. 12,* 150–153.

Borgan, O. (1984) Maximum likelihood estimation in parametric counting process models, with applications to censored survival data. *Scand. J. Statist. 11,* 1–16 (Correction 11, 275).

Doob, J.L. (1940) Regularity properties of certain families of chance variables. *Trans. Amer. Math. Soc. 47,* 455–486.

Doob, J.L. (1953) *Stochastic Processes*. John Wiley and Sons, New York.

Gill, R.D. (1984) Understanding Cox's regression model: a martingale approach. *J. Amer. Statist. Assoc.* 79, 441–447.

Gill, R.D., and Johansen, S. (1990) A survey of product-integration with a view toward application in survival analysis. *Ann. Statist.* 4, 1501–1555.

Lan, K.K.G., and Lachin, J.M. (1995) Martingales without tears. *Lifetime Data Analysis* 1, 361–375.

Rivest, L-P., and Wells, M.T. (2001) A martingale approach to the copula-graphic estimator for the survival function under dependent censoring. *J. Multiv. Anal.* 79, 138–155.

Therneau, T.M., Grambsch, P.M., and Fleming, T.R. (1990) Martingale-based residuals for survival models. *Biometrika* 77, 147–160.

General References

Abate, J., Choudhury, G.L., and Whitt, W. (1996) On the Laguerre method for numerically inverting Laplace transforms. *INFORMS Journal on Computing* 8, 413–427.

Andrews, D.F., and Herzberg, A.M. (1985) *Data: A Collection of Problems from Many Fields for the Student and Research Worker*. Springer, New York.

Archbold, J.W. (1970) *Algebra*, Fourth Edition. Pitman Publishing, London.

Atkinson, A.C. (1985) *Plots, Transformations and Regression*. Clarendon Press, Oxford.

Bowers, L., Allan, T., Simpson, A., Jones, J., and Van Der Merwe, M. (2009) Identifying key factors associated with aggression on acute inpatient psychiatric wards. *Issues in Mental Health Nursing* 30, 260–271.

Brooks, S.P., and Morgan, B.J.T. (1995) Optimisation using simulated annealing. *Statistician* 44, 241–257.

Buonaccorsi, J.P. (2010) *Measurement Error: Models, Methods and Applications*. Chapman & Hall/CRC Press, Boca Raton, FL.

Cappe, O., Moulines, E., and Ryden, T. (2005) *Inference in Hidden Markov Models*. Springer, New York.

Cheng, R.C.H., and Stevens, M.A. (1989) A goodness-of-fit test using Moran's statistic with estimated parameters. *Biometrika* 76, 385–392.

Cody, W.J., and Thacher, H.C. Jr. (1969) Chebyshev approximations for the exponential integral $E_1(x)$. *Math. Comp.* 23, 289–303.

Cook, R.D., and Weisburg, S. (1982) *Residuals and Influence in Regression*. Chapman & Hall, London.

Copas, J. B. (1983) Plotting p against x. *Appl. Statist.* 32, 25–31.

Cox, D.R. (1970) *The Analysis of Binary Data*. Methuen, London.

Cox, D.R., and Hinkley, D.V. (1974) *Theoretical Statistics*. Chapman & Hall, London.

Cox, D.R., and Miller, H.D. (1965) *The Theory of Stochastic Processes*. Methuen, London.

Cox, D.R., and Reid, N. (1987) Parameter orthogonality and approximate conditional inference (with discussion). *J. Roy. Statist. Soc.* B49, 1–39.

Crompton, R. (1922) *Just William*. Newnes, London.

Crowder, M.J. (1986) On consistency and inconsistency of estimating equations. *Econ. Theory* 2, 305–330.

Crowder, M.J. (1987) On linear and quadratic estimating functions. *Biometrika* 74, 591–597.

Crowder, M.J. (2001) Corrected p-values for tests based on estimated parameters. *Statistics and Computing* 11, 347–354.

Crowder, M.J., and Hand, D.J. (1990) *Analysis of Repeated Measures*. Chapman & Hall, London.

Crowder, M.J., Dixon, M.J., Robinson, M.E., and Ledford, A.W. (2002) Dynamic modelling and prediction of English football league matches. *The Statistician* 51, 157–168.

DeGroot, M.H. (1973) Doing what comes naturally: interpreting a tail area as a posterior probability or a likelihood ratio. *J. Amer. Statist. Assoc.* 68, 966–969.

Diggle, P.J., Heagerty, P.J., Liang, K-Y., and Zeger, S.L. (2002) *Analysis of Longitudinal Data*, Second Edition. Oxford University Press, Oxford.

Dixon, M.J., and Coles, S.G. (1997) Modelling association football scores and inefficiencies in the football betting market. *Applied Statistics* 46, 265–280.

Doyle, A.C. (1950) Silver Blaze. In *The Memoirs of Sherlock Holmes*. Penguin, London.

Efron, B., and Hinkley, D.V. (1978) Assessing the accuracy of the maximum likelihood estimator: observed versus expected Fisher information. *Biometrika* 65, 457–487.

Everitt, B.S., and Hothorn, T. (2010) *A Handbook of Statistical Analyses Using R*. Chapman & Hall/CRC Press, Boca Raton, FL.

Feller, W. (1962) *An Introduction to Probability Theory and Its Applications*, Vol. 1, Second Edition. Wiley, New York.

Fitzmaurice, G., Davidian, M., Verbeke, G., and Molenberghs, G. (2009) *Longitudinal Data Analysis*. CRC Press, Boca Raton, FL.

Fletcher, R., and Powell, M.J.D. (1963) A rapidly convergent descent method for minimisation. *Comp. J.* 6, 163–168.

Galambos, J. (1978) *The Asymptotic Theory of Extreme Order Statistics*. Wiley, New York.

Gelfand, A.E., and Smith, A.F.M. (1990) Sampling-based approaches to calculating marginal densities. *J. Amer. Statist. Assoc.* 85, 398–409.

Genz, A. (1993) A comparison of methods for numerical computation of multivariate normal integrals. *Comp. Sci. Statist.* 25, 400–405.

Geyer, C.J. (1992) Practical Markov chain Monte Carlo. *Statist. Sci.* 7, 473–511.

Gilks, W.R., Richardson, S., and Spiegelhalter, D.J. (1996) *Markov Chain Monte Carlo in Practice*. Chapman & Hall, London.

Gill, P.E., Murray, W., and Wright, M.H. (1981) *Practical Optimization*. Academic Press, London.

Hand, D.J., and Crowder, M.J. (1996) *Practical Longitudinal Data Analysis*. Chapman & Hall/CRC Press, London.

Johnson, N.L., and Kotz, S. (1972) *Distributions in Statistics: Continuous Multivariate Distributions*. Wiley, New York.

Kalbfleisch, J.D., and Sprott, D.A. (1970) Application of likelihood methods to models involving large numbers of parameters (with discussion). *J. Roy. Statist. Soc.* B 32, 175–208.

Karlin, S., and Taylor, H.M. (1975) *A First Course in Stochastic Processes*, Second Edition. Academic Press, London.

Little, R.J.A., and Rubin, D.B. (2002) *Statistical Analysis with Missing Data*, Second Edition. Wiley, New York.

Lodge, D. (1990) *Famous Cricketers*, Series No. 15: W.G. Grace. Association of Cricket Statisticians and Historians, West Bridgford, Nottingham, U.K.

MacDonald, I.L., and Zucchini, W. (1997) *Hidden Markov and Other Models for Discrete-Valued Time Series*. Chapman & Hall/CRC Press, Boca Raton, FL.

Moler, M., and Van Loan, C. (2003) Nineteen dubious ways to compute the exponential of a matrix, twenty-five years later. *SIAM Review* 45, 3–49.

Morgan, B.J.T. (1995) *Elements of Simulation*. Chapman & Hall, London.

O'Hagan, A., and Forster, J. (2004) *Kendall's Advanced Theory of Statistics*, Second Edition, Vol. 2B, Bayesian Inference. Arnold, London.

Olver, F.W.J., Lozier, D.W., Boisvert, R.F., and Clark, C.W. (2010) *NIST Handbook of Mathematical Functions*. National Institute for Standards and Technology and Cambridge University Press, Cambridge.

Rabe-Hesketh, S., and Skrondal, A. (2008) *Multilevel and Longitudinal Modeling Using Stata*, Second Edition. Stata Press, College Station, Texas.

Rayner, J., and Best, D. (1999) Smooth tests of goodness of fit: An overview. *Int. Statist. Rev.* 58, 9–17.

Ripley, B.D. (1987) *Stochastic Simulation*. John Wiley & Sons, New York.

Ross, G.J.S. (1970) The efficient use of function minimization in nonlinear maximum likelihood estimation. *Appl. Statist.* 19, 205–221.

Ross, S. (1996) *Stochastic Processes, Second Edition*. Wiley, New York.

Self, S.G., and Liang, K. (1987) Asymptotic properties of maximum likelihood estimators and likelihood ratio tests under nonstandard conditions. *J. Amer. Statist. Assoc.* 82, 605–610.

Sethuraman, J. (1965) On a characterization of the three limiting types of the extreme. *Sankhya* A 27, 357–364.

Smith, A.F.M. (1991) Bayesian computational methods. *Phil. Trans. Roy. Soc. London.* A 337, 369–386.

Smith, A.F.M., and Roberts, G.O. (1993) Bayesian computation via the Gibbs sampler and related Markov chain Monte Carlo methods. *J. Roy. Statist. Soc.* B55, 3–23.

Stewart, D., Bowers, L., and Warburton, F. (2009) Constant special observation and self-harm on acute psychiatric wards: A longitudinal analysis. *General Hospital Psychiatry* 31, 523–530.

Venables, W.N., and Ripley, B.D. (1999) *Modern Applied Statistics with S-PLUS*, Third Edition. Springer-Verlag, New York.

Walker, S.G., Damien, P., Land, P.W., and Smith, A.F.M. (1999) Bayesian nonparametric inference for random distributions and related functions (with discussion). *J. Roy. Statist. Soc.* B 61, 485–527.

Wang, Y. (2007) On fast computation of the non-parametric maximum likelihood estimate of a mixing distribution. *J. R. Statist. Soc.* 69, 185–198.

Watson, A.S., and Smith, R.L. (1985) An examination of statistical theories for fibrous materials in the light of experimental data. *J. Mat. Sci.* 20, 3260–3270.

Whittaker, E.T., and Watson, G.N. (1927) *A Course of Modern Analysis*. Cambridge University Press, Cambridge.

Wu, L. (2010) *Mixed Effects Models for Complex Data*. Chapman & Hall/CRC Press, Boca Raton, FL.

Epilogue to First Edition

What are the directions of development for the subject? I'm not normally one to be drawn into grandiose speculation — it reminds me of Peter Sellers' spoof political speech "The Future Lies Ahead." Come to think of it, it reminds me of most political speeches. However, it has been suggested that there should be a few words on this at the end of the book, so here goes. There have been some obviously identifiable key advances, such as non-parametric maximum likelihood (Kaplan and Meier, 1958), semi-parametric modelling (Cox, 1972), and martingale counting processes (Aalen, 1975). The strength of such works is demonstrated by the huge research literature that they have spawned. Also, the sequential nature of research is that there is a wealth of groundwork to be drawn on. For instance, some of what we now call non-parametric methods have been used from time immemorial in the actuarial business. So, what are the star turns of the future? One contender is the use of much more sophisticated non-parametric and semi-parametric methodology. For example, Walker et al. (1999) presented a modelling framework for the integrated hazard function based on a particular kind of stochastic process. The aim is to have a flexible prior distribution on the space of distribution functions that gives rise to tractable (conjugate) posterior forms in a Bayesian framework. The approach is mathematically advanced, but then, who would have thought that martingales and counting processes would ever become standard tools in applied statistics? In a slightly different direction there is the semi-parametric approach of Couper and Pepe (1997) in which the modern technique of *generalised additive models* is applied.

On a more personal level, if you have stayed the course, battled through thick and thin to this point, may I congratulate you. I hope that you have come to appreciate the subject and wish to go further and deeper. The immediate way ahead, I suggest, is to continue from where the present book leaves off, and a good way of doing this is to work through Fleming and Harrington (1991) Andersen et al. (1993) or both. The counting process approach has solid advantages: it provides a firm theoretical basis for modelling events unfolding over time and is particularly suited to models framed in terms of hazard rates and involving censoring. I can do no better than quote from the Preface of Fleming and Harrington (1991, p. vii): "Martingale methods can be used to obtain simple expressions for the moments of complicated statistics, to calculate and verify asymptotic distributions for test statistics and estimators, to examine the operating characteristics of non-parametric testing methods and semi-parametric censored data regression methods, and even to provide a basis for graphical diagnostics in model building with counting process data." Good studying!

Index